Design and Control of Concrete Mixtures

THIRTEENTH EDITION

by Steven H. Kosmatka and William C. Panarese

PORTLAND CEMENT PCA ASSOCIATION

5420 Old Orchard Road, Skokie, Illinois 60077-1083

An organization of cement manufacturers to improve and extend the uses of portland cement and concrete through market development, engineering, research, education, and public affairs work.

About the authors: The authors of this engineering bulletin are Steven H. Kosmatka and William C. Panarese, manager, Research and Development, and formerly manager, Construction Information Services, respectively, Portland Cement Association.

On the cover: Illustrations on the cover (clockwise from top left) are (1) compressive strength test of concrete; (2) concrete placement by bucket; (3) cross section of concrete; (4) River City, Chicago; (5) concrete mix-water temperature chart; and (6) post-tensioned concrete box girder bridge, Portland, Ore.— photo courtesy of Oregon Department of Transportation.

Thirteenth Edition print history
First printing 1988
Second printing (rev.) 1990
Third printing (rev.) 1992
Fourth printing (rev.) 1994

Library of Congress catalog card number: LC29-10592

ISBN 0-89312-087-1

Printed in the United States of America

Caution: Contact with wet (unhardened) concrete, mortar, cement, or cement mixtures can cause SKIN IRRITATION, SEVERE CHEMICAL BURNS, or SERIOUS EYE DAMAGE. Wear waterproof gloves, a long-sleeved shirt, full-length trousers, and proper eye protection when working with these materials. If you have to stand in wet cement, use waterproof boots that are high enough to keep concrete from flowing into them. Wash wet concrete, mortar, cement, or cement mixtures from your skin immediately. Flush eyes with clean water immediately after contact. Indirect contact through clothing can be as serious as direct contact, so promptly rinse out wet concrete, mortar, cement, or cement mixtures from clothing. Seek immediate medical attention if you have persistent or severe discomfort.

EB001.13T

Contents

CONTENTS (continued)

CONTENTS (continued)

Preface

Portland cement concrete is a simple material in appearance with a very complex internal nature, as this text illustrates. In contrast to its internal complexity, concrete's versatility, durability, and economy have made it the world's most used construction material. This can be seen in the variety of structures it is used in, from highways, bridges, buildings, and dams to floors, sidewalks, and even works of art. The use of concrete is unlimited and is not even earthbound, as indicated by recent interest by the National Aeronautics and Space Administration in concrete lunar structures.

Design and Control of Concrete Mixtures has been the cement and concrete industry's primary reference on concrete technology for over 70 years. Since the first edition was published in the early 1920's, it has been updated eighteen times (13 U.S. editions and five Canadian) to reflect advances in concrete technology and to meet the growing needs of architects, engineers, builders, concrete technologists, and instructors.

This fully revised and expanded 13th edition was written to provide a concise, current reference on concrete and to include many advances in concrete technology that have occurred since the last edition was published in 1979. The text is backed by over 75 years of research by the Portland Cement Association and many other organizations, and reflects requirements of the latest editions of standards, specifications, and test methods of the American Society for Testing and Materials (ASTM) and the American Concrete Institute (ACI).

Besides presenting much new information, the 13th edition is more "user friendly" than previous editions. The table of contents and index have been expanded to provide the user with easier, quicker access to the desired information. Cross referencing is extensive and references to related information are provided in expanded lists at the end of each of the fifteen chapters. Many quick-reference tables are provided and a new appendix includes frequently needed metric conversion tables and a comprehensive list of cement- and concrete-related ASTM standards.

The plan of the book remains unchanged—fifteen chapters with essentially the same titles as the 12th edition. New, updated, or expanded discussions in Chapers 1–5 include permeability and abrasion resistance of concrete; hot, blended, and expansive cements; cement strength and compound transformation; sea-

dredged aggregate, recycled concrete, thermal properties of aggregate, alkali-aggregate reactivity; and an extended discussion on the effect of various concreting practices on air content.

Chapters 6–10 have new sections on slag, fly ash, silica fume, corrosion inhibitors, superplasticizers, air detrainers, mix-design procedure including a flow chart, pumps, conveyor belts, high-energy mixers, overlays, vapor barriers, jointing, and the effect of temperature on time of set.

New topics in Chapters 11-15 include evaporation retarders, modulus of elasticity, Poisson's ratio, shear strain, chemical volume changes, carbonation, high- and low-temperature effects, cast-in-place test cylinders and other new quality control and investigative test procedures, and special types of concrete such as lightweight, heavyweight, high-strength, roller-compacted, porous, polymer-modified, and mass concrete. Most of the new or expanded material is accompanied by explanatory photos, tables, or graphs—250 illustrations in all.

The authors wish to acknowledge contributions made by many individuals and organizations who provided valuable assistance in the writing and publishing of this expanded 13th edition. A special thanks to Paul Klieger, concrete and concrete materials consultant with over 45 years of experience with PCA's Research and Development Laboratories, who reviewed the manuscript for technical accuracy; Richard C. Spring, formerly manager of publications services for PCA, who coordinated the editorial, design, production, and printing activities; Arden Orr for her editorial work; Richard C. Wagner of Wagner Design Services for his layout design and production; Robert D. Kuhart and Curtis F. Steer of Construction Technology Laboratories, Inc. (CTL), for their drafting of the many figures; Cynthia Spigelman, PCA librarian, for promptly obtaining the many reference materials; and finally the American Concrete Institute and the American Society for Testing and Materials, whose publications and articles are frequently cited.

The authors have tried to make this edition of *Design and Control of Concrete Mixtures* a concise and current reference on concrete technology. As there is always room for improvement, readers are encouraged to submit comments to improve future printings and editions of this book.

CHAPTER 1
Fundamentals of Concrete

Concrete is basically a mixture of two components: aggregates and paste. The paste, comprised of portland cement and water, binds the aggregates (sand and gravel or crushed stone) into a rocklike mass as the paste hardens because of the chemical reaction of the cement and water.*

Aggregates are generally divided into two groups: fine and coarse. Fine aggregates consist of natural or manufactured sand with particle sizes ranging up to ⅜ in.; coarse aggregates are those with particles retained on the No. 16 sieve and ranging up to 6 in. The most commonly used maximum aggregate size is ¾ in. or 1 in.

The paste is composed of portland cement, water, and entrapped air or purposely entrained air. Cement paste ordinarily constitutes about 25% to 40% of the total volume of concrete. Fig. 1-1 shows that the absolute volume of cement is usually between 7% and 15% and the water between 14% and 21%. Air content in air-entrained concrete ranges up to about 8% of the volume of the concrete, depending on the top size of the coarse aggregate.

Since aggregates make up about 60% to 75% of the total volume of concrete, their selection is important. Aggregates should consist of particles with adequate strength and resistance to exposure conditions and should not contain materials that will cause deterioration of the concrete. A continuous gradation of particle sizes is desirable for efficient use of the cement and water paste. Throughout this text, it will be assumed that suitable aggregates are being used, except where otherwise noted.

The quality of the concrete depends to a great extent upon the quality of the paste. In properly made concrete, each particle of aggregate is completely coated with paste and all of the spaces between aggregate particles are completely filled with paste, as illustrated in Fig. 1-2.

For any particular set of materials and conditions of curing, the quality of hardened concrete is determined

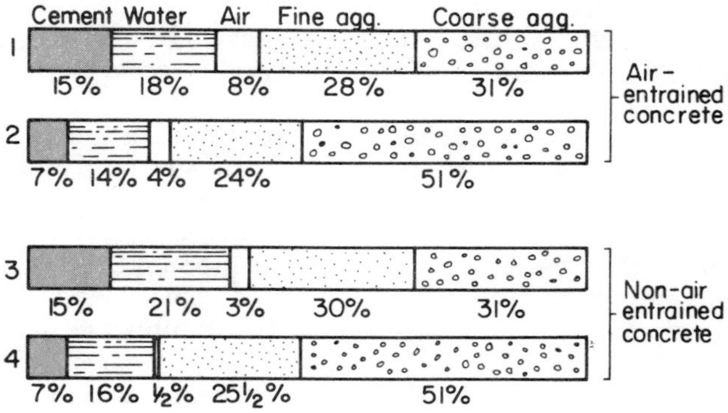

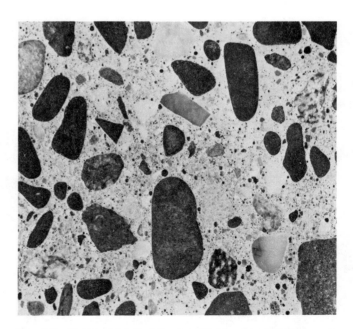

Fig. 1-1. Range in proportions of materials used in concrete, by absolute volume. Bars 1 and 3 represent rich mixes with small aggregates. Bars 2 and 4 represent lean mixes with large aggregates.

Fig. 1-2. Cross section of hardened concrete. Cement-and-water paste completely coats each aggregate particle and fills all spaces between particles.

---

*This text addresses the utilization of portland cement in the production of concrete. The term "portland cement" pertains to a calcareous hydraulic cement produced by heating the oxides of silicon, calcium, aluminum, and iron. The term "cement" used throughout the text pertains to portland cement unless otherwise stated.

by the amount of water used in relation to the amount of cement. Following are some advantages of reducing water content:

Increased compressive and flexural strength

Lower permeability, thus increased watertightness and lower absorption

Increased resistance to weathering

Better bond between successive layers and between concrete and reinforcement

Less volume change from wetting and drying

Reduced shrinkage cracking tendencies

The less water used, the better the quality of the concrete—provided it can be consolidated properly. Smaller amounts of mixing water result in stiffer mixtures; but with vibration, the stiffer mixtures can be used. For a given quality of concrete, stiffer mixtures are more economical. Thus consolidation by vibration permits improvement in the quality of concrete and in economy.

The freshly mixed (plastic) and hardened properties of concrete may be changed by adding admixtures to the concrete, usually in liquid form, during batching. Admixtures are commonly used to (1) adjust setting time or hardening, (2) reduce water demand, (3) increase workability, (4) intentionally entrain air, and (5) adjust other concrete properties. Admixtures are discussed in Chapter 6.

After completion of proper proportioning, batching, mixing, placing, consolidating, finishing, and curing, hardened concrete becomes a strong, noncombustible, durable, abrasion-resistant, and practically impermeable building material that requires little or no maintenance. Concrete is also an excellent building material because it can be formed into a wide variety of shapes, colors, and textures for use in almost unlimited number of applications.

FRESHLY MIXED CONCRETE

Freshly mixed concrete should be plastic or semifluid and generally capable of being molded by hand. A very wet concrete mixture can be molded in the sense that it can be cast in a mold, but this is not within the definition of "plastic"—that which is pliable and capable of being molded or shaped like a lump of modeling clay.

In a plastic concrete mixture all grains of sand and pieces of gravel or stone are encased and held in suspension. The ingredients are not apt to segregate during transport; and when the concrete hardens, it becomes a homogeneous mixture of all the components. Concrete of plastic consistency does not crumble but flows sluggishly without segregation.

Slump is used as a measure of the consistency of concrete. A low-slump concrete has a stiff consistency.

In construction practice, thin concrete members and heavily reinforced concrete members require workable, but never soupy, mixes for ease of placement. A plastic mixture is required for strength and for maintaining homogeneity during handling and placement.

While a plastic mixture is suitable for most concrete work, superplasticizing admixtures may be used to make concrete more flowable in thin or heavily reinforced concrete members.

Mixing

In Fig. 1-1, the five basic components of concrete are shown separately. To ensure that they are combined into a homogeneous mix requires effort and care. The sequence of charging ingredients into the mixer plays an important part in the uniformity of the finished product. The sequence, however, can be varied and still produce a quality concrete. Different sequences require adjustments in the time of water addition, the total number of revolutions of the mixer drum, and the speed of revolution. Other important factors in mixing are the size of the batch in relation to the size of the mixer drum, the elapsed time between batching and mixing, and the design, configuration, and condition of the mixer drum and blades. Approved mixers, correctly operated and maintained, ensure an end-to-end exchange of materials by a rolling, folding, and kneading action of the batch over itself as the concrete is mixed.

Workability

The ease of placing, consolidating, and finishing freshly mixed concrete is called workability. Concrete should be workable but should not segregate or bleed excessively. Bleeding is the migration of water to the top surface of freshly placed concrete caused by the settlement of the solid materials—cement, sand, and stone—within the mass. Settlement is a consequence of the combined effect of vibration and gravity.

Excessive bleeding increases the water-cement ratio near the top surface and a weak top layer with poor durability may result, particularly if finishing operations take place while bleed water is present. Because of the tendency of freshly mixed concrete to segregate and bleed, it is important to transport and place each load as close as possible to its final position. Entrained air improves workability and reduces the tendency of freshly mixed concrete to segregate and bleed.

Consolidation

Vibration sets into motion the particles in freshly mixed concrete, reducing friction between them and giving the mixture the mobile qualities of a thick fluid. The vibratory action permits use of a stiffer mixture containing a larger proportion of coarse and a smaller proportion of fine aggregate. The larger the maximum-size aggregate in concrete with a well-graded aggregate, the less volume there is to fill with paste and the less aggregate surface area there is to coat with paste; thus less water and cement are needed. With adequate consolidation, harsher as well as stiffer mixtures can be used, resulting in improved quality and economy.

If a concrete mixture is workable enough to be readily consolidated by hand rodding, there may not be an advantage in vibrating it. In fact, such mixtures may segregate when vibrated. Only by using stiffer, harsher mixtures are the full benefits of vibration realized.

Mechanical vibration has many advantages. High-frequency vibrators make it possible to economically place mixtures that are impractical to consolidate by hand under many conditions. As an example, Fig. 1-3 shows concrete of a stiff consistency (low slump) that was mechanically vibrated in forms containing closely spaced reinforcement. With hand rodding, a much wetter consistency would have been necessary.

Fig. 1-3. Concrete of a stiff consistency (low slump).

Hydration, Setting Time, Hardening

The binding quality of portland cement paste is due to the chemical reaction between the cement and water, called hydration.

Portland cement is not a simple chemical compound, it is a mixture of many compounds. Four of these make up 90% or more of the weight of portland cement: tricalcium silicate, dicalcium silicate, tricalcium aluminate, and tetracalcium aluminoferrite. In addition to these major compounds, several others play important roles in the hydration process. Different types of portland cement contain the same four major compounds, but in different proportions (see Chapter 2, "Portland Cements").

When clinker (the kiln product that is ground to make portland cement) is examined under a microscope, most of the individual cement compounds can be identified and their amounts determined. However, the smallest grains elude visual detection. The average diameter of a typical cement particle is approximately 10 μm, or about a hundredth of a millimeter. If all cement particles were average, portland cement would contain about 135 billion grains per pound, but in fact there are some 7000 billion particles per pound because of the broad range of particle sizes. The particles in a kilogram of portland cement have a surface area of about 400 square meters.

The two calcium silicates, which constitute about 75% of the weight of portland cement, react with water to form two new compounds: calcium hydroxide and *calcium silicate hydrate.* The latter is by far the most important cementing component in concrete. The engineering properties of concrete—setting and hardening, strength, and dimensional stability—depend primarily on calcium silicate hydrate gel. It is the heart of concrete.

The chemical composition of calcium silicate hydrate is somewhat variable, but it contains lime (CaO) and silicate (SiO_2) in a ratio on the order of 3 to 2. The surface area of calcium silicate hydrate is some 300 square meters per gram. The particles are so minute that they can be seen only in an electron microscope. In hardened cement paste, these particles form dense, bonded aggregations between the other crystalline phases and the remaining unhydrated cement grains; they also adhere to grains of sand and to pieces of coarse aggregate, cementing everything together. The formation of this structure is the paste's cementing action and is responsible for setting, hardening, and strength development.

When concrete sets, its gross volume remains almost unchanged, but hardened concrete contains pores filled with water and air that have no strength. The strength is in the solid part of the paste, mostly in the calcium silicate hydrate and crystalline phases.

The less porous the cement paste, the stronger the concrete. When mixing concrete, therefore, use no more water than is absolutely necessary to make the concrete plastic and workable. Even then, the water used is usually more than is required for complete hydration of the cement. The water-cement ratio (by weight) of completely hydrated cement is about 0.22 to 0.25, excluding evaporable water.

Knowledge of the amount of heat released as cement hydrates can be useful in planning construction. In winter, the heat of hydration will help protect the concrete against damage from freezing temperatures. The heat may be harmful, however, in massive structures such as dams because it may produce undesirable stresses on cooling after hardening. Type I portland cement releases a little more than half of its total heat of hydration in three days. Type III, high-early-strength cement, releases approximately the same percentage of its heat in much less than three days. Type II, a moderate-heat cement, releases less total heat than the others and more than three days are required for only half of that heat to be released. The use of Type IV, low-heat-of-hydration portland cement, should be considered where low heat of hydration is of primary importance.

Knowledge of the rate of reaction between cement and water is important because the rate determines the time of setting and hardening. The initial reaction must be slow enough to allow time for the concrete to be transported and placed. Once the concrete has been placed and finished, however, rapid hardening is desirable. Gypsum, added at the cement mill when clinker is ground, acts as a regulator of the initial rate of hydration of portland cement. Other factors that influence the rate of hydration include fineness of grinding, admixtures, amount of water added, and temperature of the materials at the time of mixing.

HARDENED CONCRETE

Moist Curing

Increase in strength with age continues as long as any unhydrated cement is still present, provided the concrete remains moist or has a relative humidity above approximately 80% and the concrete temperature remains favorable. When the relative humidity within the concrete drops to about 80% or the temperature of the concrete drops below freezing, hydration and strength gain virtually stop. Fig. 1-4 illustrates the relationship between strength gain and moist curing.

If concrete is resaturated after a drying period, hydration is resumed and strength will again increase. However, it is best to moist-cure concrete continuously from the time it is placed until it has attained the desired quality because concrete is difficult to resaturate.

The effects of concrete temperature during mixing and curing are discussed in Chapters 10, 11, and 12: "Curing Concrete," "Hot-Weather Concreting," and "Cold-Weather Concreting."

Drying Rate of Concrete

Concrete does not harden or cure by drying. Concrete (or more precisely, the cement in it) needs moisture to hydrate and harden. The drying of concrete is only indirectly related to hydration and hardening. When concrete dries, it ceases to gain strength; the fact that it is dry is no indication that it has undergone sufficient hydration to achieve the desired physical properties.

Knowledge of the rate of drying is helpful in understanding the properties or physical condition of concrete. For example, as mentioned, concrete must continue to hold enough moisture throughout the curing period in order for the cement to hydrate. Freshly cast concrete has an abundance of water, but as drying progresses from the surface inward, strength gain will continue at each depth only as long as the relative humidity at that point remains above 80%.

A common illustration of this is the surface of a concrete floor that has not had sufficient moist curing. Because it has dried quickly, concrete at the surface is weak and traffic on it creates dusting. Also, when concrete dries, it shrinks, just as wood, paper, and clay do (though not as much). Drying shrinkage is a primary cause of cracking, and the width of cracks is a function of the degree of drying.

While the surface of concrete will dry quite rapidly, it takes a much longer time for concrete in the interior to dry. Fig. 1-5 illustrates the rates of natural and artificial drying for a 6-in.-thick concrete wall or slab drying from both sides. Note that after 114 days of natural drying the concrete is still quite moist on the inside and that 850 days were required for the relative humidity at the center to drop to 50%. The figure is not typical for all concrete, only that used in the test.

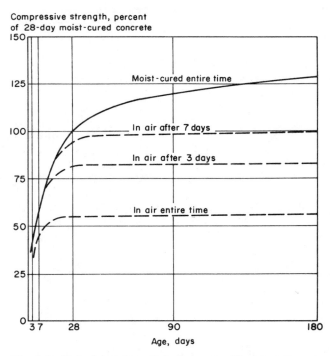

Fig. 1-4. Concrete strength increases with age as long as moisture and a favorable temperature are present for hydration of cement. Adapted from Reference 1-15, Fig. 9.

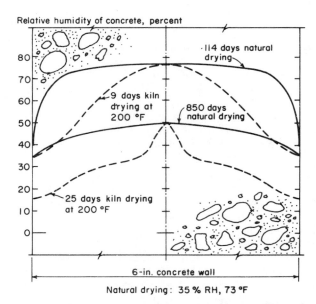

Fig. 1-5. Relative humidity distributions resulting from natural and artificial drying. The centerline humidity for the 850-day curve is based on test data, while the gradient is estimated. Reference 1-9.

The moisture content of thin concrete elements after drying in air with a relative humidity of 50% to 90% for several months is about 1% to 2% by weight of the concrete depending on the concrete's constituents, original water content, drying conditions, and the size of the concrete element (refer to Chapter 13 for more information).

Size and shape of a concrete member have an important bearing on the rate of drying. Concrete elements with large surface area in relation to volume (such as floor slabs) dry faster than large concrete volumes with relatively small surface areas (such as bridge piers).

Many other properties of hardened concrete also are affected by its moisture content; these include elasticity, creep, insulating value, fire resistance, abrasion resistance, electrical conductivity, and durability.

Strength

Compressive strength may be defined as the measured maximum resistance of a concrete or mortar specimen to axial loading. It is generally expressed in pounds per square inch (psi) at an age of 28 days and is designated by the symbol f'_c. To determine compressive strength, tests are made on specimens of mortar or concrete; in the United States, unless otherwise specified, compression tests of mortar are made on 2-in. cubes, while compression tests of concrete are made on cylinders 6 in. in diameter and 12 in. high (see Fig. 1-6).

Compressive strength of concrete is a primary physical property and one frequently used in design calculations for bridges, buildings, and other structures. Most general-use concrete has a compressive strength between 3000 psi and 5000 psi. High-strength concrete has a compressive strength of at least 6000 psi. Compressive strengths of 20,000 psi have been used in building applications.

In designing pavements and other slabs on ground, the flexural strength of concrete is generally used.

Fig. 1-6. Testing a 6x12-in. concrete cylinder in compression. The load on the test cylinder is registered on the scale.

Compressive strength can be used as an index of flexural strength, once the empirical relationship between them has been established for the materials and the size of the member involved. The flexural strength or modulus of rupture of normal-weight concrete is often approximated as 7.5 to 10 times the square root of the compressive strength.

The tensile strength of concrete is about 8% to 12% of the compressive strength and is often estimated as 5* to 7.5 times the square root of the compressive strength.**

The torsional strength for concrete is related to the modulus of rupture and the dimensions of the concrete element.†

The shear strength of concrete is about 20% of the compressive strength. The correlation between compressive strength and flexural, tensile, torsional, and shear strength varies with concrete ingredients and environment.

Modulus of elasticity, denoted by the symbol E, may be defined as the ratio of normal stress to corresponding strain for tensile or compressive stresses below the proportional limit of a material. For normal-weight concrete, E ranges from 2 to 6 million psi and can be approximated as 57,000 times the square root of the compressive strength.††

The principal factors affecting strength are water-cement ratio and age, or the extent to which hydration has progressed. Fig. 1-7 shows compressive strengths for a range of water-cement ratios at different ages. Tests were made on 6-in.-diameter cylinders that were 12 in. in height. Note that strengths increase with age and increase as the water-cement ratios decrease. These factors also affect flexural and tensile strengths and bond of concrete to steel.

The age-compressive strength relationships in Fig. 1-7 are for typical air-entrained and non-air-entrained concretes. When more precise values for concrete are required, curves should be developed for the specific materials and mix proportions to be used on the job.

For a given workability and a given amount of cement, air-entrained concrete requires less mixing water than non-air-entrained concrete. The lower water-cement ratio possible for air-entrained concrete tends to offset the somewhat lower strengths of air-entrained concrete, particularly in lean-to-medium cement content mixes.

Unit Weight

Conventional concrete, normally used in pavements, buildings, and other structures, has a unit weight in the range of 140 to 150 lb per cubic foot (pcf). The unit weight (density) of concrete varies, depending on the amount and relative density of the aggregate, the amount of air that is entrapped or purposely entrained, and the water and cement contents, which in turn are

---

*Reference 1-11.
**ACI 207.2R estimates tensile strength as $6.7\sqrt{f'_c}$.
†Torsional strength correlations are presented in Reference 1-11.
††See Section 8.5 of ACI 318.

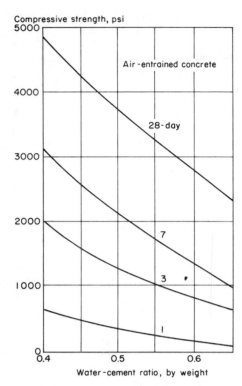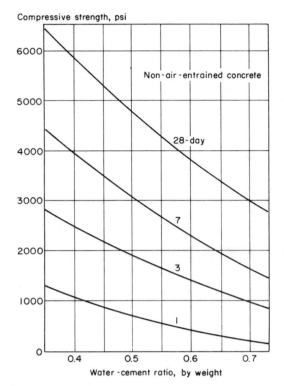

Fig. 1-7. Typical age-strength relationships of concrete based on compression tests of 6x12-in. cylinders, using Type I portland cement and moist-curing at 70°F.

influenced by the maximum-size aggregate. Values of the unit weight of fresh concrete are given in Table 1-1. In the design of reinforced concrete structures, the combination of conventional concrete and reinforcing bars is commonly assumed to weigh 150 pcf.

The weight of dry concrete equals the weight of freshly mixed concrete less the weight of evaporable water. Some of the mix water combines chemically with the cement during the hydration process, converting the cement into cement gel. Also, some of the water remains tightly held in pores and capillaries and does not evaporate under normal conditions. The amount of water that will evaporate in air at 50% relative humidity is about ½% to 3% of the concrete weight, depending on initial water content of the concrete, absorption characteristics of the aggregates, and size of the structure.

Aside from conventional concrete, there is a wide spectrum of other concretes to meet various needs,

ranging from lightweight insulating concretes with a unit weight of 15 pcf to heavyweight concrete with a unit weight of up to about 400 pcf used for counterweights or radiation shielding (see Chapter 15, "Special Types of Concrete").

Resistance to Freezing and Thawing

Concrete used in structures and pavements is expected to have long life and low maintenance. It must have good durability to resist anticipated exposure conditions. The most destructive weathering factor is freezing and thawing while the concrete is wet, particularly in the presence of deicing chemicals. Deterioration is caused by the freezing of the water in the paste, the aggregate particles, or both.

With air entrainment, concrete is highly resistant to this deterioration as shown in Fig. 1-8. During freezing,

Table 1-1. Observed Average Weight of Fresh Concrete*

Maximum size of aggregate, inches	Air content, percent	Water, pounds per cubic yard	Cement, pounds per cubic yard	Unit weight, pounds per cubic foot**				
				Specific gravity of aggregate†				
				2.55	2.60	2.65	2.70	2.75
¾	6.0	283	566	137	139	141	143	145
1½	4.5	245	490	141	143	146	148	150
3	3.5	204	408	144	147	149	152	154
6	3.0	164	282	147	149	152	154	157

*Source: Reference 1-15, Table 4.
**Air-entrained concrete with indicated air content.
†On saturated surface-dry basis.

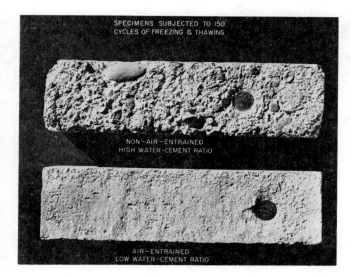

Fig. 1-8. Air-entrained concrete is highly resistant to repeated freeze-thaw cycles.

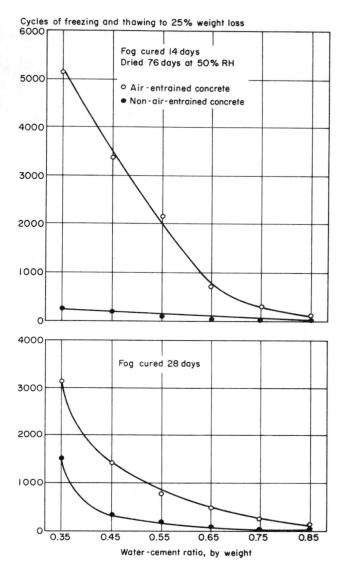

Fig. 1-9. Relationship between freeze-thaw resistance, water-cement ratio, and drying for air-entrained and non-air-entrained concretes made with Type I cement. High resistance to freezing and thawing is associated with entrained air, low water-cement ratio, and a drying period prior to freeze-thaw exposure. Reference 1-5.

the water displaced by ice formation in the paste is accommodated so that it is not disruptive; the air bubbles in the paste provide chambers for the water to enter and thus relieve the hydraulic pressure generated.

When freezing occurs in concrete containing saturated aggregate, disruptive hydraulic pressures can also be generated within the aggregate. Water displaced from the aggregate particles during the formation of ice cannot escape fast enough to the surrounding paste to relieve pressure. However, under nearly all exposure conditions, a paste of good quality (low water-cement ratio) will prevent most aggregate particles from becoming saturated. Also, if the paste is air-entrained, it will accommodate the small amounts of excess water that may be expelled from aggregates, thus protecting the concrete from freeze-thaw damage.

Fig. 1-9 illustrates, for a range of water-cement ratios, that (1) air-entrained concrete is much more resistant to freeze-thaw cycles than non-air-entrained concrete, (2) concrete with a low water-cement ratio is more durable than concrete with a high water-cement ratio, and (3) a drying period prior to freeze-thaw exposure substantially benefits the freeze-thaw resistance of air-entrained concrete but does not significantly benefit non-air-entrained concrete.* Air-entrained concrete with a low water-cement ratio and an air content of 4% to 8% will withstand a great number of cycles of freezing and thawing without distress.

Freeze-thaw durability can be determined by laboratory test procedure ASTM C 666, Standard Test Method for Resistance of Concrete to Rapid Freezing and Thawing. From the test, a durability factor is calculated that reflects the number of cycles of freezing and thawing required to produce a certain amount of deterioration. Deicer-scaling resistance can be determined by ASTM C 672, Standard Test Method for Scaling Resistance of Concrete Surfaces Exposed to Deicing Chemicals.

Permeability and Watertightness

Concrete used in water-retaining structures or exposed to weather or other severe exposure conditions must be virtually impermeable or watertight. Watertightness is often referred to as the ability of concrete to hold back or retain water without visible leakage. Permeability refers to the amount of water migration through concrete when the water is under pressure or to the ability of concrete to resist penetration of water or other substances (liquid, gas, ions, etc.). Generally,

---

*See References 1-5 and 1-6.

the same properties of concrete that make concrete less permeable also make it more watertight.

The overall permeability of concrete to water is a function of the permeability of the paste, the permeability and gradation of the aggregate, and the relative proportion of paste to aggregate. Decreased permeability improves concrete's resistance to resaturation, sulfate and other chemical attack, and chloride-ion penetration.

Permeability also affects the destructiveness of saturated freezing. Here the permeability of the paste is of particular importance because the paste envelops all constituents in the concrete. Paste permeability is related to water-cement ratio and the degree of cement hydration or length of moist curing. A low-permeability concrete requires a low water-cement ratio and an adequate moist-curing period. Air entrainment aids watertightness but has little effect on permeability. Permeability increases with drying.*

The permeability of mature hardened paste kept continuously moist ranges from 0.1×10^{-12} to 120×10^{-12} cm per sec. for water-cement ratios ranging from 0.3 to 0.7.* The permeability of rock commonly used as concrete aggregate varies from approximately 1.7×10^{-9} to 3.5×10^{-13} cm per sec. The permeability of mature, good-quality concrete is aproximately 1×10^{-10} cm per sec.

The relationship between permeability, water-cement ratio, and initial curing for 4x8-in. cylindrical concrete specimens tested after 90 days of air drying and subjected to 3000 psi of water pressure is illustrated in Fig. 1-10. The test apparatus is shown in Fig. 1-11. Although permeability values would be different for other liquids and gases, the relationship between water-cement ratio, curing period, and permeability would be similar.

Test results obtained by subjecting 1-in.-thick non-air-entrained mortar disks to 20-psi water pressure are given in Fig. 1-12. In these tests, there was no water leakage through mortar disks that had a water-cement

Fig. 1-11. Hydraulic permeability test apparatus used to obtain data illustrated in Fig. 1-10.

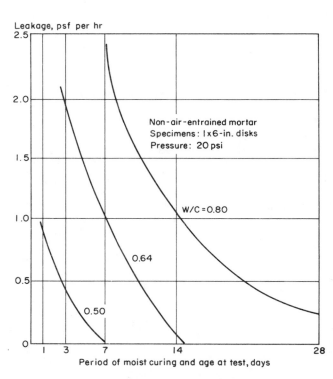

Fig. 1-12. Effect of water-cement ratio (w/c) and curing duration on permeability of mortar. Note that leakage is reduced as the water-cement ratio is decreased and the curing period increased. Reference 1-1 and PCA Major Series 227.

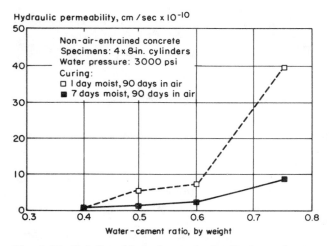

Fig. 1-10. Relationship between hydraulic (water) permeability, water-cement ratio, and initial curing on concrete specimens. Reference PCA HM1170.

ratio of 0.50 by weight or less and were moist-cured for seven days. Where leakage occurred, it was greater in mortar disks made with high water-cement ratios. Also, for each water-cement ratio, leakage was less as the length of the moist-curing period increased. In disks with a water-cement ratio of 0.80, the mortar still

*Reference 1-4.

permitted leakage after being moist-cured for one month. These results clearly show that a low water-cement ratio and a period of moist curing significantly reduce permeability.

A low water-cement ratio also reduces segregation and bleeding, further contributing to watertightness. To be watertight, concrete must also be free from cracks and honeycomb.

Occasionally, porous concrete—no-fines concrete that readily allows water to flow through—is designed for special applications. In these concretes, the fine aggregate is greatly reduced or completely removed producing a high volume of air voids. Porous concrete has been used in tennis courts, pavements, parking lots, greenhouses, and drainage structures. No-fines concrete has also been used in buildings because of its thermal insulation properties. Additional information on porous concrete is given in Chapter 15, "Special Types of Concrete."

Abrasion Resistance

Floors, pavements, and hydraulic structures are subjected to abrasion; therefore, in these applications concrete must have a high abrasion resistance. Test results indicate that abrasion resistance is closely related to the compressive strength of concrete. Strong concrete has more resistance to abrasion than does weak concrete. Since compressive strength depends on water-cement ratio and curing, a low water-cement ratio and adequate curing are necessary for abrasion resistance. The type of aggregate and surface finish or treatment used also have a strong influence on abrasion resistance. Hard aggregate is more abrasion resistant than soft aggregate and a steel-troweled surface resists abrasion more than a surface that is not troweled.

Fig. 1-13 shows results of abrasion tests on concretes of different compressive strengths and aggregate types.

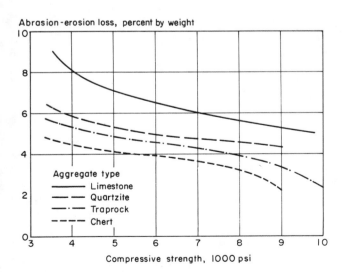

Fig. 1-13. Effect of compressive strength and aggregate type on the abrasion resistance of concrete. High-strength concrete made with a hard aggregate is highly resistant to abrasion. Reference 1-16.

Fig. 1-14 illustrates the effect hard steel troweling and surface treatments have on abrasion resistance. Abrasion tests can be conducted by rotating steel balls, dressing wheels, or disks under pressure over the surface (ASTM C 779). One type of test apparatus is pictured in Fig. 1-15. Other types of abrasion tests are also available (ASTM C 418 and C 944).

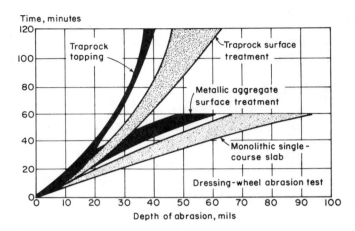

Fig. 1-14. Effect of hard steel troweling and surface treatments on the abrasion resistance of concrete. Base slab compressive strength was 6000 psi at 28 days. All slabs were steel troweled. Reference 1-12.

Fig. 1-15. Test apparatus for measuring abrasion resistance of concrete. The machine can be adjusted to use either revolving disks or dressing wheels. With a different machine, steel balls under pressure are rolled over the surface of the specimen. The tests are described in ASTM C 779, Standard Test Method for Abrasion Resistance of Horizontal Concrete Surfaces.

Volume Stability

Hardened concrete changes volume slightly due to changes in temperature, moisture, and stress. These volume or length changes may range from about 0.01% to 0.08%. Thermal volume changes of hardened concrete are about the same as those for steel.

Concrete kept continually moist will expand slightly. When permitted to dry, concrete will shrink. The primary factor influencing the amount of drying shrinkage is the water content of the freshly mixed concrete. Drying shrinkage increases directly with increases in this water content. The amount of shrinkage also depends upon several other factors, such as amounts of aggregate used, properties of the aggregate, size and shape of the concrete mass, relative humidity and temperature of the environment, method of curing, degree of hydration, and time. Cement content has little to no effect on shrinkage of concrete with cement contents between 5 and 8 bags per cu yd.

Concrete under stress will deform elastically. Sustained stress will result in additional deformation called creep. The rate of creep (deformation per unit of time) decreases with time.

The magnitude of volume changes and factors influencing them are discussed in Chapter 13, "Volume Changes of Concrete."

Control of Cracking

Two basic causes of cracks in concrete are (1) stress due to applied loads and (2) stress due to drying shrinkage or temperature changes in restrained conditions.

Drying shrinkage is an inherent, unavoidable property of concrete; therefore, properly positioned reinforcing steel is used to reduce crack widths, or joints (Fig. 1-16) are used to predetermine and control the location of cracks. Thermal stress due to fluctuations in temperature can cause cracking, particularly at an early age.

Concrete shrinkage cracks occur because of restraint. When shrinkage occurs and there is no restraint, the concrete does not crack. Restraint comes from several sources. Drying shrinkage is always greater near the surface of concrete; the moist inner portions restrain the concrete near the surface, which can cause cracking. Other sources of restraint are reinforcing steel embedded in concrete, the interconnected parts of a concrete structure, and the friction of the subgrade on which concrete is placed.

Joints are the most effective method of controlling unsightly cracking. If a sizable expanse of concrete (a wall, slab, or pavement) is not provided with properly spaced joints to accommodate drying shrinkage and temperature contraction, the concrete will crack in a random manner.*

Control joints are grooved, formed, or sawed into sidewalks, driveways, pavements, floors, and walls so that cracking will occur in these joints rather than in a

---

*Refer to Chapter 9 for more information.

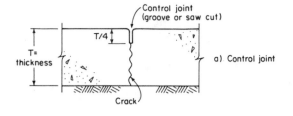

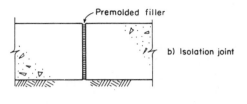

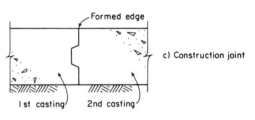

Fig. 1-16. The three basic types of joints used in concrete slab-on-ground construction.

random manner. Control joints permit movement in the plane of a slab or wall. They extend to a depth of approximately one-quarter the concrete thickness.

Isolation joints separate a slab from other parts of a structure and permit horizontal and vertical movements of the slab. They are placed at the junction of floors with walls, columns, footings, and other points where restraint can occur. They extend the full depth of the slab and include a premolded joint filler.

Construction joints occur where concrete work is concluded for the day; they separate areas of concrete placed at different times. In slabs-on-ground, construction joints usually align with and function as control or isolation joints.

REFERENCES

1-1. McMillan, F. R., and Lyse, Inge, "Some Permeability Studies of Concrete," *Journal of the American Concrete Institute, Proceedings,* vol. 26, December 1929, pages 101-142.

1-2. Powers, T. C., *The Bleeding of Portland Cement Paste, Mortar, and Concrete Treated As a Special Case of Sedimentation,* Research Department Bulletin RX002, Portland Cement Association, 1939.

1-3. Powers, T. C., and Brownyard, T. L., *Studies of the Physical Properties of Hardened Portland Cement Paste,* Research Department Bulletin RX022, Portland Cement Association, 1947.

1-4. Powers, T. C.; Copeland, L. E.; Hayes, J. C.; and Mann, H. M., *Permeability of Portland Cement Pastes,* Research Department Bulletin RX053, Portland Cement Association, 1954.

1-5. Backstrom, J. E.; Burrow, R. W.; and Witte, L. P., *Investigation into the Effect of Water-Cement Ratio on the Freezing-Thawing Resistance of Non-Air- and Air-Entrained Concrete.* Concrete Laboratory Report No. C-810, Engineering Laboratories Division, U.S. Department of the Interior, Bureau of Reclamation, Denver, November 1955.

1-6. Woods, Hubert, *Observations on the Resistance of Concrete to Freezing and Thawing,* Research Department Bulletin RX067, Portland Cement Association, 1956.

1-7. Brunauer, Stephen, and Copeland, L. E., "The Chemistry of Concrete," *Scientific American,* New York, April 1964.

1-8. Powers, T. C., *Topics in Concrete Technology,* Research Department Bulletin RX174, Portland Cement Association, 1964.

1-9. Abrams, M. S., and Orals, D. L., *Concrete Drying Methods and Their Effect on Fire Resistance,* Research Department Bulletin RX181, Portland Cement Association, 1965.

1-10. Powers, T. C., *The Nature of Concrete,* Research Department Bulletin RX196, Portland Cement Association, 1966.

1-11. Hsu, Thomas T. C., *Torsion of Structural Concrete—Plain Concrete Rectangular Sections,* Development Department Bulletin DX134, Portland Cement Association, 1968.

1-12. Brinkerhoff, C. H., "Report to ASTM C-9 Subcommittee III-M (Testing Concrete for Abrasion) Cooperative Abrasion Test Program," University of California and Portland Cement Association, 1970.

1-13. McMillam, F. R., and Tuthill, Lewis H., *Concrete Primer,* SP-1, 3rd ed., American Concrete Institute, 1973.

1-14. Kirk, Raymond E., and Othmer, Donald F., eds., "Cement," *Encyclopedia of Chemical Technology,* 3rd ed., vol. 5, John Wiley and Sons, Inc., New York, 1979, pages 163-193.

1-15. *Concrete Manual,* 8th ed., U.S. Bureau of Reclamation, Denver, revised 1981.

1-16. Liu, Tony C., "Abrasion Resistance of Concrete," *Journal of the American Concrete Institute,* September-October 1981, pages 341-350.

1-17. *ACI Manual of Concrete Practice,* American Concrete Institute, 1987.

CHAPTER 2
Portland Cements

Portland cements are hydraulic cements composed primarily of hydraulic calcium silicates. Hydraulic cements set and harden by reacting chemically with water. During this reaction, called hydration, cement combines with water to form a stonelike mass. When the paste (cement and water) is added to aggregates (sand and gravel, crushed stone, or other granular material) it acts as an adhesive and binds the aggregates together to form concrete, the world's most versatile and most used construction material.

Hydration begins as soon as cement comes in contact with water. Each cement particle forms a type of growth on its surface that gradually spreads until it links up with the growth from other cement particles or adheres to adjacent substances. This building up results in progressive stiffening, hardening, and strength development. The stiffening of concrete can be recognized by a loss of workability that usually occurs within three hours of mixing, but is dependent upon the composition and fineness of the cement, the mixture proportions, and the temperature conditions. Subsequently, the concrete sets and becomes hard.

Hydration continues as long as space for hydration products is available and moisture and temperature conditions are favorable (curing). As hydration continues, concrete becomes harder and stronger. Most of the hydration and strength development take place within the first month of concrete's life cycle, but they continue, though more slowly, for a long time; strength increases over a 50-year period have been recorded in laboratory investigations.

The invention of portland cement is generally credited to Joseph Aspdin, an English mason. In 1824, he obtained a patent for his product, which he named portland cement because it produced a concrete that resembled the color of the natural limestone quarried on the Isle of Portland, a peninsula in the English Channel. The name has endured and is used throughout the world, with many manufacturers adding their own trade or brand names. Although Aspdin was the first to prescribe a formula for portland cement and the first to have his product patented, calcareous cements had been used for many centuries. Natural cements were manufactured in Rosendale, New York, in the middle 1800's. The first recorded shipment of portland cement to the United States was in 1868 and the first portland cement made in the United States was produced at a plant in Coplay, Pa., in 1871.

MANUFACTURE OF PORTLAND CEMENT

Portland cement is produced by pulverizing clinker consisting essentially of hydraulic calcium silicates along with some calcium aluminates and calcium aluminoferrites and usually containing one or more forms of calcium sulfate (gypsum) as an interground addition.

Materials used in the manufacture of portland cement must contain appropriate proportions of calcium oxide, silica, alumina, and iron oxide components. During manufacture, analyses of all materials are made frequently to ensure a uniformly high quality cement.

Steps in the manufacture of cement are illustrated in the flow charts in Figs. 2-2 and 2-3. While the opera-

Fig. 2-1. Rotary kiln of a modern mill for manufacturing portland cement.

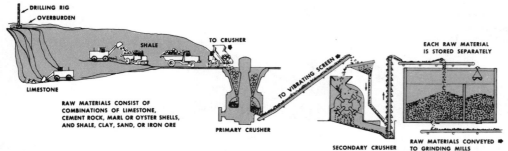

1. Stone is first reduced to 5-in. size, then to ¾ in., and stored.

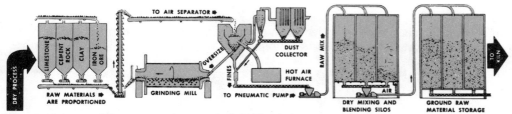

2. Raw materials are ground to powder and blended.

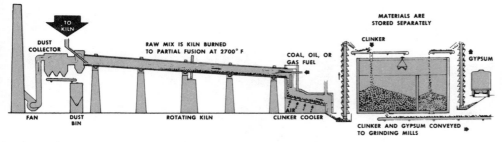

2. Raw materials are ground, mixed with water to form slurry, and blended.

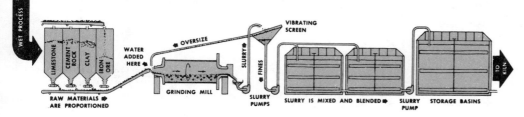

3. Burning changes raw mix chemically into cement clinker.

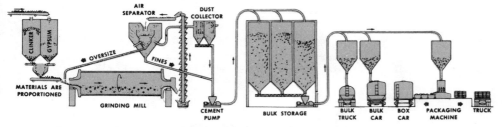

4. Clinker with gypsum is ground into portland cement and shipped.

Fig. 2-2. Steps in the manufacture of portland cement.

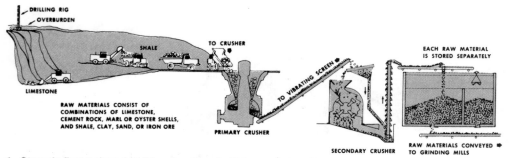

1. **Stone is first reduced to 5-in. size, then to ¾ in., and stored.**

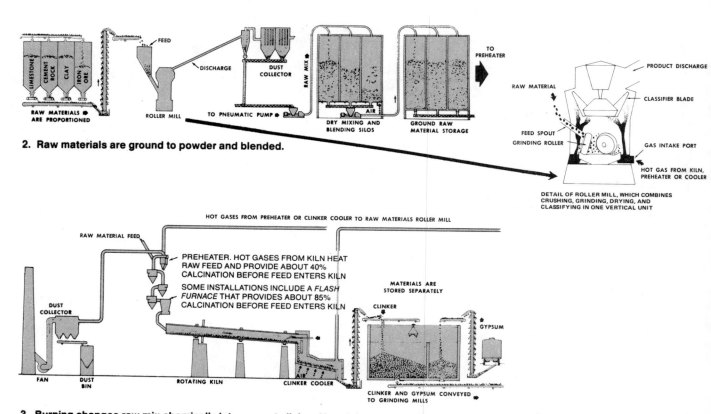

2. **Raw materials are ground to powder and blended.**

3. **Burning changes raw mix chemically into cement clinker. Note four-stage preheater, flash furnaces, and shorter kiln.**

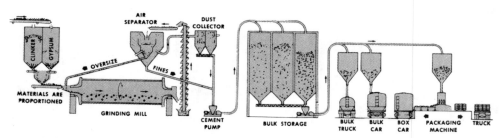

4. **Clinker with gypsum is ground into portland cement and shipped.**

Fig. 2-3. New technology in dry-process cement manufacture.

Table 2-1. Sources of Raw Materials Used in Manufacture of Portland Cement

Lime, CaO	Iron, Fe_2O_3	Silica, SiO_2	Alumina, Al_2O_3	Gypsum, $CaSO_4 \cdot 2H_2O$	Magnesia, MgO
Alkali waste	Blast-furnace flue dust	Calcium silicate	Aluminum-ore refuse*	Anhydrite	Cement rock
Aragonite*	Clay*	Cement rock	Bauxite	Calcium sulfate	Limestone
Calcite*	Iron ore*	Clay*	Cement rock	Gypsum	Slag
Cement-kiln dust	Mill scale*	Fly ash	Clay*		
Cement rock	Ore washings	Fuller's earth	Copper slag		
Chalk	Pyrite cinders	Limestone	Fly ash*		
Clay	Shale	Loess	Fuller's earth		
Fuller's earth		Marl*	Granodiorite		
Limestone*		Ore washings	Limestone		
Marble		Quartzite	Loess		
Marl*		Rice-hull ash	Ore washings		
Seashells		Sand*	Shale*		
Shale*		Sandstone	Slag		
Slag		Shale*	Staurolite		
		Slag			
		Traprock			

Note: As a generalization, probably 50% of all industrial byproducts have potential as raw materials for the manufacture of portland cement.

*Most common sources.

tions of all cement plants are basically the same, no flow diagram can adequately illustrate all plants. There is no typical portland cement manufacturing plant; every plant has significant differences in layout, equipment, or general appearance.*

Selected raw materials (Table 2-1) are crushed, milled, and proportioned in such a way that the resulting mixture has the desired chemical composition. The raw materials are generally a mixture of calcareous (calcium oxide) material, such as limestone, chalk or shells, and an argillaceous (silica and alumina) material such as clay, shale, or blast-furnace slag. Either a dry or a wet process is used. In the dry process, grinding and blending are done with dry materials. In the wet process, the grinding and blending operations are done with the materials in slurry form. In other respects, the dry and wet processes are very much alike. Fig. 2-3 illustrates important technological developments that can improve significantly the productivity and energy efficiency of dry-process plants.

After blending, the ground raw material is fed into the upper end of a kiln. The raw mix passes through the kiln at a rate controlled by the slope and rotational speed of the kiln. Burning fuel (powdered coal, oil, or gas) is forced into the lower end of the kiln where temperatures of 2600°F to 3000°F change the raw material chemically into cement clinker, grayish-black pellets predominantly the size of $1/2$-in.-diameter marbles.

The clinker is cooled and then pulverized. During this operation a small amount of gypsum is added to regulate the setting time of the cement. The clinker is ground so fine that nearly all of it passes through a No. 200 mesh (75 micron) sieve with 40,000 openings per square inch. This extremely fine gray powder is portland cement.

TYPES OF PORTLAND CEMENT

Different types of portland cement are manufactured to meet various normal physical and chemical require-

ments for specific purposes. The American Society for Testing and Materials (ASTM) Designation C 150, Standard Specification for Portland Cement, provides for eight types of portland cement as follows:

Type I	normal
Type IA	normal, air-entraining
Type II	moderate sulfate resistance
Type IIA	moderate sulfate resistance, air-entraining
Type III	high early strength
Type IIIA	high early strength, air-entraining
Type IV	low heat of hydration
Type V	high sulfate resistance

Type I

Type I portland cement is a general-purpose cement suitable for all uses where the special properties of other types are not required. It is used in concrete that is not subject to aggressive exposures, such as sulfate attack from soil or water, or to an objectionable temperature rise due to heat generated by hydration. Its uses in concrete include pavements, floors, reinforced concrete buildings, bridges, railway structures, tanks and reservoirs, pipe, masonry units, and other precast concrete products.

Type II

Type II portland cement is used where precaution against moderate sulfate attack is important, as in drainage structures where sulfate concentrations in groundwaters are higher than normal but not unusually severe (see Table 2-2). Type II cement will usually generate less heat at a slower rate than Type I. The requirement of moderate heat of hydration can be

---

*Mechanical equipment is described in Reference 2-16.

Table 2-2. Types of Cement Required for Concrete Exposed to Sulfate Attack

Sulfate exposure	Water-soluble sulfate (SO₄) in soil, percent by weight	Sulfate (SO₄) in water, ppm	Cement type
Negligible	0.00-0.10	0-150	—
Moderate*	0.10-0.20	150-1500	II, IP(MS), IS(MS), P(MS), I(PM)(MS), I(SM)(MS)
Severe	0.20-2.00	1500-10,000	V
Very severe	Over 2.00	Over 10,000	V plus pozzolan**

*Seawater.
**Pozzolan that has been determined by test or service record to improve sulfate resistance when used in concrete containing Type V cement.
Source: Adapted from Reference 2-20 and ACI 318, Table 4.5.3.

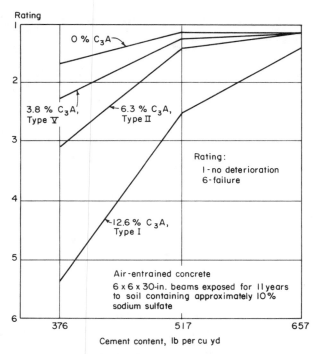

Fig. 2-4. Performance of concretes made with cements with different C_3A contents in sulfate soil. See Fig. 2-5 for the rating description. Reference 2-24.

Fig. 2-5. Range of durability represented, from left to right, by visual ratings of about 1, 4, and 6 for the sulfate-resistance tests in Fig. 2-4. Reference 2-24.

specified at the option of the purchaser. If heat-of-hydration maximums are specified, this cement can be used in structures of considerable mass, such as large piers, and heavy abutments and retaining walls. Its use will reduce temperature rise, which is especially important when concrete is placed in warm weather.

Type III

Type III portland cement provides high strengths at an early period, usually a week or less. It is chemically and physically similar to Type I cement, except that its particles have been ground finer. It is used when forms need to be removed as soon as possible or when the structure must be put into service quickly. In cold weather its use permits a reduction in the controlled curing period. Although richer mixes of Type I cement can be used to gain high early strength, Type III may provide it more satisfactorily and more economically.

Type IV

Type IV portland cement is used where the rate and amount of heat generated from hydration must be minimized. It develops strength at a slower rate than other cement types. Type IV cement is intended for use in massive concrete structures, such as large gravity dams, where the temperature rise resulting from heat generated during hardening must be minimized.

Type V

Type V portland cement is used only in concrete exposed to severe sulfate action—principally where soils or groundwaters have a high sulfate content. It gains strength more slowly than Type I cement. Table 2-2 describes sulfate concentrations requiring the use of Type V cement. The high sulfate resistance of Type V cement is attributed to a low tricalcium aluminate (C_3A) content as illustrated in Fig. 2-4. Sulfate resistance also increases with air entrainment and increasing cement contents (low water-cement ratios). Type V

cement, like other portland cements, is not resistant to acids and other highly corrosive substances.

Air-Entraining Portland Cements

Specifications for three types of air-entraining portland cement (Types IA, IIA, and IIIA) are given in ASTM C150. They correspond in composition to ASTM Types

I, II, and III, respectively, except that small quantities of air-entraining material are interground with the clinker during manufacture. These cements produce concrete with improved resistance to freeze-thaw action and to scaling caused by chemicals applied for snow and ice removal. Such concrete contains minute, well-distributed, and completely separated air bubbles.*

White Portland Cement

White portland cement is a true portland cement that differs from gray cement chiefly in color. It is made to conform to the specifications of ASTM C150, usually Type I or Type III, but the manufacturing process is controlled so that the finished product will be white. White portland cement is made of selected raw materials containing negligible amounts of iron and magnesium oxides–the substances that give cement its gray color. White portland cement is used primarily for architectural purposes such as precast curtain walls and facing panels, terrazzo surfaces, stucco, cement paint, tile grout, and decorative concrete. Its use is recommended wherever white or colored concrete or mortar is desired.

BLENDED HYDRAULIC CEMENTS

Recent concern with energy conservation has prompted the use of by-product materials in portland cement concrete. Blended hydraulic cements are produced by intimately and uniformly blending two or more types of fine materials. The primary blending materials are portland cement, ground granulated blast-furnace slag, fly ash and other pozzolans, hydrated lime, and pre-blended cement combinations of these materials. Cement kiln dust, silica fume, and other materials are undergoing research for use in blended cements. Blended hydraulic cements must conform to the requirements of ASTM C595, which recognizes five classes of blended cements as follows:

Portland blast-furnace slag cement—Type IS
Portland-pozzolan cement—Type IP and Type P
Pozzolan-modified portland cement—Type I(PM)
Slag cement—Type S
Slag-modified portland cement—Type I(SM)

Type IS

Portland blast-furnace slag cement, Type IS, may be used in general concrete construction. In producing these cements, granulated blast-furnace slag of selected quality is either interground with portland cement clinker, separately ground and blended with portland cement, or produced with a combination of intergrinding and blending. The blast-furnace slag content of this cement is between 25% and 70% by weight. Air-entrainment, moderate sulfate resistance, or moderate heat of hydration may be specified by adding the suffixes A, MS, or MH. For example, an air-entraining

portland blast-furnace slag cement that has moderate sulfate resistance would be designated as Type IS-A(MS).**

Type IP and Type P

Portland-pozzolan cements are designated as Type IP or Type P. Type IP may be used for general construction and Type P is used in construction where high early strengths are not required. These cements are manufactured by intergrinding portland cement clinker with a suitable pozzolan, by blending portland cement or portland blast-furnace slag cement and a pozzolan, or by a combination of intergrinding and blending. The pozzolan content of these cements is between 15% and 40% by weight. Laboratory tests† indicate that performance of concrete made with Type IP cement as a group is similar to that of Type I cement concrete, however strengths through 28 days can be slightly lower for the Type IP than the Type I cement. Type IP may be designated as air-entraining, moderate sulfate resistant, or with moderate heat of hydration by adding the suffixes A, MS, or MH. Type P may be designated as low heat of hydration (LH), moderate sulfate resistant (MS), or air entraining (A).

Type I(PM)

Pozzolan-modified portland cement, Type I(PM), is used in general concrete construction. The cement is manufactured by combining portland cement or portland blast-furnace slag cement and a fine pozzolan. This may be accomplished by either (1) blending portland cement with a pozzolan, (2) blending portland blast-furnace slag cement with a pozzolan, (3) intergrinding portland cement clinker and a pozzolan, or (4) a combination of intergrinding and blending. The pozzolan content is less than 15% by weight of the finished cement. Air-entrainment, moderate sulfate resistance, or moderate heat of hydration may be designated in any combination by adding the suffixes A, MS, or MH. An example of an air-entraining, moderate-heat-of-hydration Type I(PM) cement would be Type I(PM)-A(MH).

Type S

Slag cement, Type S, is used with portland cement in making concrete or with lime in making mortar, but is not used alone in structural concrete. Slag cement is manufactured by either (1) blending ground granulated blast-furnace slag and portland cement, (2) blending ground granulated blast-furnace slag and hydrated lime, or (3) a combination of blending ground granulated blast-furnace slag, portland cement, and hydrated lime. The minimum slag content is 70% of the weight of the slag cement. Air-entrainment may be designated in a

---

*See Chapter 5, "Air-Entrained Concrete."
**See References 2-11 and 2-27.
†See Reference 2-15.

17

slag cement by adding the suffix A, for example, Type S-A.

Type I(SM)

Slag-modified portland cement, Type I(SM), is used for general concrete construction. This cement is manufactured by either (1) intergrinding portland cement clinker and granulated blast-furnace slag, (2) blending portland cement and finely ground granulated blast-furnace slag, or (3) a combination of intergrinding and blending. Slag is less than 25% of the weight of the finished cement. Type I(SM) may be designated with air-entrainment, moderate sulfate resistance, or moderate heat of hydration by adding the suffixes A, MS, or MH. An example would be Type I(SM)-A(MH) for an air-entraining slag-modified portland cement with moderate heat of hydration.

Blended cements may be used in concrete construction when specific properties of other types of cements are not required. Several of the blended cements have a lower early strength gain as compared to Type I cement. Therefore, if a blended cement is diluted by the addition of more pozzolans or slags, the resulting concrete should be carefully tested for changes in strength, durability, shrinkage, permeability, and other properties. Cold placement and curing temperatures may significantly decrease strength gain and increase the time of set in high slag or pozzolan content blended cement concretes.

MASONRY CEMENTS

Masonry cements are hydraulic cements designed for use in mortar for masonry construction. They are composed of one or more of the following: portland cement, portland-pozzolan cement, portland blast-furnace slag cement, slag cement, hydraulic lime, and natural cement and, in addition, usually contain materials such as hydrated lime, limestone, chalk, calcareous shell, talc, slag, or clay. Materials are selected for their ability to impart workability, plasticity, and water retention to masonry mortars. Masonry cements meet the requirements of ASTM C 91, which classifies masonry cements as Type N, Type S, and Type M. A brief description of each type follows:

Type N masonry cement is used in ASTM C 270 Type N and Type O mortars. It may also be used with portland or blended cements to produce Type S and Type M mortars.

Type S masonry cement is used in ASTM C 270 Type S mortar. It may also be used with portland or blended cements to produce Type M mortar.

Type M masonry cement is used in ASTM C 270 Type M mortar without the addition of other cements or hydrated lime.

The workability, strength, and color of masonry cements stay at a uniform level because of manufacturing controls. In addition to mortar for masonry construction, masonry cements are used for parging and plaster (stucco); they must never be used for making concrete.

EXPANSIVE CEMENTS

Expansive cement is a hydraulic cement that expands slightly during the early hardening period after setting. It must meet the requirements of ASTM C 845 in which it is designated as Type E-1. Currently, three varieties of expansive cement are recognized and have been designated as K, M, and S, which are added as a suffix to the type. Type E-1(K) contains portland cement, anhydrous tetracalcium trialuminosulfate, calcium sulfate, and uncombined calcium oxide (lime). Type E-1(M) contains portland cement, calcium aluminate cement, and calcium sulfate. Type E-1(S) contains portland cement with a high tricalcium aluminate content and calcium sulfate.

Expansive cement may also be made of formulations other than those mentioned. The expansive properties of each type can be varied over a considerable range. Type I cement may be transformed into expansive cement by the addition of an expansive admixture at the ready mix plant.

When expansion is restrained, for example by reinforcement, expansive cement concrete (shrinkage-compensating concrete) can be used to (1) compensate for the volume decrease due to drying shrinkage, (2) induce tensile stress in reinforcement (post-tensioning), and (3) stabilize the long-term dimensions of post-tensioned concrete structures with respect to original design. One of the major advantages of using expansive cement in concrete is in the control and reduction of drying shrinkage cracks. Fig. 2-6 illustrates the length change (early expansion and drying shrinkage) history of shrinkage-compensating concrete and conventional portland cement concrete.*

---

*References 2-19 and 2-21.

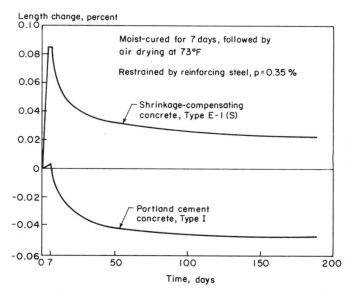

Fig. 2-6. Length-change history of shrinkage-compensating concrete containing Type E-1(S) cement and Type I portland cement concrete. Reference 2-14.

SPECIAL CEMENTS

There are special types of cement not necessarily covered by ASTM specifications, some of which contain portland cement. A few are discussed below.

Oil-Well Cements

Oil-well cements, used for sealing oil wells, are usually made from portland cement clinker or from blended hydraulic cements. Generally they must be slow-setting and resistant to high temperatures and pressures. The American Petroleum Institute Specifications for Materials and Testing for Well Cements (API Specification 10) includes requirements for nine classes of well cements (classes A through H and J). Each class is applicable for use at a certain range of well depths, temperatures, pressures, and sulfate environments. The petroleum industry also uses conventional types of portland cement with suitable cement-modifying admixtures. Expansive cements have also performed adequately as well cements.

Waterproofed Portland Cements

Waterproofed portland cement is usually made by adding a small amount of water-repellent additive such as stearate (sodium, aluminum, or other) to portland cement clinker during its final grinding. Manufactured in either white or gray color, it reduces capillary water transmission under little to no pressure but does not stop water-vapor transmission.*

Plastic Cements

Plastic cements are made by adding plasticizing agents, up to 12% by total volume, to Type I or Type II portland cement during the milling operation. Plastic cements are used for making plaster and stucco.

Regulated-Set Cements

Regulated-Set cement** is a hydraulic cement that can be formulated and controlled to produce concrete with setting times from a few minutes to an hour and with corresponding rapid early strength development of 1000 psi or more within an hour after setting. It is a modified portland cement that can be manufactured in the same kiln used to manufacture conventional portland cement. Regulated-Set cement incorporates set control (calcium sulfate) and early-strength-development components. Final physical properties of the resulting concrete are in most respects similar to comparable concretes made with portland cement.

Cements with Functional Additions

Functional additions that may be interground with cement clinker are a combination of water-reducing, retarding, and accelerating additions. These additions including set control must make a cement that meets the requirements of ASTM C 688.

AVAILABILITY OF CEMENTS

Some types of cement may not be readily available in all sections of the United States. Before specifying a type of portland cement, its availability should be determined.

Type I portland cement is usually carried in stock and is furnished when a type of cement is not specified. Type II cement is usually available, especially in parts of the United States where moderate sulfate resistance is needed. Cement Types I and II represent about 90% of the cement shipped from U.S. plants. Some cements are designated as both Type I and II, i.e., they meet specification requirements for both types. Type III cement and white cement are usually available in larger cities and represent about 4% of cement produced. Type IV cement is manufactured only when specified for particular projects (massive structures) and therefore is usually not readily available. Type V cement is available only in particular parts of the United States where it is needed to resist high sulfate environments. Air-entraining cements are sometimes difficult to obtain. Their use has been decreasing as the popularity of air-entraining admixtures has increased. Masonry cement is available in most areas.†

If a given type is not available, comparable results frequently can be obtained with one of the available types. For example, high-early-strength concrete can be made by using richer mixes (higher cement contents). Also, the effects of heat of hydration can be minimized by using lean mixes, smaller placing lifts, artificial cooling, or by adding a pozzolan to the concrete. Type V portland cement may at times be difficult to obtain, but due to the wide range of compounds permitted in specifications, a Type II cement may meet the requirements of a Type V.

Blended cements represent about 1% of the cement shipped in the United States and may not be available in many areas. When blended cements are required but are not available, similar properties may be obtained by adding pozzolans or finely ground granulated blast-furnace slag to the concrete at a ready mix plant using normal portland cement. These mixes should be thoroughly tested for time of set, strength gain, durability, and other properties prior to use in construction.

CHEMICAL COMPOUNDS IN PORTLAND CEMENT

During the burning operation in the manufacture of portland cement clinker, calcium oxide combines with the acidic components of the raw mix to form four

---

*A more complete discussion appears in Reference 2-13.
**U.S. Patent 3,628,973.
†Reference 2-25.

principal compounds that make up 90% of cement by weight. Gypsum and other materials are also present. Following are the primary compounds, their chemical formulas, and abbreviations:

Tricalcium silicate	$3CaO \cdot SiO_2$	$= C_3S$
Dicalcium silicate	$2CaO \cdot SiO_2$	$= C_2S$
Tricalcium aluminate	$3CaO \cdot Al_2O_3$	$= C_3A$
Tetracalcium aluminoferrite	$4CaO \cdot Al_2O_3 \cdot Fe_2O_3$	$= C_4AF$

In the presence of water, the four compounds hydrate to form new ones that are the infrastructure of hardened cement paste in concrete. The calcium silicates, C_3S and C_2S, which constitute about 75% of the weight of cement, hydrate to form the compounds calcium hydroxide and calcium silicate hydrate (tobermorite gel). Hydrated cement contains about 25% calcium hydroxide and 50% tobermorite gel by weight. The strength and other properties of hydrated cement are due primarily to tobermorite gel. C_3A reacts with water and calcium hydroxide to form tetracalcium aluminate hydrate. C_4AF reacts with water to form calcium aluminoferrite hydrate. C_3A, gypsum, and water may combine to form calcium sulfoaluminate hydrate. These basic compound transformations are shown in Table 2-3.*

C_3S and C_2S in clinker and cement are also referred to as alite and belite, respectively. These and other compounds may be observed and analyzed through the use of microscopical techniques (see Figs. 2-7 and 2-8).

The approximate percentage of each compound can be calculated from a chemical analysis of the cement. X-ray diffraction techniques may be used to more accurately determine compound percentages. Table 2-4 shows typical compound composition and fineness for each of the principal types of portland cement. Present knowledge of cement chemistry indicates that these compounds have the following properties:

---

*Reference 2-10. (Also see "Hydration, Setting Time, Hardening" in Chapter 1.)

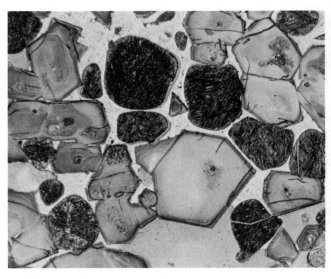

Fig. 2-7. Polished thin-section examination shows alite (C_3S) as light, angular gray crystals. The darker, rounded crystals are belite (C_2S). Magnification 400X.

Fig. 2-8. Scanning electron microscope micrograph of alite (C_3S) crystals. Magnification 3000X.

Table 2-3. Portland Cement Compound Transformations

$2(3CaO \cdot SiO_2)$ (Tricalcium silicate)	$+$ $6H_2O$ (Water)	$=$ $3CaO \cdot 2SiO_2 \cdot 3H_2O$ (Tobermorite gel)	$+$ $3Ca(OH)_2$ (Calcium hydroxide)	
$2(2CaO \cdot SiO_2)$ (Dicalcium silicate)	$+$ $4H_2O$ (Water)	$=$ $3CaO \cdot 2SiO_2 \cdot 3H_2O$ (Tobermorite gel)	$+$ $Ca(OH)_2$ (Calcium hydroxide)	
$3CaO \cdot Al_2O_3$ (Tricalcium aluminate)	$+$ $12H_2O$ (Water)	$+$ $Ca(OH)_2$ (Calcium hydroxide)	$=$ $3CaO \cdot Al_2O_3 \cdot Ca(OH)_2 \cdot 12H_2O$ (Tetracalcium aluminate hydrate)	
$4CaO \cdot Al_2O_3 \cdot Fe_2O_3$ (Tetracalcium aluminoferrite)	$+$ $10H_2O$ (Water)	$+$ $2Ca(OH)_2$ (Calcium hydroxide)	$=$ $6CaO \cdot Al_2O_3 \cdot Fe_2O_3 \cdot 12H_2O$ (Calcium aluminoferrite hydrate)	
$3CaO \cdot Al_2O_3$ (Tricalcium aluminate)	$+$ $10H_2O$ (Water)	$+$ $CaSO_4 \cdot 2H_2O$ (Gypsum)	$=$ $3CaO \cdot Al_2O_3 \cdot CaSO_4 \cdot 12H_2O$ (Calcium monosulfoaluminate hydrate)*	

*Ettringite may also develop.
Note: Table 2-3 illustrates only primary transformations and not the several minor transformations.

Table 2-4. Chemical and Compound Composition and Fineness of Some Typical Cements

Type of portland cement	Chemical composition, %						Loss on ignition, %	Insoluble residue, %	Potential compound composition, %*				Blaine fineness, m²/kg
	SiO_2	Al_2O_3	Fe_2O_3	CaO	MgO	SO_3			C_3S	C_2S	C_3A	C_4AF	
Type I	20.9	5.2	2.3	64.4	2.8	2.9	1.0	0.2	55	19	10	7	370
Type II	21.7	4.7	3.6	63.6	2.9	2.4	0.8	0.4	51	24	6	11	370
Type III	21.3	5.1	2.3	64.9	3.0	3.1	0.8	0.2	56	19	10	7	540
Type IV	24.3	4.3	4.1	62.3	1.8	1.9	0.9	0.2	28	49	4	12	380
Type V	25.0	3.4	2.8	64.4	1.9	1.6	0.9	0.2	38	43	4	9	380
White	24.5	5.9	0.6	65.0	1.1	1.8	0.9	0.2	33	46	14	2	490

*"Potential Compound Composition" refers to the maximum compound composition allowable by ASTM C 150 calculations using the chemical composition of the cement. The actual compound composition may be less due to incomplete or altered chemical reactions.
Reference 2-18.

Tricalcium silicate, C_3S, hydrates and hardens rapidly and is largely responsible for initial set and early strength. In general, the early strength of portland cement concrete is higher with increased percentages of C_3S.

Dicalcium silicate, C_2S, hydrates and hardens slowly and contributes largely to strength increase at ages beyond one week.

Tricalcium aluminate, C_3A, liberates a large amount of heat during the first few days of hydration and hardening. It also contributes slightly to early strength development. Gypsum, which is added to cement during final grinding, slows down the hydration rate of C_3A. Without gypsum, a cement with C_3A present would set rapidly. Cements with low percentages of C_3A are especially resistant to soils and waters containing sulfates.

Tetracalcium aluminoferrite, C_4AF, reduces the clinkering temperature, thereby assisting in the manufacture of cement. It hydrates rather rapidly but contributes very little to strength. Most color effects are due to C_4AF and its hydrates.

PROPERTIES OF PORTLAND CEMENT

Most specifications for portland cement place limits on both its chemical composition and physical properties. An understanding of the significance of some of the physical properties is helpful in interpreting results of cement tests. In general, tests of the physical properties of the cements should be used merely to evaluate the properties of the cement rather than the concrete. ASTM C 150 and C 595 limit the properties with respect to the type of cement. Cement should be sampled in accordance with ASTM C 183.

Fineness

Fineness of cement affects heat released and the rate of hydration. Greater cement fineness increases the rate at which cement hydrates and thus accelerates strength development. The effects of greater fineness on strength are manifested principally during the first seven days. Fineness is measured by the Wagner turbidimeter test (ASTM C 115), the Blaine air-permeability test (ASTM C 204), or the No. 325 (45μm) sieve (ASTM C 430). Approximately 85% to 95% of cement particles are smaller than 45 microns. Blaine fineness data are given in Table 2-4. Refer to the hydration discussion in Chapter 1 for more information.

Soundness

Soundness refers to the ability of a hardened paste to retain its volume after setting. Lack of soundness or delayed destructive expansion is caused by excessive amounts of hard-burned free lime or magnesia. Most specifications for portland cement limit the magnesia (periclase) content and the autoclave expansion. Since the adoption of the autoclave-expansion test (ASTM C 151) in 1943, there have been exceedingly few cases of abnormal expansion attributed to unsound cement.

Consistency

Consistency refers to the relative mobility of a freshly mixed cement paste or mortar or to its ability to flow. During cement testing, pastes are mixed to normal consistency as defined by a penetration of 10 ± 1 mm of the Vicat plunger, while mortars are mixed to obtain either a fixed water-cement ratio or to yield a flow within a prescribed range. The flow is determined on a flow table as described in ASTM C 230. Both the normal consistency method and the flow test are used to regulate water contents of pastes and mortars, respectively, to be used in subsequent tests; both allow comparing dissimilar ingredients with the same penetrability or flow.

Setting Time

To determine if a cement sets according to the time limits specified in ASTM C 150, tests are performed using either the Vicat apparatus (ASTM C 191) or a Gillmore needle (ASTM C 266). Initial set of cement paste must not occur too early; final set must not occur too late. The setting times indicate that the paste is or is not undergoing normal hydration reactions. Gypsum in the cement regulates setting time. Setting time is also affected by cement fineness, water-cement ratio, and admixtures. Setting times of concretes do not correlate directly with setting times of pastes because of water loss to air or substrate and because of temperature differences in the field as contrasted with the controlled temperature in the testing lab.

False Set

False set (ASTM C 451, Paste Method, and ASTM C 359, Mortar Method) is evidenced by a significant loss of plasticity without the evolution of much heat shortly after mixing. From a placing and handling standpoint, false-set tendencies in portland cement will cause no difficulty if the concrete is mixed for a longer time than usual or if it is remixed without additional water before it is transported or placed.

Compressive Strength

Compressive strength as specified by ASTM C 150 is that obtained from tests of standard 2-in. mortar cubes tested in accordance with ASTM C 109 (Fig. 2-9). These cubes are made and cured in a prescribed manner using a standard sand.

As indicated in Table 2-5, compressive strength is influenced by the cement type, or more precisely, the compound composition and fineness of the cement.

ASTM C 150 sets only a minimum strength requirement that is exceeded comfortably by most manufacturers. Therefore, it should not be assumed that two types of portland cement meeting the same minimum requirements will produce the same strength of mortar or concrete without modification of mix proportions.

In general, cement strengths (based on mortar-cube tests) cannot be used to predict concrete strengths with a great degree of accuracy because of the many variables in aggregate characteristics, concrete mixtures, and construction procedures. Fig. 2-10 illustrates the strength development for concrete made with various types of cement. The strength uniformity of a cement from a single source may be determined by following the procedures outlined in ASTM C 917.

Heat of Hydration

Heat of hydration is the heat generated when cement and water react. The amount of heat generated is dependent chiefly upon the chemical composition of the cement, with C_3A and C_3S being the compounds primarily responsible for high heat evolution. The water-cement ratio, fineness of the cement, and temperature of curing also are factors. An increase in the water-cement ratio, fineness, and curing temperature increases the heat of hydration.*

---
*Reference 2-7.

Table 2-5. Compressive Strength Requirements for Mortars Made with Various Types of Cement

Type of cement	Minimum compressive strength, psi				ASTM designation
	1 day	3 days	7 days	28 days	
Portland cements					C 150-85
I	—	1800	2800	4000*	
IA	—	1450	2250	3200*	
II	—	1500	2500	4000*	
	—	1000†	1700†	3200*†	
IIA	—	1200	2000	3200*	
	—	800†	1350†	2560*†	
III	1800	3500	—	—	
IIIA	1450	2800	—	—	
IV	—	—	1000	2500	
V	—	1200	2200	3000	
Blended cements					C 595-85
I(SM), IS, I(PM), IP	—	1800	2800	3500	
I(SM)-A, IS-A, I(PM)-A, IP-A	—	1450	2250	2800	
IS(MS), IP(MS)	—	1500	2500	3500	
IS-A(MS), IP-A(MS)	—	1200	2000	2800	
S	—	—	600	1500	
SA	—	—	500	1250	
P	—	—	1500	3000	
PA	—	—	1250	2500	
Expansive cement					C 845-80
E-1	—	—	2100	3500	
Masonry cements					C 91-83a
N	—	—	500	900	
S	—	—	1300	2100	
M	—	—	1800	2900	

*Optional requirement.
†Applicable when the optional heat of hydration or chemical limit on the sum of C_3S and C_3A is specified.
Note: When low or moderate heat of hydration is specified for blended cements (ASTM C 595), the strength requirement is 80% of the value shown.

Fig. 2-9. Test for compressive strength of cement mortar using a 2-in.-cube specimen.

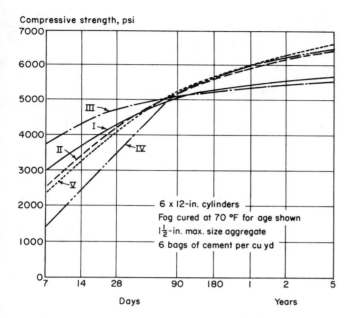

Fig. 2-10. Rates of compressive strength development for concrete made with various types of cement. Reference 2-20.

In certain structures, such as those with considerable mass, the rate and amount of heat generated are important. If this heat is not rapidly dissipated, a significant rise in concrete temperature can occur. This may be undesirable since, after hardening at an elevated temperature, nonuniform cooling of the concrete to ambient temperature may create undesirable stresses due to thermal contraction and restraint conditions. On the other hand, a rise in concrete temperature caused by heat of hydration is often beneficial in cold weather since it helps maintain favorable curing temperatures. The heat of hydration is tested in accordance with ASTM C 186.

The approximate amounts of heat generated during the first seven days, based on 100% for Type I normal portland cement are as follows:

Type II	moderate	80% to 85%
Type III	high early strength	up to 150%
Type IV	low heat of hydration	40% to 60%
Type V	sulfate resistant	60% to 75%

Loss on Ignition

Loss on ignition of portland cement is determined by heating a cement sample of known weight to 900°C to 1000°C until a constant weight is obtained. The weight loss of the sample is then determined. Normally, a high loss on ignition is an indication of prehydration and carbonation, which may be caused by improper and prolonged storage or adulteration during transport and transfer. The test for loss on ignition is performed in accordance with ASTM C 114.

Specific Gravity

The specific gravity of portland cement is generally about 3.15. Portland-blast-furnace-slag and portland-pozzolan cements may have specific gravity values of about 2.90. The specific gravity of a cement, which is determined by ASTM C 188, is not an indication of the cement's quality; it's principal use is in mixture proportioning calculations.

Weight of Cement

Most portland cements are shipped in bulk by rail, truck, or barge. Pneumatic loading and unloading of the transport vehicle is the most popular means of handling bulk cement.

Bulk cement is measured by the ton (2000 lb)—smaller quantities are bagged. In the United States a bag of portland cement weighs 94 lb and has a volume of about 1 cu ft when freshly packed. The weight of masonry cement is printed on the bag.

The actual density of bulk portland cement can vary considerably depending on how it is handled and stored. Portland cement that is fluffed may weigh only 52 pcf, whereas if it is consolidated by vibration, the same cement may weigh as much as 103 pcf.* For this reason, good practice has decreed that bulk cement must be weighed for each batch of concrete produced.

STORAGE OF CEMENT

Portland cement is a moisture-sensitive material; if kept dry, it will retain its quality indefinitely. Portland cement stored in contact with damp air or moisture sets more slowly and has less strength than portland cement that is kept dry. The relative humidity in a warehouse or shed used to store bagged cement should be as low as possible. All cracks and openings in walls and roofs should be closed. Cement bags should not be stored on damp floors but should rest on pallets. Bags should be stacked close together to reduce air circulation but should never be stacked against outside walls. Bags to be stored for long periods should be covered with tarpaulins or other waterproof covering. Stack the bags so that the first in are the first out.

On small jobs where a shed is not available, bags should be placed on raised wooden platforms at least 4 to 6 in. above the ground. Waterproof coverings should fit over the pile and extend over the edges of the platform to prevent rain from reaching the cement and the platform. Rain-soaked platforms can damage the bottom bags of cement.

Cement stored for long periods may develop what is called warehouse pack. This can usually be corrected by rolling the bags on the floor. At the time of use, cement should be free-flowing and free of lumps. If lumps do not break up easily, the cement should be

---

*Reference 2-9.

tested before it is used in important work. Standard strength tests or loss-on-ignition tests should be made whenever the quality of a cement is doubtful.

Ordinarily, cement does not remain in storage long, but it can be stored for long periods without deterioration. Bulk cement should be stored in weathertight concrete or steel bins or silos. Dry low-pressure aeration or vibration should be used in bins or silos to make the cement flow better and avoid bridging. Due to fluffing of cement, silos may hold only about 80% of rated capacity.

HOT CEMENT

When cement clinker is pulverized in the grinding mill, the friction generates heat. Freshly ground cement is therefore hot when placed in storage silos at the cement plant and this heat dissipates slowly. Therefore, during summer months when demand is high, cement may still be hot when delivered to a ready mix plant or jobsite. Tests* have shown that the effect of hot cement on the workability and strength development of concrete is not significant. The temperatures of the mixing water and aggregates play a much greater role in determining the concrete temperature.

―――――
*Reference 2-4.

REFERENCES

2-1. McMillan, F. R.; Tyler, I. L.; Hansen, W. C.; Lerch, W.; Ford, C. L.; and Brown, L. S., *Long-Time Study of Cement Performance in Concrete,* Research Department Bulletin RX026, Portland Cement Association, 1948.

2-2. Gonnerman, H. F., and Lerch, William, *Changes in Characteristics of Portland Cement As Exhibited by Laboratory Tests Over the Period 1904 to 1950,* Research Department Bulletin RX039, Portland Cement Association, 1952.

2-3. Bogue, R. H., *The Chemistry of Portland Cement,* Reinhold Publishing Corporation, New York, 1955.

2-4. Lerch, William, *Hot Cement and Hot Weather Concrete Tests,* TA015T, Portland Cement Association, 1955.

2-5. Brunauer, S., *Some Aspects of the Physics and Chemistry of Cement,* Research Department Bulletin RX080, Portland Cement Association, 1957.

2-6. Gonnerman, H. F., *Development of Cement Performance Tests and Requirements,* Research Department Bulletin RX093, Portland Cement Association, 1958.

2-7. Copeland, L. E.; Kantro, D. L.; and Verbeck, George, *Chemistry of Hydration of Portland Cement,* Research Department Bulletin RX153, Portland Cement Association, 1960.

2-8. Clausen, C. F., *Cement Materials,* Research Department Report MP-95, Portland Cement Association, 1960.

2-9. Toler, H. R., *Flowability of Cement,* Research Department Report MP-106, Portland Cement Association, October, 1963.

2-10. Brunauer, Stephen, and Copeland, L. E., "The Chemistry of Concrete," *Scientific American,* Scientific American, Inc., New York, April 1964.

2-11. Klieger, Paul, and Isberner, Albert W., *Laboratory Studies of Blended Cements—Portland Blast-Furnace Slag Cements,* Research Department Bulletin RX218, Portland Cement Association, 1967.

2-12. Verbeck, G. J., *Field and Laboratory Studies of the Sulphate Resistance of Concrete,* Research Department Bulletin RX227, Portland Cement Association, 1967.

2-13. Lea, F. M., *The Chemistry of Cement and Concrete,* 3rd ed., Chemical Publishing Co., Inc., New York, 1971.

2-14. Pfeifer, Donald W., and Perenchio, W. F., *Reinforced Concrete Pipe Made with Expansive Cements,* Research and Development Bulletin RD015W, Portland Cement Association, 1973.

2-15. Perenchio, William F., and Klieger, Paul, *Further Laboratory Studies of Portland-Pozzolan Cements,* Research and Development Bulletin RD041T, Portland Cement Association, 1976.

2-16. Duda, Walter H., *Cement Data Book, International Process Engineering in the Cement Industry,* 2nd ed., Macdonald & Evans, London, 1977.

2-17. Russell, H. G., *Performance of Shrinkage-Compensating Concretes in Slabs,* Research and Development Bulletin RD057D, Portland Cement Association, 1978.

2-18. Kirk, Raymond E., and Othmer, Donald F., eds., "Cement," *Encyclopedia of Chemical Technology,* 3rd ed., vol. 5, John Wiley & Sons, Inc., New York, 1979, pages 163-193.

2-19. *Cedric Willson Symposium on Expansive Cements,* SP-64, American Concrete Institute, Detroit, 1980.

2-20. *Concrete Manual,* 8th ed., U.S. Bureau of Reclamation, Denver, revised 1981.

2-21. *Standard Practice for the Use of Shrinkage-Compensating Concrete,* ACI 223-83, ACI Committee 223 Report, American Concrete Institute, Detroit, 1983.

2-22. Young, J. Francis, *Hydraulic Cements for Concrete,* E3-83, American Concrete Institute, 1983.

2-23. Bhatty, M. S. Y., "Use of Cement-Kiln Dust in Blended Cements," *World Cement,* vol. 15, no. 4, Palladian Publications Ltd., London, 1984.

2-24. Stark, David, *Longtime Study of Concrete Durability in Sulfate Soils,* Research and Development Bulletin RD086T, Portland Cement Association, 1984.

2-25. *The U.S. Cement Industry, an Economic Report,* SP016G, Portland Cement Association, 1984.

2-26. *Guide to the Selection and Use of Hydraulic Cements,* ACI 225R-85, ACI Committee 225 Report, American Concrete Institute, 1985.

2-27. Frohnsdorff, Geoffrey, ed., *Blended Cements,* STP 897, American Society for Testing and Materials, Philadelphia, 1986.

2-28. Campbell, Donald H., *Microscopical Examination and Interpretation of Portland Cement and Clinker,* SP030T, Portland Cement Association, 1986.

CHAPTER 3
Mixing Water for Concrete

Almost any natural water that is drinkable and has no pronounced taste or odor can be used as mixing water for making concrete. However, some waters that are not fit for drinking may be suitable for concrete.

Six typical analyses of city water supplies and seawater are shown in Table 3-1. These waters approximate the composition of domestic water supplies for most of the cities over 20,000 population in the United States and Canada. Water from any of these sources is suitable for making concrete. A water source comparable in analysis to any of the waters in the table is probably satisfactory for use in concrete.

Water of questionable suitability can be used for making concrete if mortar cubes (ASTM C 109) made with it have 7-day strengths equal to at least 90% of companion specimens made with drinkable or distilled water. In addition, ASTM C 191 tests should be made to ensure that impurities in the mixing water do not adversely shorten or extend the setting time of the cement. Acceptable criteria for water to be used in concrete is given in ASTM C 94 and American Association of State Highway and Transportation Officials (AASHTO) T 26* (see Tables 3-2 and 3-3).

Excessive impurities in mixing water not only may affect setting time and concrete strength, but also may cause efflorescence, staining, corrosion of reinforcement, volume instability, and reduced durability. Therefore, certain optional limits may be set on chlorides, sulfates, alkalies, and solids in the mixing water

Table 3-2. Acceptance Criteria for Questionable Water Supplies (ASTM C 94)

	Limits	Test method
Compressive strength, minimum percentage of control at 7 days	90	C 109*
Time of set, deviation from control, hr:min.	from 1:00 earlier to 1:30 later	C 191*

*Comparisons should be based on fixed proportions and the same volume of test water compared to control mix using city water or distilled water.

or appropriate tests can be performed to determine the effect the impurity has on various properties. Some impurities may have little effect on strength and setting time, yet they can adversely affect durability and other properties.

Water containing less than 2000 parts per million (ppm) of total dissolved solids can generally be used satisfactorily for making concrete. Water containing more than 2000 ppm of dissolved solids should be tested for its effect on strength and time of set.

A résumé of the effects of certain impurities in mixing water on the quality of normal concrete follows.**

---

*Reference 3-3.

**Additional information on the effects various impurities in mix water have on concrete is in Reference 3-1. Over 100 different compounds and ions are discussed.

Table 3-1. Typical Analyses of City Water Supplies and Seawater, parts per million

	Analysis No.						
Chemicals	1	2	3	4	5	6	Seawater*
Silica (SiO_2)	2.4	0.0	6.5	9.4	22.0	3.0	—
Iron (Fe)	0.1	0.0	0.0	0.2	0.1	0.0	—
Calcium (Ca)	5.8	15.3	29.5	96.0	3.0	1.3	50-480
Magnesium (Mg)	1.4	5.5	7.6	27.0	2.4	0.3	260-1410
Sodium (Na)	1.7	16.1	2.3	183.0	215.0	1.4	2190-12,200
Potassium (K)	0.7	0.0	1.6	18.0	9.8	0.2	70-550
Bicarbonate (HCO_3)	14.0	35.8	122.0	334.0	549.0	4.1	—
Sulfate (SO_4)	9.7	59.9	5.3	121.0	11.0	2.6	580-2810
Chloride (Cl)	2.0	3.0	1.4	280.0	22.0	1.0	3960-20,000
Nitrate (NO_3)	0.5	0.0	1.6	0.2	0.5	0.0	—
Total dissolved solids	31.0	250.0	125.0	983.0	564.0	19.0	35,000

*Different seas contain different amounts of dissolved salts.

Table 3-3. Chemical Limits for Wash Water used as Mixing Water (ASTM C 94)

Chemical	Maximum concentration,* ppm	Test method**
Chloride, as Cl		ASTM D 512
Prestressed concrete or concrete in bridge decks	500†	
Other reinforced concrete in moist environments or containing aluminum embedments or dissimilar metals or with stay-in-place galvanized metal forms	1,000†	
Sulfate, as SO_4	3,000	ASTM D 516
Alkalies, as ($Na_2O + 0.658 K_2O$)	600	
Total solids	50,000	AASHTO T 26

*Wash water reused as mixing water in concrete can exceed the listed concentrations of chloride and sulfate if it can be shown that the concentration calculated in the total mixing water, including mixing water on the aggregates and other sources, does not exceed the stated limits.
**Other test methods that have been demonstrated to yield comparable results can be used.
†For conditions allowing use of $CaCl_2$ accelerator as an admixture, the chloride limitation may be waived by the purchaser.

Alkali Carbonate and Bicarbonate

Carbonates and bicarbonates of sodium and potassium have different effects on the setting times of different cements. Sodium carbonate can cause very rapid setting, bicarbonates can either accelerate or retard the set. In large concentrations these salts can materially reduce concrete strength. When the sum of the dissolved salts exceeds 1000 ppm, tests for their effect on setting time and 28-day strength should be made. The possibility of aggravated alkali-aggregate reactions should also be considered.

Chloride

Concern over a high chloride content in mixing water is chiefly due to the possible adverse effect of chloride ions on the corrosion of reinforcing steel or prestressing strands. Chloride ions attack the protective oxide film formed on the steel by the highly alkaline (pH > 12.5) chemical environment present in concrete. The water soluble chloride ion level at which steel reinforcement corrosion begins in concrete is about 0.15% by weight of cement.* Of the total chloride-ion content in concrete, only about 50% to 85% is water soluble; the rest becomes chemically combined in cement reactions.**

Chlorides can be introduced into concrete with the separate mix ingredients—admixtures, aggregates, cement, and mixing water—or through exposure to deicing salts, seawater, or salt-laden air in coastal environments. Placing an acceptable limit on chloride content for any one ingredient, such as mixing water, is difficult considering the several possible sources of chloride ions in concrete. An acceptable limit in the

concrete depends primarily upon the type of structure and the environment to which it is exposed during its service life.

A high dissolved solids content of a natural water is usually due to a high content of sodium chloride or sodium sulfate. Both can be tolerated in rather large quantities. Concentrations of 20,000 ppm of sodium chloride are generally tolerable in concrete that will be dry in service and has low potential for corrosive reactions. Water used in prestressed concrete or in concrete that is to have aluminum embedments should not contain deleterious amounts of chloride ion. The contribution of chlorides from ingredients other than water should also be considered. Calcium chloride admixtures should be used with caution.

The American Concrete Institute's Building Code Requirements for Reinforced Concrete, ACI 318, limits water soluble chloride ion content in concrete to the following percentages by weight of cement:†

Prestressed concrete	0.06%
Reinforced concrete exposed to chloride in service	0.15%
Reinforced concrete that will be dry or protected from moisture in service	1.00%
Other reinforced concrete construction	0.30%

Sulfate

Concern over a high sulfate content in mix water is due to possible expansive reactions and deterioration by sulfate attack, especially in areas where the concrete will be exposed to high sulfate soils or water. Although mixing waters containing 10,000 ppm of sodium sulfate have been used satisfactorily, the limit in Table 3-3 should be considered unless special precautions are taken.

Other Common Salts

Carbonates of calcium and magnesium are not very soluble in water and are seldom found in sufficient concentration to affect the strength of concrete. Bicarbonates of calcium and magnesium are present in some municipal waters. Concentrations up to 400 ppm of bicarbonate in these forms are not considered harmful.

Magnesium sulfate and magnesium chloride can be present in high concentrations without harmful effects on strength. Good strengths have been obtained with concentrations up to 40,000 ppm of magnesium chloride. Concentrations of magnesium sulfate should be less than 25,000 ppm.

---

*Reference 3-5.
**Reference 3-4.
†ACI 222R suggests that the maximum acid-soluble chloride content, ASTM C 114, for prestressed concrete and reinforced concrete be 0.08% and 0.20%, respectively. ACI 318 and 222R do not address chloride limits for concrete without reinforcement. Refer to Chapter 14 for tests to determine chloride content. Also see Reference 3-6 and Chapter 6, "Corrosion Inhibitors."

Iron Salts

Natural groundwaters seldom contain more than 20 to 30 ppm of iron; however, acid mine waters may carry rather large quantities. Iron salts in concentrations up to 40,000 ppm do not usually affect strengths adversely.

Miscellaneous Inorganic Salts

Salts of manganese, tin, zinc, copper, and lead in mixing water can cause a significant reduction in strength and large variations in setting time. Of these, salts of zinc, copper, and lead are the most active. Salts that are especially active as retarders include sodium iodate, sodium phosphate, sodium arsenate, and sodium borate. All can greatly retard both set and strength development when present in concentrations of a few tenths percent by weight of the cement. Generally, concentrations of these salts up to 500 ppm can be tolerated in mixing water.

Another salt that may be detrimental to concrete is sodium sulfide; even the presence of 100 ppm warrants testing.

Seawater

Seawater containing up to 35,000 ppm of dissolved salts is generally suitable as mixing water for *unreinforced concrete*. About 78% of the salt is sodium chloride, and 15% is chloride and sulfate of magnesium. Although concrete made with seawater may have higher early strength than normal concrete, strengths at later ages (after 28 days) may be lower. This strength reduction can be compensated for by reducing the water-cement ratio.

Seawater is not suitable for use in making steel reinforced concrete and it should not be used in prestressed concrete due to the risk of corrosion of the reinforcement, particularly in warm and humid environments.

Sodium or potassium in salts present in seawater used for mix water can combine with alkali-reactive aggregates in the same manner as alkalies in cement. Thus, seawater should not be used as mix water for concrete with known potentially alkali-reactive aggregates, even when the alkali content of the cement is low.

Seawater used for mix water also tends to cause efflorescence and dampness on concrete surfaces exposed to air and water.* Marine-dredged aggregates are discussed in Chapter 4.

Acid Waters

Acceptance of acid mixing water should be based on the concentration (in parts per million) of acids in the water. Occasionally, acceptance is based on the pH, which is a measure of the hydrogen-ion concentration.** The pH value is an intensity index and is not the best measure of potential acid or base reactivity.

Generally, mixing waters containing hydrochloric, sulfuric, and other common inorganic acids in concentrations as high as 10,000 ppm have no adverse effect on strength. Acid waters with pH values less than 3.0 may create handling problems and should be avoided if possible.

Alkaline Waters

Waters with sodium hydroxide concentrations of 0.5% by weight of cement do not greatly affect concrete strength provided quick set is not induced. Higher concentrations, however, may reduce concrete strength.

Potassium hydroxide in concentrations up to 1.2% by weight of cement has little effect on the concrete strength developed by some cements, but the same concentration when used with other cements may substantially reduce the 28-day strength.

The possibility for increased alkali-aggregate reactivity should be considered.

Wash Water

The U.S. Environmental Protection Agency and state agencies forbid discharging into the nation's waterways untreated wash water used in reclaiming sand and gravel from returned concrete or mixer washout operations. It is permissible, however, to reuse wash water as mixing water in concrete if the wash water satisfies the limits in Tables 3-2 and 3-3.

Industrial Wastewater

Most waters carrying industrial wastes have less than 4000 ppm of total solids. When such water is used as mixing water in concrete, the reduction in compressive strength is generally not greater than about 10%-15%. Wastewaters such as those from tanneries, paint factories, coke plants, and chemical and galvanizing plants may contain harmful impurities. It is best to test any wastewater that contains even a few hundred parts per million of unusual solids.

Waters Carrying Sanitary Sewage

A typical sewage may contain about 400 ppm of organic matter. After the sewage is diluted in a good disposal system, the concentration is reduced to about 20 ppm or less. This amount is too low to have any significant effect on strength.

Organic Impurities

The effect of organic substances in natural waters on the setting time of portland cement or the ultimate strength of concrete is a problem of considerable complexity. Highly colored waters, waters with a noticeable odor, or those in which green or brown algae are

---

*Reference 3-1.
**The pH of neutral water is 7.0; values below 7.0 indicate acidity and those above 7.0 alkalinity.

visible should be regarded with suspicion and tested accordingly.

Sugar

Small amounts of sucrose, as little as 0.03% to 0.15% by weight of cement, usually retard the setting of cement. The upper limit of this range varies with different cements. The 7-day strength may be reduced while the 28-day strength may be improved. Sugar in quantities of 0.25% or more by weight of cement may cause rapid setting and a substantial reduction in 28-day strength. Each type of sugar influences setting time and strength differently.

Less than 500 ppm of sugar in mix water generally has no adverse effect on strength, but if the concentration exceeds this amount, tests for setting time and strength should be made.

Silt or Suspended Particles

About 2000 ppm of suspended clay or fine rock particles can be tolerated in mixing water. Higher amounts might not affect strength but might influence other properties of some concrete mixtures. Before use, muddy water should pass through settling basins or be otherwise clarified to reduce the amount of silt and clay added to the mix. When cement fines are returned to the concrete in reused wash water, 50,000 ppm can be tolerated.

Oils

Various kinds of oil are occasionally present in mixing water. Mineral oil (petroleum) not mixed with animal or vegetable oils probably has less effect on strength development than other oils. However, mineral oil in concentrations greater than 2.5% by weight of cement may reduce strength by more than 20%.

Algae

Water containing algae is unsuited for making concrete because the algae can cause excessive reduction in strength either by influencing cement hydration or by causing a large amount of air to be entrained in the concrete. Algae may also be present on aggregates, in which case the bond between the aggregate and cement paste is reduced.

REFERENCES

3-1. Steinour, H. H., *Concrete Mix Water—How Impure Can It Be?*, Research Department Bulletin RX119, Portland Cement Association, 1960.

3-2. McCoy, W. J., "Mixing and Curing Water for Concrete," *Significance of Tests and Properties of Concrete and Concrete-Making Materials*, STP 169-A, American Society for Testing and Materials, Philadelphia, 1966, pages 515-521.

3-3. *Method of Test for Quality of Water to Be Used in Concrete*, T26-79, American Association of State Highway and Transportation Officials, Washington, D.C., 1979.

3-4. *Guide to Durable Concrete*, ACI 201.2R-77, Reaffirmed 1982, ACI Committee 201 Report, American Concrete Institute, Detroit.

3-5. *Corrosion of Metals in Concrete*, ACI 222R-85, ACI Committee 222 Report, American Concrete Institute, 1985.

3-6. Gaynor, Richard D., *Understanding Chloride Percentages*, NRMCA Publication Number 173, National Ready Mixed Concrete Association, Silver Spring, Maryland, 1985.

CHAPTER 4
Aggregates for Concrete

The importance of using the right type and quality of aggregates cannot be overemphasized since the fine and coarse aggregates generally occupy 60% to 75% of the concrete volume (70% to 85% by weight) and strongly influence the concrete's freshly mixed and hardened properties, mixture proportions, and economy. Fine aggregates generally consist of natural sand or crushed stone with most particles smaller than 0.2 in. Coarse aggregates consist of one or a combination of gravels or crushed aggregate with particles predominantly larger than 0.2 in. and generally between ⅜ and 1½ in.* Some natural aggregate deposits, sometimes called bank gravel, consist of gravel and sand that can be readily used in concrete after minimal processing. Natural gravel and sand are usually dug or dredged from a pit, river, lake, or seabed. Crushed aggregate is produced by crushing quarry rock, boulders, cobbles, or large-size gravel. Crushed air-cooled blast-furnace slag is also used as fine or coarse aggregate. The aggregates are usually washed and graded at the pit or plant. Some variation in the type, quality, cleanliness, grading, moisture content, and other properties is expected. Close to half of the coarse aggregates used in portland cement concrete in the United States are gravels; most of the remainder are crushed stones.

Naturally occurring concrete aggregates are a mixture of rocks and minerals (see Table 4-1). A mineral is a naturally occurring solid substance with an orderly internal structure and a chemical composition that ranges within narrow limits. Rocks (classified as igneous, sedimentary, or metamorphic, depending on origin) are generally composed of several minerals. For example, granite contains quartz, feldspar, mica, and a few other minerals; most limestones consist of calcite, dolomite, and minor amounts of quartz, feldspar, and clay. Weathering and erosion of rocks produce particles of stone, gravel, sand, silt, and clay.

Recycled concrete, or crushed waste concrete, is a feasible source of aggregates and an economic reality where good aggregates are scarce. Conventional crushing equipment can be used, and new equipment is being developed to reduce noise and dust.

Aggregates must conform to certain standards for optimum engineering use: they must be clean, hard, strong, durable particles free of absorbed chemicals, coatings of clay, and other fine materials in amounts

Fig. 4-1. Stockpiles of fine and coarse aggregate at a ready mix plant formed with a clamshell bucket using the cast-and-spread stockpile method.

that could affect hydration and bond of the cement paste. Aggregate particles that are friable or capable of being split are undesirable. Aggregates containing any appreciable amounts of shale or other shaly rocks, soft and porous materials, and certain types of chert should be especially avoided since they have low resistance to weathering and can cause surface defects such as popouts.

Identification of the constituents of an aggregate cannot alone provide a basis for predicting the behavior of aggregates in service. Visual inspection will often disclose weaknesses in coarse aggregates. Service records are invaluable in evaluating aggregates. In the absence of a record of performance, the aggregates should be tested before they are used in concrete.

The most commonly used aggregates such as sand, gravel, crushed stone, and air-cooled blast-furnace slag produce wet or freshly mixed normal-weight concrete

---

*ASTM C 125 further defines fine and coarse aggregate.

weighing about 135 to 160 lb per cubic foot. Aggregates of expanded shale, clay, slate, and slag are used to produce structural lightweight concrete with a freshly mixed unit weight ranging from about 90 to 120 lb per cubic foot. Other lightweight materials such as pumice, scoria, perlite, vermiculite, and diatomite are used to produce insulating lightweight concretes ranging in weight from about 15 to 90 lb per cubic foot. Heavyweight materials such as barite, limonite, magnetite, ilmenite, hematite, iron, and steel punchings or shot are used to produce heavyweight concrete and radiation-shielding concrete (ASTM C 637 and C 638). Only normal-weight aggregates are discussed in this chapter. See Chapter 15 for special types of aggregates and concretes.

Normal-weight aggregates should meet the requirements of ASTM C 33. This specification limits the permissible amounts of deleterious substances and states the requirements for aggregate characteristics. However, the fact that aggregates satisfy ASTM C 33 requirements does not necessarily assure defect-free concrete. Compliance is determined by using one or more of the several ASTM standard tests cited in the following sections and tables.

For adequate consolidation of concrete, the desirable amount of air, water, cement, and fine aggregate (i.e., the mortar fraction) is about 50% to 65% by absolute volume (45% to 60% by weight). Rounded aggregate, such as gravel, requires slightly lower values while crushed aggregate requires slightly higher values. Fine-aggregate content is usually 35% to 45% by weight or volume of the total aggregate content.

Table 4-1. Rock and Mineral Constituents in Aggregates

Minerals	Igneous rocks	Metamorphic rocks
Silica	Granite	Marble
Quartz	Syenite	Metaquartzite
Opal	Diorite	Slate
Chalcedony	Gabbro	Phyllite
Tridymite	Peridotite	Schist
Cristobalite	Pegmatite	Amphibolite
Silicates	Volcanic glass	Hornfels
Feldspars	Obsidian	Gneiss
Ferromagnesian	Pumice	Serpentinite
Hornblende	Tuff	
Augite	Scoria	
Clay	Perlite	
Illites	Pitchstone	
Kaolins	Felsite	
Chlorites	Basalt	
Montmorillonites		
Mica	**Sedimentary rocks**	
Zeolite	Conglomerate	
Carbonate	Sandstone	
Calcite	Quartzite	
Dolomite	Graywacke	
Sulfate	Subgraywacke	
Gypsum	Arkose	
Anhydrite	Claystone, siltstone,	
Iron sulfide	argillite, and shale	
Pyrite	Carbonates	
Marcasite	Limestone	
Pyrrhotite	Dolomite	
Iron oxide	Marl	
Magnetite	Chalk	
Hematite	Chert	
Goethite		
Ilmenite		
Limonite		

For brief descriptions, see "Standard Descriptive Nomenclature of Constituents of Natural Mineral Aggregates" (ASTM C 294).

CHARACTERISTICS OF AGGREGATES

The important characteristics of aggregates for concrete are listed in Table 4-2 and most are discussed in this section.

Grading

The grading is the particle-size distribution of an aggregate as determined by a sieve analysis (ASTM C 136). The aggregate particle size is determined by using wire-mesh sieves with square openings. The seven standard ASTM C 33 sieves for fine aggregate have openings ranging from the No. 100 sieve (150 μm) to $\frac{3}{8}$ in. The 13 standard sieves for coarse aggregate listed in Table 4-3 have openings ranging from 0.046 in. to 4 in. Tolerances for the sizes of openings in sieves are listed in ASTM E 11.

Size numbers (grading sizes) for coarse aggregate apply to the amounts of aggregate (by weight) in percentages that pass through an assortment of sieves (Fig. 4-2). For highway construction, ASTM D 448 lists the 13 size numbers in ASTM C 33 (Table 4-3) plus six more coarse aggregate size numbers. Fine aggregate or sand has only one range of particle sizes.

The grading and grading limits are usually expressed as the percentage of material passing each sieve. Fig.

4-3 shows these limits for fine aggregate and for one size of coarse aggregate.

There are several reasons for specifying grading limits and maximum aggregate size. The grading and maximum size of aggregate affect relative aggregate proportions as well as cement and water requirements, workability, pumpability, economy, porosity, shrinkage, and durability of concrete. Variations in grading can seriously affect the uniformity of concrete from batch to batch. Very fine sands are often uneconomical; very coarse sands and coarse aggregate can produce harsh, unworkable mixes. In general, aggregates that do not have a large deficiency or excess of any size and give a smooth grading curve will produce the most satisfactory results.

The effect of a collection of sizes in reducing the total volume of voids between aggregates is illustrated by the simple method shown in Fig. 4-4. The beaker on the left is filled with large aggregate particles of uniform size and shape; the middle beaker is filled with an equal volume of small aggregate particles of uniform size and shape. The beaker on the right is filled with particles of both sizes. Below each beaker is a graduate with the amount of water required to fill the voids in that beaker. Note that when the beakers are filled with one particle size of equal volume, the void

Table 4-2. Characteristics and Tests of Aggregates

Characteristic	Significance	Test designation*	Requirement or item reported
Resistance to abrasion and degradation	Index of aggregate quality; wear resistance of floors, pavements	ASTM C 131 ASTM C 535 ASTM C 779	Maximum percentage of weight loss. Depth of wear and time
Resistance to freezing and thawing	Surface scaling, roughness, loss of section, and unsightliness	ASTM C 666 ASTM C 682	Maximum number of cycles or period of frost immunity; durability factor
Resistance to disintegration by sulfates	Soundness against weathering action	ASTM C 88	Weight loss, particles exhibiting distress
Particle shape and surface texture	Workability of fresh concrete	ASTM C 295 ASTM D 3398	Maximum percentage of flat and elongated pieces
Grading	Workability of fresh concrete; economy	ASTM C 117 ASTM C 136	Minimum and maximum percentage passing standard sieves
Bulk unit weight or bulk density	Mix design calculations; classification	ASTM C 29	Compact weight and loose weight
Specific gravity	Mix design calculations	ASTM C 127, fine aggregate ASTM C128, coarse aggregate	—
Absorption and surface moisture	Control of concrete quality	ASTM C 70 ASTM C 127 ASTM C 128 ASTM C 566	—
Compressive and flexural strength	Acceptability of fine aggregate failing other tests	ASTM C 39 ASTM C 78	Strength to exceed 95% of strength achieved with purified sand
Definitions of constituents	Clear understanding and communication	ASTM C 125 ASTM C 294	—
Aggregate constituents	Determine amount of deleterious and organic materials	ASTM C 40 ASTM C 87 ASTM C 117 ASTM C 123 ASTM C 142 ASTM C 295	Maximum percentage of individual constituents
Resistance to alkali reactivity and volume change	Soundness against volume change	ASTM C 227 ASTM C 289 ASTM C 295 ASTM C 342 ASTM C 586	Maximum length change, constituents and amount of silica, and alkalinity

*The majority of the tests and characteristics listed are referenced in ASTM C 33. Reference 4-22 presents additional test methods and properties of concrete influenced by aggregate properties.

Table 4-3. Grading Requirements for Coarse Aggregates (ASTM C 33)

Size number	Nominal size (sieves with square openings)	Amounts finer than each laboratory sieve (square-openings), weight percent passing								
		4 in. (100 mm)	3½ in. (90 mm)	3 in. (75 mm)	2½ in. (63 mm)	2 in. (50 mm)	1½ in. (37.5 mm)	1 in. (25.0 mm)	¾ in. (19.0 mm)	½ in. (12.5 mm)
1	3½ to 1½ in.	100	90 to 100	—	25 to 60	—	0 to 15	—	0 to 5	—
2	2½ to 1½ in.	—	—	100	90 to 100	35 to 70	0 to 15	—	0 to 5	—
3	2 to 1 in.	—	—	—	100	90 to 100	35 to 70	0 to 15	—	0 to 5
357	2 in. to No. 4	—	—	—	100	95 to 100	—	35 to 70	—	10 to 30
4	1½ to ¾ in.	—	—	—	—	100	90 to 100	20 to 55	0 to 15	—
467	1½ in. to No. 4	—	—	—	—	100	95 to 100	—	35 to 70	—
5	1 to ½ in.	—	—	—	—	—	100	90 to 100	20 to 55	0 to 10
56	1 to ⅜ in.	—	—	—	—	—	100	90 to 100	40 to 85	10 to 40
57	1 in to No. 4	—	—	—	—	—	100	95 to 100	—	25 to 60
6	¾ to ⅜ in.	—	—	—	—	—	—	100	90 to 100	20 to 55
67	¾ in. to No. 4	—	—	—	—	—	—	100	90 to 100	—
7	½ in. to No. 4	—	—	—	—	—	—	—	100	90 to 100
8	⅜ in. to No. 8	—	—	—	—	—	—	—	—	100

Fig. 4-2. Making a sieve analysis test of coarse aggregate in the laboratory.

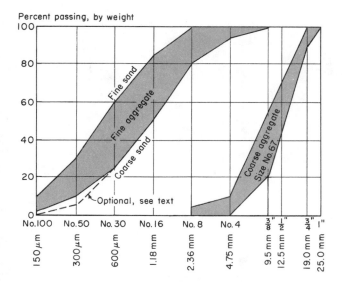

Fig. 4-3. Curves indicate the limits specified in ASTM C 33 for fine aggregate and for one typically used size number (grading size) of coarse aggregate.

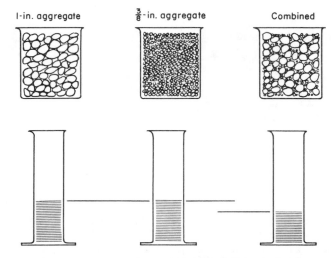

Fig. 4-4. The level of liquid in the graduates, representing voids, is constant for equal absolute volumes of aggregates of uniform but different size. When different sizes are combined, the void-content decreases. The illustration is not to scale.

content is constant, regardless of the particle size. When the two aggregate sizes are combined, the void content is decreased. If this operation were repeated with several additional sizes, a further reduction in voids would occur. The cement paste requirement for concrete is proportional to the void content of the combined aggregates.

During the early years of concrete technology it was sometimes assumed that the smallest percentage of voids (greatest density of aggregates) was the most

suitable for concrete. At the same time, limits were placed on the amount and size of the smallest particles. It is now known that, even on this restricted basis, this is not the best target for the mix designer. However, production of satisfactory, economical concrete requires aggregates of low void content, but not the lowest. Voids in aggregates can be tested according to ASTM C 29.

In reality, the amount of cement paste required is greater than the volume of voids between the aggregates. This is illustrated in Fig. 4-5. Sketch A represents large aggregates alone, with all particles in contact. Sketch B represents the dispersal of aggregates in a

3/8 in. (9.5 mm)	No. 4 (4.75 mm)	No. 8 (2.36 mm)	No. 16 (1.18 mm)
—	—	—	—
—	—	—	—
—	0 to 5	—	—
0 to 5	—	—	—
10 to 30	0 to 5	—	—
0 to 5	—	—	—
0 to 15	0 to 5	—	—
—	0 to 10	0 to 5	—
0 to 15	0 to 5	—	—
25 to 55	0 to 10	0 to 5	—
40 to 70	0 to 15	0 to 5	—
85 to 100	10 to 30	0 to 10	0 to 5

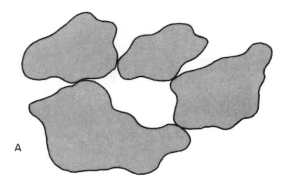

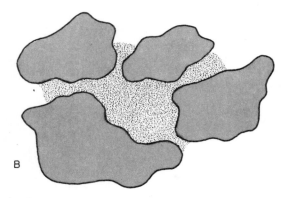

Fig. 4-5. Illustration of the dispersion of aggregates in cohesive concrete mixtures.

matrix of paste. The amount of paste is necessarily greater than the void content of A in order to provide workability to the concrete, and the amount is influenced by the workability and cohesiveness of the paste.

Fine-Aggregate Grading

Requirements of ASTM C 33 permit a relatively wide range in fine-aggregate grading, but specifications by other organizations are sometimes more restrictive. The most desirable fine-aggregate grading depends on the type of work, the richness of mix, and the maximum size of coarse aggregate. In leaner mixes or when small-size coarse aggregates are used, a grading that approaches the maximum recommended percentage passing each sieve is desirable for workability. In general, if the water-cement ratio is kept constant and the ratio of fine-to-coarse aggregate is chosen correctly, a wide range in grading can be used without measurable effect on strength. The greatest economy will sometimes be achieved by adjusting the concrete mixture to suit the gradation of the local aggregates. The more uniform the grading, the greater the economy.

Fine-aggregate grading within the limits of ASTM C 33 is generally satisfactory for most concretes. The ASTM C 33 limits with respect to sieve size are as follows:

Sieve size	Percent passing by weight
⅜ in. (9.5 mm)	100
No. 4 (4.75 mm)	95 to 100
No. 8 (2.36 mm)	80 to 100
No. 16 (1.18 mm)	50 to 85
No. 30 (600 µm)	25 to 60
No. 50 (300 µm)	10 to 30
No. 100 (150 µm)	2 to 10

These specifications permit the minimum percentages (by weight) of material passing the No. 50 and No. 100 sieves to be reduced to 5% and 0% respectively, provided

1. The aggregate is used in air-entrained concrete containing more than 400 lb of cement per cubic yard and having an air content of more than 3%.
2. The aggregate is used in concrete containing more than 500 lb of cement per cubic yard when the concrete is not air-entrained.
3. An approved mineral admixture is used to supply the deficiency in material passing these two sieves.

Other requirements of ASTM C 33 are

1. The fine aggregate must not have more than 45% retained between any two consecutive standard sieves.
2. The fineness modulus must be not less than 2.3 nor more than 3.1, nor vary more than 0.2 from the typical value of the aggregate source. If this value is exceeded, the fine aggregate should be rejected unless suitable adjustments are made in proportions of fine and coarse aggregate.

The amounts of fine aggregate passing the No. 50 and No. 100 sieves affect workability, surface texture, and bleeding of concrete. Most specifications allow 10% to 30% to pass the No. 50 sieve. The lower limit may be sufficient for easy placing conditions or where concrete is mechanically finished such as in pavements. However, for hand-finished concrete floors or where a smooth surface texture is desired, fine aggregate with at least 15% passing the No. 50 sieve and 3% or more passing the No. 100 sieve should be used.

The fineness modulus (FM) of either fine or coarse aggregate according to ASTM C 125 is obtained by adding the cumulative percentages by weight retained on each of a specified series of sieves and dividing the sum by 100. The specified sieves for determining FM are No. 100, No. 50, No. 30, No. 16, No. 8, No. 4, ⅜ in., ¾ in., 1½ in., 3 in., and 6 in. FM is an index of the fineness of an aggregate—the higher the FM, the coarser the aggregate. Different aggregate grading may have the same FM. FM of fine aggregate is useful in estimating proportions of fine and coarse aggregates in concrete mixtures. An example of determining the FM of a fine aggregate with an assumed sieve analysis follows:

Sieve size	Percentage of individual fraction retained, by weight*	Cumulative percentage passing, by weight	Cumulative percentage retained, by weight
⅜″	0	100	0
No. 4	2	98	2
No. 8	13	85	15
No. 16	20	65	35
No. 30	20	45	55
No. 50	24	21	79
No. 100	18	3	97
Pan	3	0	–
Total	100		283

Fineness modulus
= 283 ÷ 100 =
2.83

Coarse-Aggregate Grading

The coarse aggregate grading requirements of ASTM C 33 (see Table 4-3) permit a wide range in grading and a variety of grading sizes. The grading for a given maximum-size coarse aggregate can be varied over a moderate range without appreciable effect on cement and water requirements if the proportion of fine aggregate to total aggregate produces concrete of good workability. Mixture proportions should be changed to produce workable concrete if wide variations occur in the coarse-aggregate grading. Since variations are difficult to anticipate, it is often more economical to maintain uniformity in handling and manufacturing coarse aggregate to reduce variations in gradation.

The maximum size of coarse aggregate used in concrete has a bearing on economy. Usually more water and cement is required for small-size aggregates than for large sizes. The water and cement required for a slump of approximately 3 in. is shown in Fig. 4-6 for a wide range of coarse-aggregate sizes. Fig. 4-6 shows that, for a given water-cement ratio, the amount of cement required decreases as the maximum size of coarse aggregate increases. The increased cost of obtaining or handling aggregates larger than 2 in. may offset the savings in using less cement. Furthermore, aggregates of different maximum sizes may give slightly different concrete strengths for the same water-cement ratio. In some instances, at the same water-cement ratio, the concrete with the smaller maximum-size aggregate has higher compressive strength. This is especially true of high-strength concrete. The optimum maximum size of coarse aggregate for higher-strength ranges depends on relative strength of the cement paste, cement-aggregate bond, and strength of the aggregate particles (see "High-Strength Concrete" in Chapter 15).

The terminology used to specify size of coarse aggregate must be chosen carefully. As stated, particle size is determined by size of sieve and applies to the aggregate passing that sieve and not passing the next smaller sieve. When speaking of an assortment of particle sizes, the *size number* (or grading size) of the gradation is used. The size number applies to the collective amount of aggregate that passes through an assortment

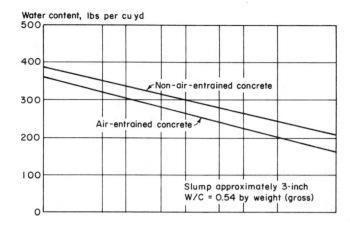

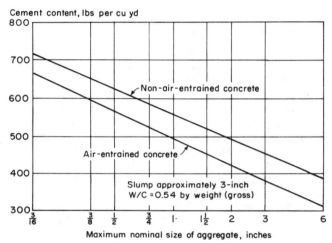

Fig. 4-6. Cement and water contents in relation to maximum size of aggregates, for air-entrained and non-air-entrained concrete. Less cement and water are required in mixes having large, coarse aggregate. Reference 4-16.

of sieves (Fig. 4-2). As shown in Fig. 4-3 and Table 4-3, the amount of aggregate passing the respective sieves is given in percentages and is also termed a sieve analysis.

Because of past usage, there is sometimes confusion about what is meant by the maximum size of aggregate. ASTM C 125 and ACI 116 define this term and distinguish it from nominal maximum size of aggregate. The maximum size of an aggregate is the smallest sieve that *all* of a particular aggregate must pass through. The nominal maximum size of an aggregate is the smallest sieve size through which the major portion of the aggregate must pass. The nominal maximum-size sieve may retain 5% to 15% of the aggregate depending on the size number. For example, size number 67 aggregate has a maximum size of 1 in. and a nominal maximum size of ¾ in. Ninety to one hundred percent of this aggregate must pass the ¾ in. sieve and all of the particles must pass the 1-in. sieve.

---

*Percentage of material retained between consecutive sieves. For example, 13% of the sample is retained between the No. 4 and No. 8 sieves.

The maximum size of aggregate that can be used generally depends on the size and shape of the concrete member and the amount and distribution of reinforcing steel. The maximum size of aggregate particles generally should not exceed

1. One-fifth the narrowest dimension of a concrete member
2. Three-fourths the clear spacing between reinforcing bars
3. One-third the depth of slabs

These requirements may be waived if, in the judgment of the engineer, the mixture possesses sufficient workability that the concrete can be placed without honeycomb or voids.

Gap-Graded Aggregates

In gap-graded aggregates certain particle sizes are omitted. For cast-in-place concrete, typical gap-graded aggregates consist of only one size of coarse aggregate with all the particles of fine aggregate able to pass through the voids in the compacted coarse aggregate. Gap-graded mixes are used to obtain uniform textures in exposed-aggregate concrete. They are also used in normal structural concrete because of possible improvements in density, permeability, shrinkage, creep, strength, consolidation, and to permit the use of local aggregate gradations.*

For an aggregate of ¾-in. maximum size, the No. 4 to ⅜-in. particles can be omitted without making the concrete unduly harsh or subject to segregation. In the case of 1½-in. aggregate, usually the No. 4 to ¾-in. sizes are omitted.

Care must be taken in choosing the percentage of fine aggregate in a gap-graded mix. A wrong choice can result in concrete that is likely to produce segregation or honeycomb because of an excess of coarse aggregate or concrete with a low density and high water requirement because of an excess of fine aggregate. Fine aggregate is usually 25% to 35% by volume of the total aggregate. The lower percentage is used with rounded aggregates and the higher with crushed material. For a smooth off-the-form finish, a somewhat higher percentage of fine aggregate to total aggregate may be used than for an exposed-aggregate finish, but both use a lower fine aggregate content than continuously graded mixes. Fine aggregate content depends upon cement content, type of aggregate, and workability.

Air entrainment is usually required for workability since low-slump, gap-graded mixes use a low fine-aggregate percentage and produce harsh mixes without entrained air.

Segregation of gap-graded mixes must be prevented by restricting the slump to the lowest value consistent with good consolidation. This may vary from zero to 3 in. depending on the thickness of the section, amount of reinforcement, and height of casting. Close control of grading and water content is also required because variations might cause segregation. If a stiff mixture is required, gap-graded aggregates may produce higher strengths than normal aggregates used with comparable cement contents. Because of their low fine-aggregate volumes and low water-cement ratios, gap-graded mixtures might be considered unworkable for cast-in-place construction. When properly proportioned, however, these concretes are readily consolidated with vibration.

Particle Shape and Surface Texture

The particle shape and surface texture of an aggregate influence the properties of freshly mixed concrete more than the properties of hardened concrete. Rough-textured, angular, elongated particles require more water to produce workable concrete than do smooth, rounded, compact aggregates. Hence, aggregate particles that are angular require more cement to maintain the same water-cement ratio. However, with satisfactory gradation, both crushed and noncrushed aggregates (of the same rock types) generally give essentially the same strength for the same cement factor.** Angular or poorly graded aggregates can also be more difficult to pump.

The bond between cement paste and a given aggregate generally increases as particles change from smooth and rounded to rough and angular. This increase in bond is a consideration in selecting aggregates for concrete where flexural strength is important or where high compressive strength is needed.

Void contents of compacted coarse or fine aggregate can be used as an index of differences in the shape and texture of aggregates of the same grading. The mixing water and mortar requirement tends to increase as aggregate void content increases. Voids between aggregate particles increase with aggregate angularity. See "Unit Weight and Voids" in this chapter.

Aggregate should be relatively free of flat and elongated particles. Flat and elongated aggregate particles should be avoided or at least limited to about 15% by weight of the total aggregate. This requirement is equally important for coarse and for crushed fine aggregate, since fine aggregate made by crushing stone often contains flat and elongated particles. Such aggregate particles require an increase in mixing water and thus may affect the strength of concrete, particularly in flexure, if water-cement ratio is not maintained.

Unit Weight and Voids

The unit weight (also called unit mass or bulk density) of an aggregate is the weight of the aggregate required to fill a container of a specified unit volume. The volume referred to here is occupied by both aggregates and the voids between aggregate particles. The approximate bulk density of aggregate commonly used in normal-weight concrete ranges from about 75 to 110 lb per cubic foot. The void content between particles affects mortar requirements in mix design (see preceding section, "Particle Shape and Surface Texture" and "Grading"). Void contents range from about 30% to 45% for coarse aggregates to about 40% to 50% for fine

---

*References 4-4 and 4-6.
**Also see "High-Strength Concrete," Chapter 15.

aggregate. Angularity increases void content; larger sizes of well-graded aggregate and improved grading decreases void content (Fig. 4-4). Methods of determining the bulk density of aggregates and void content are given in ASTM C 29. Three methods are described for consolidating the aggregate in the container depending on the maximum size of the aggregate: rodding, jigging, and shoveling.

Specific Gravity

The specific gravity (relative density) of an aggregate is the ratio of its weight to the weight of an equal absolute volume of water (water displaced on immersion). It is used in certain computations for mixture proportioning and control, such as the absolute volume occupied by the aggregate. It is not generally used as a measure of aggregate quality, though some porous aggregates that exhibit accelerated freeze-thaw deterioration do have low specific gravities. Most natural aggregates have specific gravities between 2.4 and 2.9 with corresponding mass densities of 150 pcf and 181 pcf.*

Test methods for determining specific gravities for coarse and fine aggregates are described in ASTM C 127 and C 128, respectively. The specific gravity of an aggregate may be determined on an ovendry basis and a saturated surface-dry (SSD) basis. Both the ovendry and saturated surface-dry specific gravities may be used in concrete mixture proportioning calculations. Ovendry aggregates do not contain any absorbed or free water. They are dried in an oven to constant weight. Saturated surface-dry aggregates are aggregates in which the pores in each aggregate particle are filled with water and no excess water is on the particle surface.

Absorption and Surface Moisture

The absorption and surface moisture of aggregates should be determined according to ASTM C 70, C 127, C 128, and C 566 so that the net water content of the concrete can be controlled and correct batch weights determined. The internal structure of an aggregate particle is made up of solid matter and voids that may or may not contain water.

The moisture conditions of aggregates are shown in Fig. 4-7. They are designated as

1. **Ovendry**—fully absorbent
2. **Air dry**—dry at the particle surface but containing some interior moisture, thus still somewhat absorbent
3. **Saturated surface dry (SSD)**—neither absorbing water from nor contributing water to the concrete mixture
4. **Damp or wet**—containing an excess of moisture on the surface (free water)

The amount of water used for the concrete mixture at the jobsite must be adjusted for the moisture conditions of the aggregates in order to meet the designated water requirement. If the water content of the concrete mixture is not kept constant, the compressive strength,

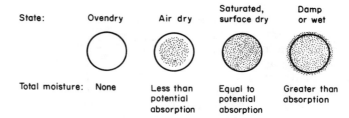

Fig. 4-7. Moisture conditions of aggregates.

workability, and other properties will vary from batch to batch. Coarse and fine aggregate will generally have absorption levels (moisture contents at SSD) in the range of about 0.2% to 4% and 0.2% to 2%, respectively. Free-water contents will usually range from 0.5% to 2% for coarse aggregate and 2% to 6% for fine aggregate. The maximum water content of drained coarse aggregate is usually less than that of fine aggregate. Most fine aggregates can maintain a maximum drained moisture content of about 3% to 8% whereas coarse aggregates can maintain only about 1% to 6%.

Bulking is the increase in total volume of moist fine aggregate over the same weight dry. Surface tension in the moisture holds the particles apart, causing an increase in volume. Bulking of fine aggregates occurs when they are shoveled or otherwise moved in a damp condition, even though they may have been fully consolidated beforehand. Fig. 4-8 illustrates how the

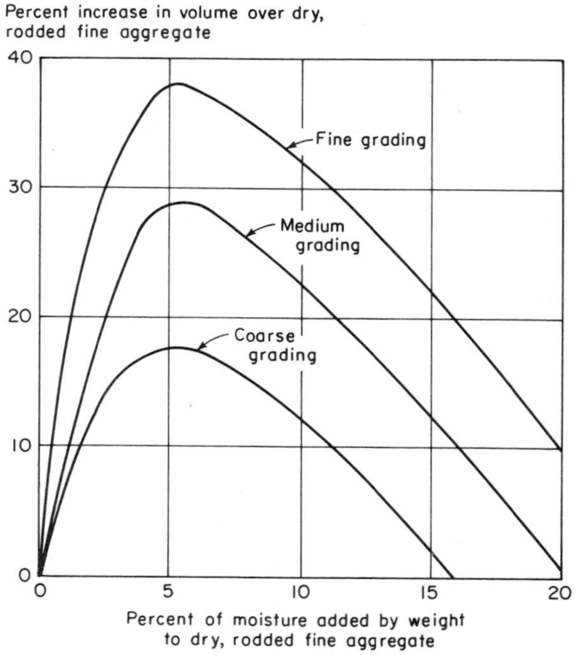

Percent increase in volume over dry, rodded fine aggregate

Fig. 4-8. Surface moisture on fine aggregate can cause considerable bulking, the amount of which varies with the amount of moisture and the aggregate grading. Reference: PCA Major Series 172 and PCA ST20.

---

*The density of the aggregate used in mixture proportioning computations (not including voids between particles) is determined by multiplying the specific gravity of the aggregate times the density of water. A value of 62.4 pcf is often used for the density of water.

amount of bulking of fine aggregates varies with moisture content and grading; fine gradings bulk more than coarse gradings for a given amount of moisture. Fig. 4-9 shows similar information in terms of weight for a particular fine aggregate. Since most fine aggregates are delivered in a damp condition, wide variations can occur in batch quantities if the batching is done according to volume. For this reason, good practice has long favored weighing the aggregate and adjusting for moisture content when proportioning concrete.

Resistance to Freezing and Thawing

The freeze-thaw resistance of an aggregate, an important characteristic in exterior concrete, is related to its porosity, absorption, permeability, and pore structure. An aggregate particle may absorb so much water (to critical saturation) that it cannot accommodate the expansion and hydraulic pressure that occurs during the freezing of water. The result is expansion of the aggregate and possible disintegration of the concrete if enough of the offending particles are present. If a problem particle is near the surface of the concrete, it can cause a popout. Popouts generally appear as conical fragments that break out of the concrete surface. The offending aggregate particle is usually at the bottom of the void. Generally it is coarse rather than fine aggregate particles with higher porosity values and

medium-sized pores (0.1 to 5 μm) that are easily saturated and cause concrete deterioration and popouts. Larger pores do not usually become saturated or cause concrete distress, and water in very fine pores may not freeze readily.

At any freezing rate, there may be a critical particle size above which a particle will fail if frozen when critically saturated. This critical size is dependent upon the rate of freezing and the porosity, permeability, and tensile strength of the particle. For fine-grained aggregates with low permeability (cherts for example), the critical particle size may be within the range of normal aggregate sizes. It is higher for coarse-grained materials or those with capillary systems interrupted by numerous macropores (voids too large to hold moisture by capillary action). For these aggregates the critical particle size may be sufficiently large to be of no consequence, even though the absorption may be high. If potentially vulnerable aggregates are used in concrete subjected to periodic drying while in service, they may never become sufficiently saturated to cause failure.

D-cracking is a freeze-thaw deterioration of concrete pavements that has been observed in some pavements after three or more years of service. D-cracks are closely spaced crack formations parallel to transverse and longitudinal joints that later multiply outward from the joints toward the center of the pavement panel (Fig. 4-10). D-cracking is a function of the pore properties of certain types of aggregate particles and the environment in which the pavement is placed. Due to the natural accumulation of water under pavements in the base and subbase structures, the aggregate may eventually become saturated. Then with freezing and thawing cycles, cracking of the concrete starts in the saturated aggregate (Fig. 4-11) at the bottom of the slab and progresses upward until it reaches the wearing

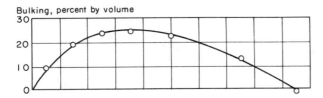

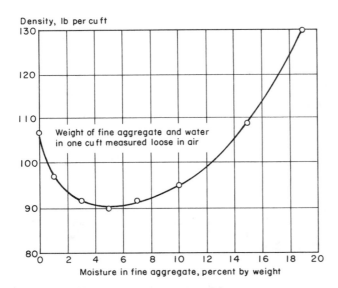

Fig. 4-9. The bulk weight per cubic foot and volume increase is compared with the moisture content for a particular sand. Reference: PCA Major Series 172.

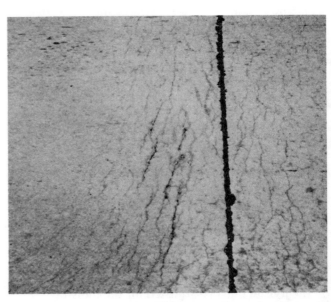

Fig. 4-10. Severe D-cracking along a transverse joint caused by failure of carbonate coarse aggregate. Reference 4-11.

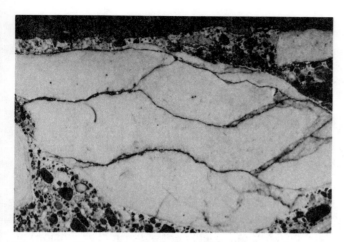

Fig. 4-11. Fractured carbonate aggregate particle as a source of distress in D-cracking (magnification 2.5X). Reference 4-11.

surface. This problem can be reduced either by selecting aggregates that perform better in freeze-thaw cycles or, where marginal aggregates must be used, by reducing the maximum particle size. Also, installation of effective drainage facilities for carrying free water from the pavement subbase may be helpful.

The performance of aggregates under exposure to freezing and thawing can be evaluated in two ways: past performance in the field and laboratory freeze-thaw tests of concrete specimens. If aggregates from the same source have previously given satisfactory service when used in concrete, they might be considered suitable. Aggregates not having a service record can be considered acceptable if they perform satisfactorily in air-entrained concretes subjected to freeze-thaw tests. In ASTM C 666 freeze-thaw tests, concrete specimens made with the aggregate in question are subjected to alternate cycles of freezing and thawing in water. Deterioration is measured by (1) the reduction in the dynamic modulus of elasticity, (2) linear expansion, and (3) weight loss of the specimens. An expansion failure criterion of 0.035% in 350 freeze-thaw cycles or less is used by a number of state highway departments to help indicate whether or not an aggregate is susceptible to D-cracking. Different aggregate types may influence the criteria levels and empirical correlations between laboratory freeze-thaw tests. Field service records should be made to select the proper criterion.

Specifications may require that resistance to weathering be demonstrated by a sodium sulfate or magnesium sulfate test (ASTM C 88). The test consists of a number of immersion cycles for a sample of the aggregate in a sulfate solution (to create a pressure through salt-crystal growth in the aggregate pores similar to that produced by freezing water). The sample is then ovendried and the percentage of weight loss calculated. Unfortunately, this test is sometimes misleading. Aggregates behaving satisfactorily in the test might produce concrete with low freeze-thaw resistance; conversely, aggregates performing poorly might produce concrete with adequate resistance. This is attributed,

at least in part, to the fact that the aggregates in the test are not confined by cement paste (as they would be in concrete) and the mechanisms of attack are not the same as in freezing and thawing. The test is most reliable for stratified rocks with porous layers or weak bedding planes.

Wetting and Drying Properties

Weathering due to wetting and drying can also affect the durability of aggregates. The expansion and contraction coefficients of rocks vary with temperature and moisture content. If alternate wetting and drying occurs, severe strain develops in some aggregates, and with certain types of rock this can cause a permanent increase in volume of the concrete and eventual breakdown. Clay lumps and other friable particles can degrade rapidly with repeated wetting and drying. Popouts can also develop due to the moisture-swelling characteristics of certain aggregates, especially clay balls and shales. While no specific tests are available to determine this tendency, an experienced petrographer can often be of assistance in determining this potential for distress.

Abrasion and Skid Resistance

The abrasion resistance of an aggregate is often used as a general index of its quality. Abrasion resistance is essential when the aggregate is to be used in concrete subject to abrasion, as in heavy-duty floors or pavements. Low abrasion resistance of an aggregate may increase the quantity of fines in the concrete during mixing and consequently may increase the water requirement.

The most common test for abrasion resistance is the Los Angeles abrasion test (rattler method) performed in accordance with ASTM C 131 or C 535. In this test, a specified quantity of aggregate is placed in a steel drum containing steel balls, the drum is rotated, and the percentage of material worn away is measured. Specifications often set an upper limit on this weight loss. However, a comparison of the results of aggregate abrasion tests with the abrasion resistance of concrete made with the same aggregate do not generally show a clear correlation. Weight loss due to impact is often as much as that due to abrasion. The wear resistance of concrete is determined more accurately by abrasion tests of the concrete itself (see Chapter 1).

To provide good skid resistance on pavements, the siliceous particle content of the fine aggregate should be at least 25%. For specification purposes, the siliceous particle content is considered equal to the insoluble residue content after treatment in hydrochloric acid under standardized conditions (ASTM D 3042). Certain manufactured sands produce slippery pavement surfaces and should be investigated for acceptance before use.

Strength and Shrinkage

The strength of an aggregate is rarely tested and generally does not influence the strength of normal-strength concrete as much as paste strength and paste-aggregate bond. However, aggregate strength does become important in high-strength concrete. Aggregate tensile strengths range from 300 to 2300 psi and compressive strengths from 10,000 to 40,000 psi.*

Different aggregate types have different compressibility, modulus of elasticity, and moisture-related shrinkage characteristics that can influence like properties of the concrete. Aggregates with high absorption properties may have high shrinkage properties on drying. Quartz and feldspar aggregates along with limestone, dolomite, and granite are considered low-shrinkage aggregates; and aggregates with sandstone, shale, slate, hornblende, and graywacke are often associated with high shrinkage in concrete.*

Resistance to Acid and Other Corrosive Substances

Portland cement concrete is durable in most natural environments; however, concrete in service is occasionally exposed to substances that will attack it.

Most acidic solutions will slowly or rapidly disintegrate portland cement concrete depending on the type and concentration of acid. Certain acids, such as oxalic acid, are harmless. Weak solutions of some acids have insignificant effects. Although acids generally attack and leach away the calcium compounds of the cement paste, they may not readily attack certain aggregates, such as siliceous aggregates. Calcareous aggregates often react readily with acids. However, the sacrificial effect of calcareous aggregates is often a benefit over siliceous aggregate in mild acid exposures or in areas where water is not flowing. With calcareous aggregate, the acid attacks the entire exposed concrete surface uniformly, reducing the rate of attack on the paste and preventing loss of aggregate particles at the surface. Calcareous aggregates also tend to neutralize the acid, especially in stagnant locations. Acids can also discolor concrete. Siliceous aggregate should be avoided when strong solutions of sodium hydroxide are present, as these attack this type of aggregate.

Acid rain (often with a pH of 4 to 4.5) can slightly etch concrete surfaces, usually without affecting the performance of exposed concrete structures. Extreme acid rain or strong acid water conditions may warrant special concrete designs or precautions, especially in submerged areas. Continuous replenishment in acid with a pH of less than 4 is considered highly aggressive to buried concrete, such as pipe.** Concrete continuously exposed to liquid with a pH of less than 3 should be protected in a similar manner as recommended for concrete exposed to dilute acid solutions.† Natural waters usually have a pH of more than 7 and seldom less than 6. Waters with a pH greater than 6.5 may be aggressive if they contain bicarbonates. Carbonic acid solutions with concentrations between 0.9 and 3 parts per million are considered to be destructive to concrete.†

A low water-cement ratio, low permeability, and low-to-moderate cement content can increase the acid or corrosion resistance of concrete. A low permeability resulting from a low water-cement ratio helps keep the corrosive agent from penetrating into the concrete. Low-to-moderate cement contents (5 to 7 bags per cubic yard) result in less available paste to attack. The use of sacrificial calcareous aggregates should be considered where indicated.

Certain acids, gases, salts, and other substances that are not mentioned here also can distintegrate concrete. Acids and other chemicals that severely attack portland cement concrete should be prevented from coming in contact with the concrete by using protective coatings.††

Fire Resistance and Thermal Properties

The fire resistance and thermal properties (conductivity, diffusivity, and coefficient of thermal expansion) of concrete depend to some extent on the mineral constituents of the aggregates used. Manufactured and some naturally occurring lightweight aggregates are more fire resistant than normal-weight aggregates due to their insulating properties and high-temperature stability. Concrete containing a calcareous coarse aggregate performs better under fire exposure than a concrete containing quartz or siliceous aggregate such as granite or quartzite. At about 1060°F, quartz expands 0.85% causing disruptive expansion.‡ The coefficient of thermal expansion of aggregates ranges from 1×10^{-6}°F to 9×10^{-6}°F. For more information refer to Chapter 13 for temperature-induced volume changes and to Chapter 15 for thermal conductivity and mass concrete considerations.

HARMFUL MATERIALS AND ALKALI REACTIVITY

Harmful substances that may be present in aggregates include organic impurities, silt, clay, shale, iron oxide, coal, lignite, and certain lightweight and soft particles (Table 4-4). In addition, rocks and minerals such as some cherts, strained quartz,‡‡ and certain dolomitic limestones are alkali reactive (see Table 4-5). Gypsum and anhydrite may cause sulfate attack. Certain aggregates, such as some shales, will cause popouts by swelling simply by absorbing water or by freezing of the water present (Fig. 4-12). Most specifications limit the permissible amounts of these substances in aggregates. The performance history of an aggregate should be a determining factor in setting the limits for harmful

---

*Reference 4-22.
**Reference 4-32, pages 496-497.
†Reference 4-23.
††The effects of various substances on concrete and guidelines for protective treatments and coatings are discussed in References 4-23 and 4-28. References 4-19 and 4-22 also present additional information on chemical attack.
‡References 4-18 and 4-22.
‡‡Reference 4-20.

Fig. 4-12. A popout is the breaking away of a small fragment of concrete surface due to internal pressure that leaves a shallow, typically conical depression.

Table 4-4. Harmful Materials in Aggregates

Substances	Effect on concrete	Test designation
Organic impurities	Affect setting and hardening, may cause deterioration	ASTM C 40 ASTM C 87
Materials finer than No. 200 (80-μm) sieve	Affect bond, increase water requirement	ASTM C 117
Coal, lignite, or other lightweight materials	Affect durability, may cause stains and popouts	ASTM C 123
Soft particles	Affect durability	
Clay lumps and friable particles	Affect workability and durability, may cause popouts	ASTM C 142
Chert of less than 2.40 specific gravity	Affects durability, may cause popouts	ASTM C 123 ASTM C 295
Alkali-reactive aggregates	Abnormal expansion, map cracking, popouts	ASTM C 227 ASTM C 289 ASTM C 295 ASTM C 342 ASTM C 586

Table 4-5. Some Potentially Harmful Reactive Minerals, Rocks, and Synthetic Materials

Alkali-silica reactive substances*		Alkali-carbonate reactive substances**
Andesites	Opal	Calcitic dolomites
Argillites	Opaline shales	Dolomitic limestones
Certain siliceous limestones and dolomites	Phyllites	Fine-grained dolomites
	Quartzites	
	Quartzoses	
Chalcedonic cherts	Cherts	
Chalcedony	Rhyolites	
Cristobalite	Schists	
Dacites	Siliceous shales	
Glassy or cryptocrystalline volcanics	Strained quartz and certain other forms of quartz	
Granite gneiss	Synthetic and natural silicious glass	
Graywackes		
Metagraywackes	Tridymite	

*Several of the rocks listed (granite gneiss and certain quartz formations for example) react very slowly and may not show evidence of harmful degrees of reactivity until the concrete is over 20 years old.
**Only certain sources of these materials have shown reactivity.

substances. ASTM test methods for detecting such substances qualitatively or quantitatively are listed in Table 4-4.

Aggregates are potentially harmful if they contain compounds known to react chemically with portland cement concrete and produce (1) significant volume changes of the paste, aggregates, or both, (2) interference with normal hydration of cement, and (3) otherwise harmful byproducts.

Organic impurities may delay setting and hardening of concrete, may reduce strength gain, and in unusual cases may cause deterioration. Organic impurities such as peat, humus, and organic loam may not be as detrimental but should be avoided.

Materials finer than the No. 200 sieve, especially silt and clay, may be present as loose dust and may form a coating on the aggregate particles. Even thin coatings of silt or clay on gravel particles can be harmful because they may weaken the bond between the cement paste and aggregate. If certain types of silt or clay are present in excessive amounts, water requirements may have to be increased significantly.

Coal or lignite—or other low-density materials such as wood or fibrous materials—in excessive amounts will affect the durability of concrete. If these impurities occur at or near the surface, they might disintegrate, pop out, or cause stains. Potentially harmful chert in coarse aggregate can be identified by ASTM C 123.

Soft particles in coarse aggregate are objectionable because they can affect durability and wear resistance of concrete and cause popouts. If friable, they could break up during mixing and thereby increase the amount of water required. Where abrasion resistance is critical, such as in heavy-duty floors, testing may indicate that further investigation or another aggregate source is warranted.

Clay lumps present in concrete may absorb some of the mixing water, cause popouts in hardened concrete, and affect durability and wear resistance. They can also break up during mixing and thereby increase the mixing-water demand.

Aggregates can occasionally contain particles of iron oxide and iron sulfide that result in unsightly stains on exposed concrete surfaces. The aggregate should meet the staining requirements of ASTM C 330 when tested according to ASTM C 641, and the quarry face and stockpiles should not show evidence of staining.

As an additional aid in identifying staining particles, the aggregate can be immersed in a lime slurry. If staining particles are present, a blue-green gelatinous precipitate will form within 5 to 10 minutes; this will rapidly change to a brown color on exposure to air and light. The reaction should be complete within 30 minutes. If no brown gelatinous precipitate is formed when a suspect aggregate is placed in the lime slurry, there is little likelihood of any reaction taking place in concrete. These tests should be required when aggregates with no record of successful prior use are used in architectural concrete.

Alkali-Aggregate Reactivity

Chemically stable aggregates in concrete do not react chemically with cement in a harmful manner. However, aggregates with certain mineral constituents (such as certain forms of silica or carbonates) will react with alkalies (sodium oxide and potassium oxide) in cement, particularly when the concrete is subject to a warm, moist environment. Certain aggregates also contain potentially harmful leachable alkalies. In order of concern and occurrence are alkali-silica reactivity, alkali-carbonate reactivity, and other cement-aggregate reactions. The alkali-silica reaction generally forms reaction products that can cause excessive expansion and cracking or popouts in concrete (Figs. 4-13 and 4-14). Alkali-silica gel in voids and cracks, aggregate reaction rims, and microcracks are commonly present with alkali-silica reactivity. Although many carbonate rocks react with cement hydration products, very rarely do they produce expansive reactions. Expansive alkali-carbonate reactivity is suspect only in extremely fine grained dolomitic limestones with large amounts of calcite, clay, silt, or dolomite rhombs found in a matrix of clay and fine calcite (see Table 4-5). Alkali reactivity is influenced by the amount, type, and particle size of reactive material as well as soluble alkali and water content of the concrete.

Alkali reactivity can be significantly reduced by keeping the concrete as dry as possible and the reactivity can be virtually stopped if the internal relative humidity of the concrete is kept below 80%. In most cases, however, this condition is difficult to achieve and maintain. Warm seawater, due to the presence of dissolved alkalies, can particularly aggravate alkali reactivity.*

Blended cements or mineral admixtures are beneficial when used in concrete that otherwise would be affected by deleterious alkali-silica reactivity. Pozzolans produce additional calcium silicate hydrates which retain some of the available alkali in the concrete, thereby reducing the amount of alkali available to react with reactive aggregate. Pozzolans with high water-soluble alkali contents should be avoided as they may increase reactivity. The quantity and type of mineral admixture should be determined by test for a particular concrete mixture (see PCA IS415T and ASTM C 441).**

Limestone sweetening (the popular term for replacing approximately 30% of the reactive sand-gravel aggregate with crushed limestone) is effective in preventing deterioration in some sand-gravel aggregate concretes. Smaller amounts of limestone dust may also provide adequate protection. Potential expansion of these cement-aggregate combinations can be determined by measuring the length change of mortar bars during storage under prescribed temperature and moisture conditions in accordance with ASTM C 342.

If the above options are not available, alkali-silica reactivity can be controlled by limiting the alkali content of the concrete when reactive aggregates are used. This and other options are discussed in Reference 4-34.

Field service records generally provide the best information for selection of aggregates. If an aggregate has no service record and is suspected of being potentially reactive, laboratory tests should be made. Three ASTM tests for identifying alkali-reactive aggregates are discussed in the following paragraphs. In addition, an ASTM C 295 petrographic examination performed by a qualified petrographer can be helpful in identifying potentially reactive aggregates.

ASTM C 227, a mortar-bar test, is used to determine the potentially expansive alkali-silica reactivity of cement-aggregate combinations. In this test, the expansion developed in small mortar bars during storage under prescribed temperature and moisture conditions is measured. Mortar-bar tests can be used for either fine or coarse aggregates. Generally three to six months must elapse before conclusions can be drawn.

Fig. 4-13. Deterioration of concrete from alkali-aggregate reactivity.

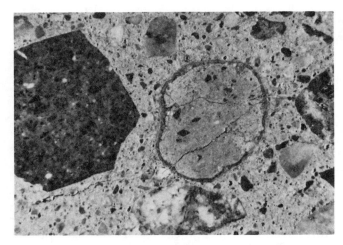

Fig. 4-14. Polished section view of an alkali reactive aggregate. Observe the alkali-silica reaction rim around the reactive aggregate and the crack formation.

---

*Reference 4-21
**Reference 4-25, 4-29, 4-32, 4-33, and 4-34.

ASTM C 289, a quick chemical test, is used for identifying potentially reactive siliceous aggregates. It can be completed in two or three days. Conclusions are based on the degree of reaction that occurs between a sodium hydroxide solution and a crushed specimen of the aggregate in question. This test is not completely dependable.

ASTM P 214 (Proposed Test Method for Accelerated Detection of Potential Deleterious Expansion of Mortar Bars Due to Alkali-Silica Reaction, 1991) is a 14-day mortar bar test. Reference 4-34 provides detailed guidance to determining which aggregates are potentially reactive using P 214, along with petrography and a concrete prism test. Reference 4-34 also uses P 214 and C 441 to determine the effectiveness of mineral admixtures and blended cements to control reactivity.

ASTM C 586, the rock-cylinder test, is used to determine potentially expansive carbonate rock aggregates (alkali-carbonate reactivity). Length changes are measured on a sample that has been immersed in a sodium hydroxide solution. Expansive tendencies are usually observable after immersion for 28 days. Restrictions on alkali content of cement and the use of pozzolanic inhibitors are not as effective as for the alkali-silica reaction.

It must be pointed out that in quite a few acceptable commercial aggregates, a portion of the aggregate will have potentially reactive minerals. It is accepted engineering procedure to use past service records coupled with laboratory test data and petrographic examination to make a judgment whether, for the intended concrete exposure, there is enough risk of deleterious expansion to justify switching to another aggregate source or to adopt other precautions. Again, service records for field performance are the most important criteria. A petrographic analysis (ASTM C 856) on hardened concrete will identify alkali reactivity in concrete that has experienced deterioration due to these reactions.*

AGGREGATE BENEFICIATION

Aggregate processing consists of (1) basic processing—crushing, screening, and washing—to obtain proper gradation and cleanliness, and (2) beneficiation—upgrading quality by processing methods such as heavy media separation, jigging, rising-current classification, and crushing.

In heavy media separation, aggregates are passed through a heavy medium comprised of finely ground heavy minerals and water proportioned to have a specific gravity less than that of the aggregate particles but greater than that of the deleterious particles. The heavier particles sink and the lighter particles float. This process can be used when acceptable and harmful particles have distinguishable specific gravities.

Jigging separates particles with small differences in specific gravity by pulsating water current. Upward pulsations of water through a jig (a box with a perforated bottom) move the lighter material into a layer on top of the heavier material. The top layer is then removed.

Rising-current classification separates particles with large differences in specific gravities. Light materials, such as wood and lignite, are floated away in a rapidly upward moving stream of water.

Crushing removes soft and friable particles from coarse aggregates. This process is sometimes the only means of making material suitable for use. Unfortunately, with any process some acceptable material is always lost and removal of harmful particles may be difficult or expensive.

HANDLING AND STORING AGGREGATES

Aggregates should be handled and stored in a way that minimizes segregation and degradation and prevents contamination by deleterious substances. Stockpiles should be built up in thin layers of uniform thickness to minimize segregation. The most economical and acceptable method of forming aggregate stockpiles is the truck-dump method, which discharges the loads in a way that keeps them tightly joined. The aggregate is then reclaimed with a front-end loader. The loader or reclaimer should remove slices from the edges of the pile from top to bottom so that every slice will contain a portion of each horizontal layer.

When the aggregates are not delivered by truck, acceptable and least expensive results are obtained by forming the stockpile in layers with a clamshell bucket (cast-and-spread method, Fig. 4-1) or, in the case of aggregates not subject to degradation, by spreading the aggregates with a rubber-tire dozer and reclaiming with a front-end loader. By spreading the material in thin layers, segregation is minimized. Whether aggregates are handled by truck, bucket loader, clamshell, or conveyor, stockpiles should not be built up in high, cone-shaped piles since this results in segregation. However, if circumstances necessitate construction of a conical pile or if a stockpile has segregated, gradation variations can be minimized when the pile is reclaimed if the aggregates are loaded by continually moving around the circumference of the pile to blend sizes rather than by starting on one side and working straight through the pile.

Crushed aggregates segregate less than rounded (gravel) aggregates and larger-size aggregates segregate more than smaller sizes. To avoid segregation of coarse aggregates, size fractions can be stockpiled and batched separately. Proper stockpiling procedures, however, should eliminate the need for this requirement. Specifications provide a range in the amount of aggregates permitted in any size fraction partly because of segregation in stockpiling and batching operations.

Washed aggregates should be stockpiled in sufficient time before use so that they can drain to a uniform moisture content. Damp fine material has less tenden-

*Reference 4-30.

cy to segregate than dry material. When dry fine aggregate is dropped from buckets or conveyors, the wind can blow out the fines. This should be avoided if possible.

Bulkheads or dividers should be used to avoid contamination of the stockpiles. Partitions between bin compartments should be high enough to prevent intermingling of materials. Storage bins should preferably be circular or nearly square. Their bottoms should slope not less than 50 degrees from the horizontal on all sides to a center outlet. The material should fall vertically over the outlet into the bin. Chuting the material into a bin at an angle and against the bin sides will cause it to segregate. Baffle plates or dividers will help minimize segregation. Bins should be kept as full as possible since this reduces the breakage of aggregate particles and the tendency to segregate.*

MARINE-DREDGED AGGREGATE

Marine-dredged aggregate from tidal estuaries and sand and gravel from the seashore can be used with caution in limited concrete applications when other aggregate sources are not available. Aggregates obtained from seabeds have two problems: (1) seashells and (2) salt.

Seashells may be present in the aggregate source. These shells are hard material that can produce good quality concrete; however, a higher cement content may be required. Due to the angularity of the shells, additional cement paste is required to obtain the desired workability. Aggregate containing complete shells (uncrushed) should be avoided as their presence will result in voids in the concrete and lower the compressive strength.

Marine-dredged aggregates often contain salt from the seawater. The primary salts are sodium chloride and magnesium sulfate and the amount of salt on the aggregate is often not more than about 1% of the weight of the mixing water. The highest salt content occurs in sands located just above the high-tide level. Use of these aggregates with drinkable mix water often contributes less salt to the mixture than the use of seawater with salt-free aggregates. Marine aggregates can be an appreciable source of chlorides. The presence of these chlorides may affect the concrete by (1) altering the time of set, (2) increasing drying shrinkage, (3) significantly increasing the risk of corrosion of steel reinforcement, and (4) causing efflorescence. Generally, marine aggregates containing large amounts of chloride should not be used in reinforced concrete.

Marine-dredged aggregates can be washed with fresh water to reduce the salt content. There is no maximum limit on the salt content of coarse or fine aggregate; however, the chloride limits presented in Chapter 7 should be followed.

RECYCLING OLD CONCRETE

In recent years, the concept of using old concrete pavements, buildings, and other structures as a source of aggregate has been demonstrated on several projects, resulting in both material and energy savings. The procedure involves (1) breaking up and removing the old concrete, (2) crushing in primary and secondary crushers, (3) removing reinforcing steel and embedded items, (4) grading and washing, and (5) finally stockpiling the resulting coarse and fine aggregate. Dirt, gypsum board, wood, and other foreign materials should be prevented from contaminating the final product.

Recycled concrete is simply old concrete that has been crushed to produce aggregate. Recycled concrete is primarily used in pavement reconstruction. It has been satisfactorily used as an aggregate in granular subbases, lean-concrete subbases, soil-cement, and in new concrete as the only source of aggregate or as a partial replacement of new aggregate. Recycled concrete aggregate generally has a higher absorption and a lower specific gravity than conventional aggregate. The particle shape is similar to crushed rock as shown in Fig. 4-15.

The sulfate content of recycled concrete should be determined to assess the possibility of deleterious sulfate reactivity. The water-soluble chloride content should also be determined where applicable.

New concrete made from recycled concrete aggregate generally has good workability, durability, and resistance to saturated freeze-thaw action. The compressive strength will vary with the compressive strength of the original concrete and the water-cement ratio of the new concrete. The compressive strength may be increased by using a higher cement content and replacing

Fig. 4-15. Recycled-concrete aggregate.

---

*Recommended methods of handling aggregates are discussed at length in References 4-7, 4-9, and 4-16.

some of the recycled concrete aggregate with conventional aggregate. The new concrete will also have a lower density. As with any new aggregate source, recycled concrete aggregate should be tested for durability, gradation, and other properties.

Concrete trial mixtures should be made to check the new concrete's quality and to determine the proper mixture proportions. One major problem with using recycled concrete is the variability in the properties of the old concrete that will in turn affect the properties of the new concrete. This can partially be avoided by frequent monitoring of the properties of the old concrete that is being recycled. Adjustments in the mixture proportions may then need to be made.

REFERENCES

4-1. *Symposium on Mineral Aggregates,* ASTM STP83, American Society for Testing and Materials, Philadelphia, 1948.

4-2. Brown, L. S., *Some Observations on the Mechanics of Alkali-Aggregate Reaction,* Research Department Bulletin RX054, Portland Cement Association, 1955.

4-3. Verbeck, George, and Landgren, Robert, *Influence of Physical Characteristics of Aggregates on Frost Resistance of Concrete,* Research Department Bulletin RX126, Portland Cement Association, 1960.

4-4. Houston, B. J., *Investigation of Gap-Grading of Concrete Aggregates; Review of Available Information,* Technical Report No. 6-593, Report 1, Waterways Experiment Station, U.S. Army Corps of Engineers, Vicksburg, Mississippi, February 1962.

4-5. Hadley, David W., *Alkali Reactivity of Dolomitic Carbonate Rocks,* Research Department Bulletin RX176, Portland Cement Association, 1964.

4-6. Litvin, Albert, and Pfeifer, Donald W., *Gap-Graded Mixes for Cast-in-Place Exposed Aggregate Concrete,* Development Department Bulletin DX090, Portland Cement Association, 1965.

4-7. Matthews, C. W., "Stockpiling of Materials," *Rock Products,* series of 21 articles, Maclean Hunter Publishing Company, Chicago, August 1965 through August 1967.

4-8. "Tests and Properties of Concrete Aggregates," Part III, *Significance of Tests and Properties of Concrete and Concrete-Making Materials,* STP 169-A, American Society for Testing and Materials, 1966, pages 379-512.

4-9. *Effects of Different Methods of Stockpiling and Handling Aggregates,* NCHRP Report 46, Transportation Research Board, Washington, D.C., 1967.

4-10. Stark, David, and Klieger, Paul, *Effect of Maximum Size of Coarse Aggregate on D-Cracking in Concrete Pavements,* Research and Development Bulletin RD023P, Portland Cement Association, 1974.

4-11. Stark, David, *Characteristics and Utilization of Coarse Aggregates Associated with D-Cracking,* Research and Development Bulletin RD047P, Portland Cement Association, 1976.

4-12. Buck, Alan D., "Recycled Concrete as a Source of Aggregate," *ACI Journal,* American Concrete Institute, Detroit, May 1977, pages 212-219.

4-13. *Acid Rain, Research Summary,* EPA-600/8-79-028, U.S. Environmental Protection Agency, Washington, D.C., October 1979.

4-14. Ray, Gordon K., "Quarrying Old Pavements to Build New Ones," *Concrete Construction,* Concrete Construction Publications, Inc., Addison, Illinois, October 1980, pages 725-729.

4-15. *Recycling D-Cracked Pavement in Minnesota,* PL146P, Portland Cement Association, 1980.

4-16. *Concrete Manual,* 8th ed., U.S. Bureau of Reclamation, Denver, revised 1981.

4-17. Stark, D. C., *Alkali-Silica Reactivity: Some Reconsiderations,* Research and Development Bulletin RD076T, Portland Cement Association, 1981.

4-18. *Guide for Determining the Fire Endurance of Concrete Elements,* ACI 216R-81, ACI Committee 216 Report, American Concrete Institute, 1981.

4-19. *Guide to Durable Concrete,* ACI 201.2R-77, reaffirmed 1982, ACI Committee 201 Report, American Concrete Institute.

4-20. Buck, Alan D.; Mather, Katharine, *Reactivity of Quartz at Normal Temperatures,* Technical Report SL-84-12, Structures Laboratory, Waterways Experiment Station, U.S. Army Corps of Engineers, July 1984.

4-21. Buck, Alan D.; Mather, Katharine; and Mather, Bryant, *Cement Composition and Concrete Durability in Sea Water,* Technical Report SL-84-21, Structures Laboratory, Waterways Experiment Station, U.S. Army Corps of Engineers, December 1984.

4-22. *Guide for Use of Normal Weight Aggregates in Concrete,* ACI 221R-84, ACI Committee 221 Report, American Concrete Institute, 1984.

4-23. *A Guide to the Use of Waterproofing, Dampproofing, Protective, and Decorative Barrier Systems for Concrete,* ACI 515.1R-79, revised 1985, ACI Committee 515, American Concrete Institute.

4-24. "Popouts: Causes, Prevention, Repair," *Concrete Technology Today,* PL852B, Portland Cement Association, June 1985.

4-25. Bhatty, Muhammad S. Y., "Mechanism of Pozzolanic Reactions and Control of Alkali-Aggregate Expansion" *Cement, Concrete, and Aggregates,* American Society for Testing and Materials, Winter 1985.

4-26. *Final Report, U.S. EPA Workshop on Acid Deposition Effects on Portland Cement Concrete and Related Materials,* Atmospheric Sciences Research Laboratory, U.S. Environmental Protection Agency, Research Triangle Park, North Carolina, February 1986.

4-27. *New Pavement from Old—Michigan Recycles Concrete,* PL215P, Portland Cement Association, 1986.

4-28. *Effects of Substances on Concrete and Guide to Protective Treatments,* IS001T, Portland Cement Association, 1986.

4-29. *Alkalies in Concrete,* STP 930, American Society for Testing and Materials, 1986.

4-30. Kosmatka, Steven H., "Petrographic Analysis of Concrete," *Concrete Technology Today,* PL862B, Portland Cement Association, July 1986.

4-31. Kong, Hendrik, and Orbison, James G., "Concrete Deterioration Due to Acid Precipitation," *ACI Materials Journal,* American Concrete Institute, March-April 1987.

4-32. *Concrete Durability, Katharine and Bryant Mather International Conference,* SP100, American Concrete Institute, 1987.

4-33. Buck, Alan D., and Mather, Katharine, *Methods for Controlling Effects of Alkali-Silica Reaction in Concrete,* Technical Report SL-87-6, Structures Laboratory, Waterways Experiment Station, U.S. Army Corps of Engineers, 1987.

4-34. *Guide Specification for Concrete Subject to Alkali-Silica Reactions,* IS415T, Portland Cement Association, 1994.

CHAPTER 5
Air-Entrained Concrete

One of the greatest advances in concrete technology was the development of air-entrained concrete in the mid-1930's. Today air entrainment is recommended for nearly all concretes, principally to improve their resistance to freezing when exposed to water and de-icing chemicals. However, there are other important benefits of entrained air in both freshly mixed and hardened concrete.

Air-entrained concrete is produced by using either an air-entraining cement or adding an air-entraining agent that stabilizes bubbles formed during the mixing process. The air-entraining agent enhances the incorporation of bubbles of various sizes by lowering the surface tension of the mixing water.

Anionic air-entraining agents* are hydrophobic (repel water) and are electrically charged. The negative electric charge is attracted to positively charged cement grains, which aids in stabilizing bubbles. The air-entraining agent forms a tough water-repelling film—similar to a soap film—with sufficient strength and elasticity to contain and stabilize the air bubbles and prevent them from coalescing. The hydrophobic film also keeps water out of the bubbles. The stirring and kneading action of mechanical mixing disperses the air bubbles. The fine aggregate particles also act as a three-dimensional grid to aid in holding the bubbles in the mixture.

Unlike entrapped air voids, which occur in all concretes and are largely a function of aggregate characteristics, intentionally entrained air bubbles are extremely small in size and are between 10 to 1000 μm in diameter. Entrapped voids are 1000 μm (1 mm) or larger. The majority of the entrained air voids in normal concrete are between 10 μm and 100 μm in diameter. As shown in Fig. 5-1, the bubbles are not interconnected and are well distributed. Non-air-entrained concrete with a 1-in. maximum-size aggregate has an air content of approximately 1½%. This same mixture air entrained for severe frost exposure would require an air content of about 6%.

PROPERTIES OF AIR-ENTRAINED CONCRETE

The primary concrete properties influenced by air entrainment are presented in the following sections. A

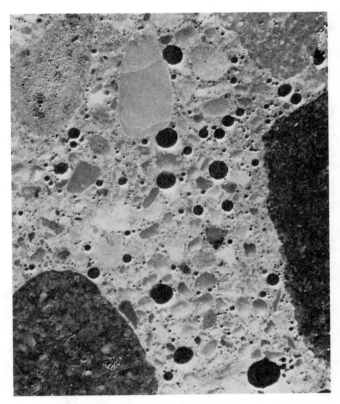

Fig. 5-1. Polished section of air-entrained concrete as seen through a microscope.

brief summary of other properties not discussed below is presented in Table 5-1.

Freeze-Thaw Resistance

The resistance of hardened concrete to freezing and thawing in a moist condition is significantly improved by the use of intentionally entrained air, even when various deicers are involved. Convincing proof of the improvement in durability effected by air entrainment is shown in Figs. 5-2 and 5-3.

---

*Nonionic admixtures are also available.

Table 5-1. Effect of Entrained Air on Concrete Properties

Properties	Effect
Abrasion	Little effect; increased strength increases abrasion resistance
Absorption	Little effect
Alkali-silica reactivity	Expansion decreases with increased air
Bleeding	Reduced significantly
Bond to steel	Decreased
Compressive strength	Reduced approximately 2% to 6% per percentage point increase in air; harsh or lean mixes may gain strength
Creep	Little effect
Deicer scaling	Significantly reduced
Fatigue	Little effect
Flexural strength	Reduced approximately 2% to 4% per percentage point increase in air
Freeze-thaw resistance	Significantly improved resistance to water-saturated freeze-thaw deterioration
Heat of hydration	No significant effect
Modulus of elasticity (static)	Decreases with increased air approximately 105,000 to 200,000 psi per percentage point of air
Permeability	Little effect; reduced water-cement ratio reduces permeability
Scaling	Significantly reduced
Shrinkage (drying)	Little effect
Slump	Increases with increased air approximately 1 in. per ½ to 1 percentage point of air
Specific heat	No effect
Sulfate resistance	Significantly improved
Temperature of wet concrete	No effect
Thermal conductivity	Decreases 1% to 3% per percentage point increase of air
Thermal diffusivity	Decreases about 1.6% per percentage point increase of air
Unit weight	Decreases with increased air
Water demand of wet concrete for equal slump	Decreases with increased air; approximately 5 to 10 pounds per cubic yard per percentage point of air
Watertightness	Increases slightly; reduced water-cement ratio increases watertightness
Workability	Increases with increased air

Note: The table information may not apply to all situations.

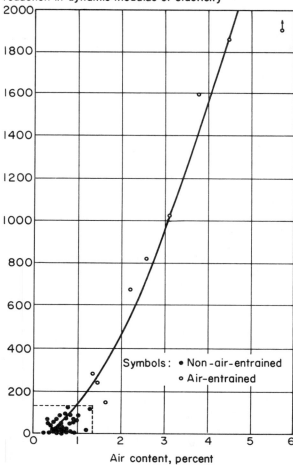

Fig. 5-2. Effect of entrained air on the resistance of concrete to freezing and thawing in laboratory tests. Concretes were made with cements of different fineness and composition and with various cement contents and water-cement ratios. References 5-12 and 5-22.

As the water in moist concrete freezes, it produces osmotic and hydraulic pressures in the capillaries and pores of the cement paste and aggregate. If the pressure exceeds the tensile strength of the paste or aggregate, the cavity will dilate and rupture. The accumulative effect of successive freeze-thaw cycles and disruption of paste and aggregate eventually cause significant expansion and deterioration of the concrete. Deterioration is visible in the form of cracking, scaling,* and crumbling.

Hydraulic pressures are caused by the 9% expansion of water upon freezing, in which growing ice crystals displace unfrozen water. If a capillary is above critical saturation (91.7% filled with water), hydraulic pressures result as freezing progresses. At lower water contents, no hydraulic pressure should exist. At critical saturation, all of the capillary void space would become filled with ice upon freezing, theoretically with no development of pressure.

Osmotic pressures develop from differential concentrations of alkali solutions in the paste. As ice develops, it creates an adjacent high-alkali solution. The high-alkali solution, through the mechanism of osmosis, draws water from lower alkali solutions in the pores. This osmotic transfer of water continues until equilibrium in the fluids' alkali concentration is achieved. Osmotic pressure is considered a minor factor, if pres-

*Scaling is the flaking or peeling off of surface mortar or concrete, usually caused by freezing and thawing of concrete in a damp or water-saturated environment (Fig. 5-3). Several factors can aggravate scaling (see Chapter 9).

Fig. 5-3. Effect of weathering on slabs on ground at the Long-Time Study outdoor test plot, Project 10C, PCA, Skokie, Illinois. Specimen at top is air-entrained, specimen at bottom exhibiting severe scaling is non-air-entrained. Both were made with 564 lb of Type I portland cement per cubic yard. Periodically 0.9 lb of calcium chloride was applied per square yard of specimen per winter. Specimens were 20 years old when photographed. Reference 5-23.

ent at all, in aggregate frost action, whereas it may be dominant in certain cement pastes. Osmotic pressures, as described above, are considered to be a major factor in "salt scaling."

Capillary ice (or any ice in large voids or cracks) draws water from pores to advance its growth. Also, since most pores in cement paste and some aggregates are too small for ice crystals to form, water attempts to migrate to locations where it can freeze.

Entrained air voids act as empty chambers in the paste for the freezing and migrating water to enter, thus relieving the pressures described above and preventing damage to the concrete. Upon thawing, most of the water returns to the capillaries due to capillary action

and pressure from air compressed in the bubbles. Thus the bubbles are ready to protect the concrete from the next cycle of freezing.*

The pressure developed by water as it expands during freezing depends largely upon the distance the water must travel to the nearest air void for relief. Therefore, the voids must be spaced close enough to reduce the pressure below that which would exceed the tensile strength of the concrete.

The spacing and size of air voids are important factors contributing to the effectiveness of air entrainment in concrete. ASTM C 457 describes a means of evaluating the air-void system in hardened concrete. Most authorities consider the following air-void characteristics as representative of a system with adequate freeze-thaw resistance:**

1. Calculated spacing factor, $\bar{L}$, (average maximum distance from any point in cement paste to the edge of the nearest air void)—less than 0.008 in.
2. Specific surface, α, (surface area of the air voids)— 600 sq in. per cubic inch of air-void volume, or greater.
3. Number of voids per linear inch of traverse, n,—at least one and a half to two times greater than the numerical value of the percentage of air in the concrete.

Current field control practice involves only the measurement of air volume in freshly mixed concrete. Although measurement of air volume alone does not permit full evaluation of the important characteristics of the air-void system, air-entrainment is generally considered effective for freeze-thaw resistance when the volume of air in the mortar fraction of the concrete (material passing the No. 4 sieve) is about 9 ± 1%.† The total required concrete air content for durability increases as the coarse-aggregate size is reduced and the exposure conditions become more severe (see "Recommended Air Contents" in this chapter and see also Chapter 7).

Freeze-thaw resistance is also significantly increased with the use of good quality aggregate, a low water-cement ratio (0.50 or less), a minimum cement content of 564 lb per cu yd,†† and proper finishing and curing techniques.‡ Concrete elements should be properly drained and kept as dry as possible as greater degrees of saturation increase the likelihood of distress due to freeze-thaw cycles. Concrete that is dry or contains only a small amount of moisture is essentially not

---

*References 5-12, 5-17, 5-22, and 5-26.
**For more information, refer to References 5-8, 5-11, 5-19, 5-21, 5-25, 5-26, and 5-27.
†Reference 5-11. Note that for equal admixture dosage rates per pound of cement the air content of ASTM C 185 standard mortar would be about 19% due to the standard aggregate's properties. The air content of concrete with ¾-in. maximum-size aggregate would be about 6%. The relationship between air content of standard mortar and concrete is illustrated in Reference 5-5.
††References 5-32 and 5-38.
‡A minimum compressive strength of 3500 psi to 4000 psi is desirable before air-entrained concrete is exposed to repeated, saturated freeze-thaw cycling.

affected by even a large number of cycles of freezing and thawing. Refer to the section on "Resistance to Deicers and Salts" and "Recommended Air Contents" in this chapter and to Chapter 7 for mixture design considerations.

Resistance to Deicers and Salts

Deicing chemicals used for snow and ice removal can cause and aggravate surface scaling. The damage is primarily a physical action. Deicer scaling of inadequately air-entrained or non-air-entrained concrete during freezing is believed to be primarily caused by a buildup of osmotic and hydraulic pressures in excess of the normal hydraulic pressures produced when water in concrete freezes. These pressures become critical and, unless entrained air voids are present to act as relief valves, scaling results. The hygroscopic (moisture absorbing) properties of salts also attract water and keep the concrete more saturated, increasing the potential for freeze-thaw deterioration. Properly designed and placed air-entrained concrete will withstand deicers for many years.

Studies have also shown that the formation of salt crystals in concrete may contribute to concrete scaling and deterioration similar to the crumbling of rocks by salt weathering. The entrained air voids in concrete allow space for salt crystals to grow, thus relieving internal stress similar to the way the voids relieve stress from freezing water in concrete.*

Deicers can have many effects on concrete and the immediate environment. Sodium chloride, calcium chloride, and urea are the most frequently used deicers. In absence of freezing, sodium chloride has little to no chemical effect on concrete but will damage plants and corrode metal. Calcium chloride in weak solutions generally has little chemical effect on concrete and vegetation but does corrode metal. Studies have shown that concentrated calcium chloride solutions can chemically attack concrete. The reaction is accelerated with increased temperature.** Urea does not chemically damage concrete, vegetation, or metal. The use of deicers containing ammonium nitrate and ammonium sulfate should be strictly prohibited as they rapidly attack and disintegrate concrete.

The extent of scaling depends upon the amount of deicer used and the frequency of application. Relatively low concentrations (on the order of 2% to 4% by weight) of deicer produce more surface scaling than higher concentrations or the absence of deicer.†

Deicers can reach concrete surfaces in ways other than direct application, such as splashing by vehicles and dripping from the undersides of vehicles. Scaling is more severe in poorly drained areas because the deicer solution is retained on the concrete surface during freezing and thawing.

Air entrainment is effective in preventing surface scaling and is recommended for all concretes that may come in any contact with deicing chemicals. To provide adequate durability and scale resistance in severe exposure with deicers present, air-entrained concrete should be composed of durable materials and have (1) a low water-cement ratio (maximum 0.45), (2) a slump of 4 in. or less,†† (3) a cement content of 564 lb per cu yd or more, (4) proper finishing after bleed water has evaporated from the surface, (5) adequate drainage with a slope of 1/8 in. per linear foot or more, (6) a minimum of 7 days moist curing at or above 50°F, (7) a minimum compressive strength of 4000 psi at 28 days, and (8) a minimum 30-day drying period after moist curing if concrete is placed in the fall and will be exposed to freeze-thaw cycles and deicers when saturated. Recommended air contents are shown in Table 5-4.

When temperatures fall to near 40°F, a 14-day curing period may be necessary when using Type I, normal cement. However, the required curing period can be reduced to seven days by using a Type III high-early-strength cement or a higher cement content in the concrete.

Air drying. The resistance of air-entrained concrete to freeze-thaw cycles and deicers is greatly increased by air drying after initial moist curing. Air drying removes excess moisture from the concrete which in turn reduces the internal stress caused by freeze-thaw conditions and deicers. Water-saturated concrete will deteriorate faster than an air-dried concrete when exposed to moist freeze-thaw cycling and deicers. Concrete placed in the spring or summer has an adequate drying period. Concrete placed in the fall season, however, often does not dry out enough before deicers are used. This is especially true of fall paving cured by membrane-forming compounds. These membranes remain intact until worn off by traffic and thus adequate drying may not occur before the onset of winter. Curing methods that allow drying at the completion of the curing period are preferable for fall paving on all projects where deicers will be used. Concrete placed in the fall should be allowed at least 30 days for drying after the moist-curing period. The exact length of time for sufficient drying to take place may vary with climate and weather conditions.

Treatment of scaled surfaces. If surface scaling (an indication of an inadequate air-void system or poor finishing practices) should develop during the first frost season, or if the concrete is of poor quality, a breathable surface treatment can be applied to the dry concrete to help protect the concrete against further damage. Treatment often consists of a penetrating sealer made with boiled linseed oil,‡ breathable methacrylate, or other materials. Nonbreathable formulations should be avoided as they can cause delamination.

The effect of mix design, surface treatment, curing, or other variables on resistance to surface scaling can be evaluated by ASTM C 672.

---

*References 5-31 and 5-37.
**Reference 5-30.
†Reference 5-18.
††A superplasticizer may be used to increase slump and workability without increasing the water-cement ratio.
‡See Reference 5-24.

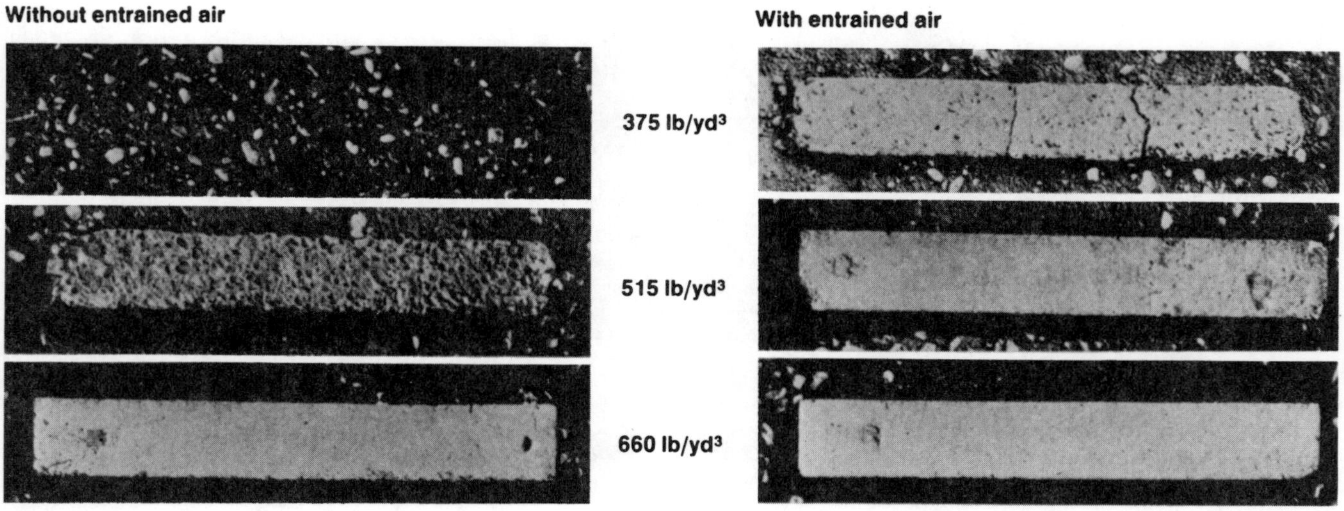

Without entrained air

With entrained air

375 lb/yd³

515 lb/yd³

660 lb/yd³

Fig. 5-4. Effect of entrained air and cement content on performance of concrete specimens (Type II cement) exposed to a sulfate soil. Without entrained air the specimens made with lesser amounts of cement deteriorated badly. Specimens made with the most cement, lowest water-cement ratio, were further improved by air entrainment. Numbers indicate cement content. Specimens were 5 years old when photographed. References 5-4 and 5-22.

Sulfate Resistance

Sulfate resistance of concrete is improved by air entrainment, as shown in Figs. 5-4 and 5-5, when advantage is taken of the reduction in water-cement ratio. Air-entrained concrete made with a low water-cement ratio and an adequate cement factor with a low tricalcium aluminate cement will be resistant to attack from sulfate soils and waters.

Resistance to Alkali-Silica Reactivity

The expansive disruption caused by alkali-silica reactivity is reduced through the use of air-entrainment.[*] Cement alkalies react with the silica of reactive aggregates to form expansive reaction products, causing the concrete to expand. Excessive expansion will disrupt and deteriorate concrete. As shown·in Fig. 5-6, the expansion of mortar bars made with reactive materials is reduced as the air content is increased.

Strength

When the air content is maintained constant, strength varies inversely with the water-cement ratio. Fig. 5-7 shows a typical relationship between 28-day compressive strength and water-cement ratio for concrete that has the recommended percentages of entrained air. As air content is increased, a given strength generally can be maintained by holding to a constant voids-cement ratio[**]; this may, however, necessitate some increase in cement content in richer mixes.

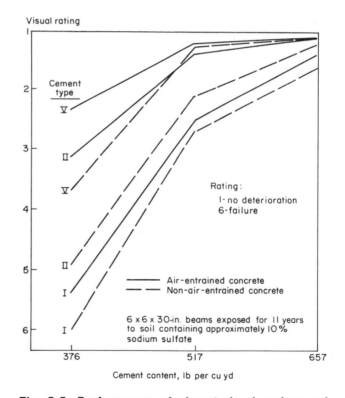

Fig. 5-5. Performance of air-entrained and non-air-entrained concretes exposed to sulfate soil with respect to cement type and content. Sulfate resistance is increased with the use of Types II and V cements, a higher cement content, lower water-cement ratio, and air entrainment. See Fig. 2-5 and Reference 5-36.

*Reference 5-7.
**Air plus water.

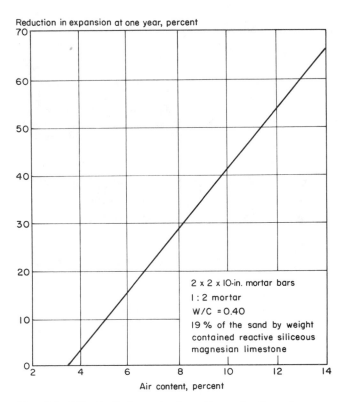

Reduction in expansion at one year, percent

2 x 2 x I0-in. mortar bars
I : 2 mortar
W/C = 0.40
19 % of the sand by weight
contained reactive siliceous
magnesian limestone

Air content, percent

Fig. 5-6. Effect of air content on the reduction of expansion due to alkali-silica reaction. Reference 5-7.

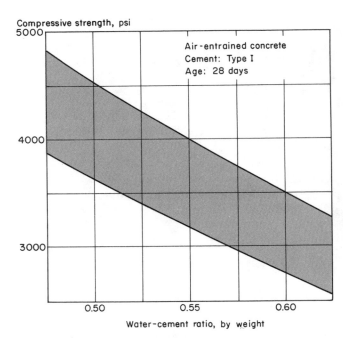

Compressive strength, psi

Air-entrained concrete
Cement: Type I
Age: 28 days

Water-cement ratio, by weight

Fig. 5-7. Typical relationship between 28-day compressive strength and water-cement ratio for a wide variety of air-entrained concretes using Type I cement.

Air-entrained as well as non-air-entrained concrete can readily be proportioned to provide similar moderate strengths. Both generally contain the same amount of coarse aggregate. When the cement content and slump are held constant, air entrainment reduces the sand and water requirements as illustrated in Fig. 5-8. Thus, air-entrained concretes can have lower water-cement ratios than non-air-entrained concretes, which minimizes the reductions in strength that generally accompany air entrainment. At constant water-cement ratios, increases in air will proportionally reduce strength. However, some reductions in strength may be tolerable in view of other benefits of air such as improved workability. Reductions in strength become more significant only in higher-strength (higher cement content) mixes as illustrated in Fig. 5-9. In lower-cement-content, harsh mixes, strength is generally increased by entrainment of air in proper amounts due to the reduced water-cement ratio and improved workability. For moderate-strength concrete, each percentile of entrained air reduces the compressive strength about 2% to 6%.* Actual strength varies and is affected by the cement source, admixtures, and other concrete ingredients.**

Attainment of high strength with air-entrained concrete may be difficult at times. Even though a reduction in mixing water is associated with air entrainment, mixtures with high cement contents require more mixing water than lower-cement-content mixtures; hence, the increase in strength expected from the additional cement is offset somewhat by the additional water.

Workability

Entrained air improves the workability of concrete. It is particularly effective in lean (low cement content) mixes that otherwise might be harsh and difficult to work. In one study, an air-entrained mix made with natural aggregate, 3% air, and a 1½-in. slump had about the same workability as a non-air-entrained concrete with 1% air and 3-in. slump, even though less cement was required for the air-entrained mix.† Workability of mixes with angular and poorly graded aggregates is similarly improved. Because of this improved workability, water and sand content can be reduced significantly (Fig. 5-8). A volume of air-entrained concrete requires less water than an equal volume of non-air-entrained concrete of the same consistency and maximum aggregate size. Freshly mixed concrete containing entrained air is cohesive, looks and feels fatty or workable, and can be handled and finished with ease. Entrained air also reduces segregation and bleeding in freshly mixed and placed concrete.

---

*References 5-11, 5-19, and 5-33.
**Reference 5-2 illustrates that concrete having the same water-cement ratio and aggregate grading is reduced in 28-day compressive strength approximately 200 psi for each percentage point increase in air content for 3000 to 5000 psi concrete.
†Reference 5-2.

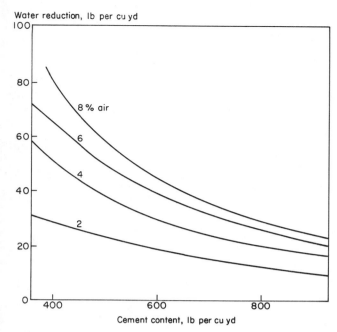

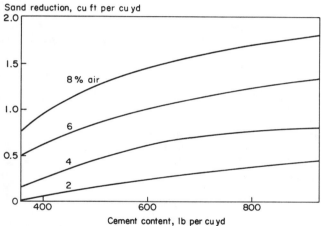

Fig. 5-8. Reduction of water and sand content obtained at various levels of air and cement contents. Reference 5-20.

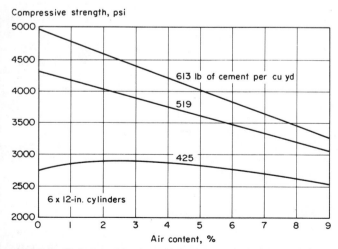

Fig. 5-9. Relationship between air content and 28-day compressive strength for concrete at three constant cement contents. Water content was reduced with increased air content to maintain a constant slump. Reference 5-2.

AIR-ENTRAINING MATERIALS

The entrainment of air in concrete can be accomplished by adding an air-entraining admixture at the mixer, by using an air-entraining cement, or by a combination of these methods. Regardless of the method used, adequate control and monitoring is required to ensure the proper air content at all times.

Numerous commercial air-entraining admixtures, manufactured from a variety of materials, are available. Most air-entraining admixtures generally consist of one or more of the following materials: wood resin (Vinsol resin), sulfonated hydrocarbons, fatty and resinous acids, and synthetic materials. Air-entraining admixtures are usually liquids and should not be allowed to freeze. Such admixtures, added at the mixer, should conform to ASTM C 260.

Air-entraining cements should comply with the specifications in ASTM C 150 and C 595. To produce such cements, acceptable air-entraining additions conforming to ASTM C 226 are interground with the cement clinker during manufacture. Air-entraining cements generally provide an adequate amount of entrained air to meet most job conditions; however, a specified air content may not necessarily be obtained in the concrete. If an insufficient volume of air is entrained, it may also be necessary to add an air-entraining agent at the mixer.

Each of these methods has certain advantages. On jobs where careful control is not practical, air-entraining cements are especially useful to ensure that a significant portion of the required air content will always be obtained. They eliminate the possibility of human or mechanical error that can occur when adding an admixture during batching. With air-entraining admixtures, the volume of entrained air can be readily adjusted to meet job conditions by changing the amount of admixture added at the mixer.

Variations in air content can be expected with variations in aggregate proportions and gradation, mixing time, temperature, and slump. The order of batching and mixing concrete ingredients when using an air-entraining admixture has a significant influence on the amount of air entrained; therefore, consistency in batching is needed to maintain adequate control.

When entrained air is excessive, it can be reduced by using one of several defoaming (air-detraining) agents such as tributyl phosphate, dibutyl phthalate, octyl alcohol, water-insoluble esters of carbonic acid and boric acid, and silicones.* Only the smallest possible dosage of defoaming agent should be used to reduce the air content to the specified limits. Excessive amounts might have adverse effects on concrete properties.

*Reference 5-33, addendum, p. 106.

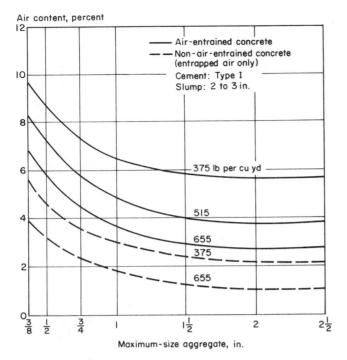

Fig. 5-10. Relationship between aggregate size, cement content, and air content of concrete. The air-entraining admixture dosage per unit of cement was constant for air-entrained concrete. PCA Major Series 336.

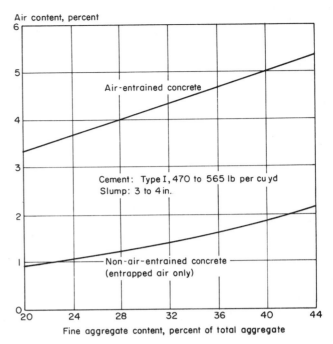

Fig. 5-11. Relationship between percentage of fine aggregate and air content of concrete. PCA Major Series 336.

FACTORS AFFECTING AIR CONTENT

Cement

As cement content increases, the air content decreases for a set dosage of air-entraining admixture per unit of cement within the normal range of cement contents (see Fig. 5-10). In going from 400 to 600 lb of cement per cubic yard, the dosage rate may double to maintain a constant air content. However, studies indicate that when this is done the air-void spacing factor generally decreases with an increase in cement content; and for a given air content the specific surface increases, thus improving durability.

An increase in cement fineness will result in a decrease in the amount of air entrained. Type III cement, a very fine material, may require twice as much air-entraining agent as a Type I cement of normal fineness.

High-alkali cements may entrain more air than low-alkali cements with the same amount of air-entraining material. A low-alkali cement may require 20% to 40% (occasionally up to 70%) more air-entraining agent than a high-alkali cement to achieve an equivalent air content. Precautions are therefore necessary when using more than one cement source in a batch plant to ensure that proper admixture requirements are determined for each cement.*

Coarse Aggregate

The size of coarse aggregate has a pronounced effect on the air content of both air-entrained and non-air-

entrained concrete, as shown in Fig. 5-10. There is little change in air content when the size of aggregate is increased above 1½ in. For smaller aggregate sizes the air content increases sharply with a constant admixture dosage rate as aggregate size decreases because of the larger increase in mortar volume as aggregate size decreases below 1½ in.

Fine Aggregate

The fine-aggregate content of a mix affects the percentage of entrained air. As shown in Fig. 5-11, increasing the amount of fine aggregate causes more air to be entrained for a given amount of air-entraining cement or admixture (more air is also entrapped in non-air-entrained concrete).

Fine-aggregate particles passing the No. 30 to No. 100 sieves entrain more air than either very fine or coarser particles. Appreciable amounts of material passing the No. 100 sieve will result in a significant reduction of entrained air.

Fine aggregates from different sources may entrain different amounts of air even though they have identical gradations. This may be due to differences in shape and surface texture or contamination by small amounts of organic materials.

Mixing Water and Slump

An increase in the mixing water makes more water available for the generation of air bubbles, thereby

---

*Reference 5-28.

increasing the air content as slumps increase up to about 6 or 7 inches. An increase in the water-cement ratio from 0.4 to 1.0 can increase the air content by four percentage points. A portion of the air increase is due to the relationship between slump and air content. (Air content increases with slump even when the water-cement ratio is held constant.) The spacing factor, $\bar{L}$, of the air-void system increases, i.e., the voids become coarser at higher water-cement ratios, thereby reducing concrete freeze-thaw durability.[*]

The addition of one gallon of water per cubic yard of concrete can increase the slump by approximately one inch. A 1-in. increase in slump increases the air content by approximately one-half to one percentage point for concretes with a low-to-moderate slump and constant air-entraining admixture dosage. However, this approximation is greatly affected by concrete temperature, slump, and the type and amount of cement and admixtures present in the concrete. A low-slump concrete with a high dosage of water-reducing and air-entraining admixtures can undergo large increases in slump and air content with a small addition of water. On the other hand, a very fluid concrete mixture (8- to 10-in. slump) may lose air with the addition of water. Refer to Tables 5-1 and 5-2 for more information.

The mixing water used may also affect air content. Algae-contaminated water increases air content. Highly alkaline wash water from truck mixers can also cause problems. The effect of water hardness in most municipal water supplies is generally insignificant; however, very hard water may decrease the air content in concrete.

Slump and Vibration

The effect of slump and vibration on the air content of concrete is shown in Fig. 5-12. For a constant amount of air-entraining admixture, air content increases as slump increases up to about 6 or 7 in. and then begins to decrease with further increases in slump. At all slumps, however, even 15 seconds of vibration will cause a considerable reduction in air content. Prolonged vibration of concrete should be avoided.

The greater the slump, air content, and vibration time, the larger the percentage of reduction in air content during vibration (see Fig. 5-12). However, if vibration is properly applied, little of the intentionally entrained air is lost. Air lost during handling and moderate vibration consists mostly of large bubbles that are usually undesirable from the standpoint of strength and durability. The average size of the air voids is reduced and the air-void spacing factor remains relatively constant.

Internal vibrators reduce air content more than external vibrators. The air loss due to vibration increases as the volume of concrete is reduced or the vibration frequency is significantly increased. Lower vibration frequencies (8000 vpm) also have less effect on spacing factors and air contents than high vibration frequencies (14,000 vpm). High frequencies can significantly increase spacing factors and decrease air contents after 20 seconds of vibration.[**]

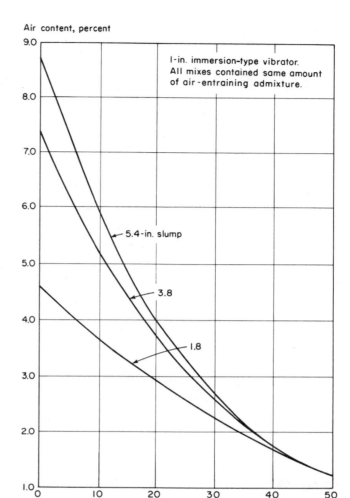

Fig. 5-12. **Relationship between slump, duration of vibration, and air content of concrete. Reference 5-6.**

Concrete Temperature

Temperature of the concrete affects air content, as shown in Fig. 5-13. Less air is entrained as the temperature of the concrete increases, particularly as slump is increased. This effect is especially important during hot-weather concreting when the concrete might be quite warm. A decrease in air content can be offset when necessary by increasing the quantity of air-entraining admixture.

In cold-weather concreting, the air-entraining admixture may lose some of its effectiveness because hot water is used during batching. To offset this loss, such admixtures should be added to the batch after the temperature of the concrete has equilibrated.

Although increased concrete temperature during mixing generally reduces air volume, the spacing factor and specific surface are only slightly affected.

[*]Reference 5-38.
[**]References 5-6 and 5-38.

Table 5-2. Effect of Mixture Design and Concrete Constituents on Control of Air Content in Concrete

Type of constituent	Effects on		Corrective action
	Air content	Air-void system	
Accelerators	Calcium chloride increases air content. Other types have little effect	Unknown	Decrease AEA* when calcium chloride is used
Cement composition	Higher fineness (Type III) requires more AEA. Alkali increases air content	Effects not well defined	Use 50% to 100% more AEA for Type III. Decrease AEA dosage 20% to 40% for high alkali
Cement contaminants	Oxidized oils increase air. Unoxidized oils decrease air	Little apparent effect	Obtain certification on cement. Test for contaminants if problems develop
Cement content in mix design	Decreases with increase in cement	Smaller voids and greater number with increasing cement content	Increase AEA 50% for 200 lb per cubic yard increase in cement. Increase AEA 10 times or more for very rich, low-slump mixtures
Coarse aggregate	Decreases as maximum size of aggregate increases. Crusher fines on coarse aggregate decreases air content	Little effect	No action needed as required air decreases with increase in aggregate size. Hold percentage fines below 4%
Fine aggregate	Increases with increase in sand content. Organic impurities may increase or decrease air content	Surface texture may affect specific surface of voids	Decrease AEA as sand content increases. Check sand with ASTM C 40 prior to acceptance
Fly ash	High loss on ignition or carbon decrease air content. Fineness of ash may have effect	Little effect	Increase AEA. May need up to 5 times more with high-carbon ash. Foam Index test is useful check procedure. Air reduction with long mixing times (90 minutes) can be significant with high-carbon ash. Add more AEA
Mix-water contaminants	Truck mixer wash water decreases air. Extreme water hardness may decrease air. Algae increases air	Unknown	Test water supplies for algae and other contaminants prior to acceptance
Pigments	Carbon-black and black iron oxide based pigments may absorb AEA, depress air content	Unknown	Prequalification of pigment with job materials
Slump	Increases ½ to 1 percentage point per 1-in. slump increase for slumps up to 6 or 7 in., then higher slumps result in decreased air	Becomes coarser with higher slumps	Reduce AEA dosage
Superplasticizers (high-range water reducers)	Melamine-based materials may decrease air or have little effect. Naphthalene and lignosulfonate-based materials increase air content. Highly fluid mixtures may lose air	Produces coarser air-void systems. Spacing factors increase	Use less AEA with naphthalenes. Specify 1% to 2% higher air content if possible
Water content in mix design	Increases with increase in water content about ½ to 1 percentage point per gallon of water. Fluid mixes show loss of air	Becomes coarser at high water content	Decrease AEA accordingly
Water reducers, retarders	Lignosulfonates increase air. Other types have less effect	Spacing factors increase at higher dosages	Decrease AEA 50% to 90% for lignosulfonates especially at lower temperatures. Decrease AEA 20% to 40% for other types. Do not mix admixtures prior to batching

*Air-entraining admixture.
Reference 5-33.
Note: The table information may not apply to all situations.

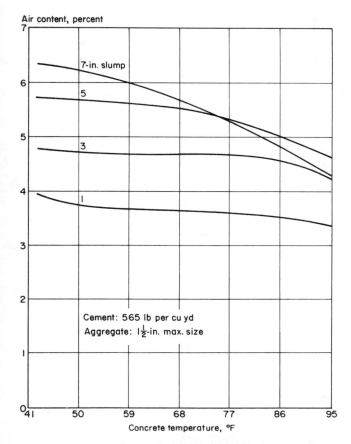

Air content, percent

7-in. slump
5
3
1

Cement: 565 lb per cu yd
Aggregate: 1½-in. max. size

Concrete temperature, °F

Fig. 5-13. Relationship between temperature, slump, and air content of concrete. PCA Major Series 336 and Reference 5-22.

Admixtures and Coloring Agents

Fly ash, coloring agents such as carbon black, or other finely divided materials usually decrease the amount of air entrained for a given amount of admixture.* This is especially true of materials with increasing percentages of carbon.

Water-reducing and set-retarding admixtures generally increase the efficiency of air-entraining admixtures 50% to 100%; therefore, when these are used, less air-entraining admixture will usually give the desired air content. Also, the time of addition of these admixtures into the mix affects the amount of entrained air, delayed additions generally increasing air content.

Set retarders may increase the air-void spacing in concrete. Some water-reducing or set-retarding admixtures are not compatible with some air-entraining agents. If they are added together to the mixing water before being introduced into the mixer, a precipitate may form. This will settle out and result in large reductions in entrained air. The fact that individual admixtures interact in this manner does not mean that they will not be fully effective if they are dispensed separately into a batch of concrete.

Superplasticizers (high-range water reducers) may increase or decrease the air content of a concrete mixture based on the admixture's chemical formulation

and the slump of the concrete. Naphthalene-based superplasticizers tend to increase the air content while melamine-based materials may decrease or have little effect on air content. Normal air loss in flowing concrete during mixing and transport is about 1½ percentage points.**

Superplasticizers also affect the air-void system characteristics of hardened concrete by increasing the general size of the entrained air voids. This results in a higher-than-normal spacing factor, occasionally higher than what may be considered desirable for freeze-thaw durability. However, tests on superplasticized concrete with slightly higher spacing factors have indicated that superplasticized concretes have very good freeze-thaw durability. This may be due to the reduced water-cement ratio often associated with superplasticized concretes.

A small quantity of calcium chloride is sometimes used in cold weather to accelerate the hardening of concrete. It can be used successfully with air-entraining admixtures if it is added separately in solution form to the mix water. Calcium chloride will slightly increase air content. However, if calcium chloride comes in direct contact with some air-entraining admixtures, a chemical reaction can take place that makes the admixture less effective. Nonchloride accelerators may increase or decrease air content, depending upon the individual chemistry of the admixture, but they generally have little effect on air content.

Mixing Action

Mixing action is one of the most important factors in the production of entrained air in concrete. Uniform distribution of entrained air voids is essential to produce scale-resistant concrete; nonuniformity might result from inadequate dispersion of the entrained air during mixing. In production of ready mixed concrete, it is especially important that adequate and consistent mixing be maintained at all times.

The amount of entrained air varies with the type and condition of the mixer, the amount of concrete being mixed, and the rate and duration of mixing. The amount of air entrained in a given mixture will decrease appreciably as the mixer blades become worn or if hardened concrete is allowed to accumulate in the drum or on the blades. Because of differences in mixing action and time, concretes made in a stationary mixer and those made in a transit mixer may differ significantly in amounts of air entrained. The air content may increase or decrease when the size of the batch departs significantly from the rated capacity of the mixer. Little air is entrained in very small batches in a large mixer; however, the air content increases as the mixer capacity is approached.

Fig. 5-14 shows the effect of mixing speed and duration on the air content of freshly mixed concretes made in a transit mixer. Generally, more air is entrained as the speed of mixing is increased up to about 20 rpm, beyond which air entrainment decreases. In the tests

*Reference 5-5.
**Reference 5-33.

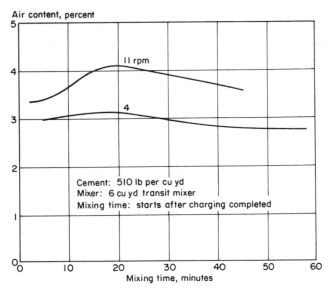

Fig. 5-14. Relationship between mixing time and air content of concrete. PCA Major Series 336.

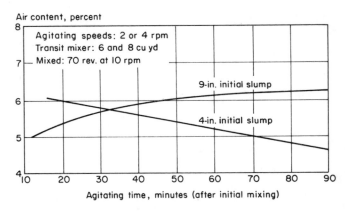

Fig. 5-15. Relationship between agitating time, air content, and slump of concrete. PCA Major Series 336.

from which the data in Fig. 5-14 were derived, the air content reached an upper limit during mixing and a gradual decrease in air content occurred with prolonged mixing. Mixing time and speed on different mixes will have different effects on air content. Significant amounts of air can be lost during mixing with certain mixes and types of mixing equipment.*

Fig. 5-15 shows the effect of continued mixer agitation on air content. The changes in air content with prolongated agitation can be explained by the relationship between slump and air content. For high-slump concretes, the air content increases with continued agitation as the slump decreases to about 6 or 7 in. Prolonged agitation will decrease slump further and decrease air content. For initial slumps lower than 6 in., both the air content and slump decrease with continued agitation. When concrete is retempered (the addition of water and remixing to restore original slump), the air content is increased; however, after 4 hours, retempering is ineffective in increasing air content. Prolonged mixing or agitation of concrete is accompanied by a progressive reduction in slump.

Transporting and Handling

Generally some air, approximately 1 to 2 percentage points, is lost during transportation of the concrete from the mixer to the jobsite. The air content during transport is influenced by several variables including haul time, amount of agitation or vibration during transport, temperature, slump, amount of retempering and concrete ingredients.

Once at the jobsite, the concrete air content remains essentially constant during handling by chute discharge, crane and bucket, wheelbarrow, power buggy, and shovel. However, concrete pumping and long-distance conveyor-belt handling can cause some loss of air, especially with high-air-content mixes. Concrete pumps have been known to cause a loss of up to 2½ percentage points of air.**

Premature Finishing

Proper screeding, floating, and general finishing practices should not affect the air content. However, premature finishing operations may reduce the amount of entrained air in the surface region—thus making the concrete surface vulnerable to scaling. Overfinishing also can cause a loss of entrained air at the surface.

TESTS FOR AIR CONTENT

Four methods for determining the air content in freshly mixed concrete are available. Although they measure only air volume and not air-void characteristics, it has been shown by laboratory tests that these methods are generally indicative of the adequacy of the air-void system.

Acceptance tests for air content of freshly mixed concrete should be made regularly for routine control purposes. Samples should be obtained and tested in accordance with ASTM C 172. Also, because of the effects of handling, placing, and vibration, samples for air-content testing should be taken frequently after the concrete has been placed and consolidated.

Following are methods for determining the air content of freshly mixed concrete:

1. Pressure method (ASTM C 231, Standard Test Method for Air Content of Freshly Mixed Concrete by the Pressure Method)—practical for field-testing all concretes except those made with highly porous and lightweight aggregates.

2. Volumetric method (ASTM C 173, Standard Test Method for Air Content of Freshly Mixed Concrete by the Volumetric Method)—practical for field-testing all concretes, but particularly useful for concretes made with lightweight and porous aggregates. When aggregates larger than 2 in. are used, they

---

*See References 5-9 and 5-22 for more information.
**Reference 5-33.

Table 5-3. Effect of Production Procedures, Construction Practices, and Environment on Control of Air Content in Concrete

Variable	Effects	Corrective action
Admixture metering	Accuracy, reliability of metering system will affect uniformity of air content	Avoid manual-dispensing gravity-feed system, timers. Positive displacement devices preferred. Establish frequent maintenance and calibration program
Batching sequence	Simultaneous batching lowers air	Avoid slurry-mix addition of AEA*
	Late addition of AEA raises air	Do not batch AEA onto cement. Maintain uniformity in batching sequence
Consolidation	Air content decreases under prolonged vibration or at high frequencies	Do not overvibrate. Avoid high-frequency vibrators. Avoid multiple passes of vibratory screeds
Finishing	Air content reduced in surface layer by excessive finishing	Avoid finishing with bleed water still on surface. Avoid overfinishing. Do not sprinkle water on surface prior to finishing
Haul time	Long hauls reduce air, especially in hot weather	Optimize delivery schedules. Maintain concrete temperatures in recommended ranges
Mixer capacity	Air increases as capacity is approached	Run mixer close to full capacity, avoid overloading, clean mixer frequently
Mixing speed	Air increases up to approximately 20 rpm. Decreases at higher speeds	Avoid high drum speeds
Mixing time	Central mixers—air increases up to 90 seconds. Truck mixers—air increases up to 10 minutes. Air decreases after optimum time is reached	Establish optimum mixing time for particular mixer. Avoid overmixing
Retempering	Air content increases after retempering. Ineffective beyond 4 hours	Retemper only enough to restore workability. Avoid addition of excess water
Temperature	Air content decreases with increase in temperature	Increase AEA dosage as temperature increases
Transport	Some air (1% to 2%) normally lost during transport. Air lost in pumping and on belt conveyors, especially at higher air contents	Avoid high air contents in pumped concrete. Do not use aluminum pipelines, dump trucks

*Air-entraining admixture.
Reference 5-33.
Note: The table information may not apply to all situations.

should be removed by hand and the effect of their removal calculated in arriving at the total air content.

3. Gravimetric method (ASTM C 138, Standard Test Method for Unit Weight, Yield, and Air Content [Gravimetric] of Concrete)—requires accurate knowledge of relative density and absolute volumes of concrete ingredients. It is impractical for a field test but can be used satisfactorily in the laboratory.

4. Chace air indicator (AASHTO T199, Standard Method of Test for Air Content of Freshly Mixed Concrete by the Chace Indicator)—is a very simple and inexpensive way to check the approximate air content of freshly mixed concrete. It is a pocket-size device that tests a mortar sample from the concrete. *This test is not a substitute, however, for the more accurate pressure, volumetric, and gravimetric methods.*

The foam-index test can be used to measure the relative air-entraining admixture requirement for concretes containing fly ash-cement combinations.*

The air-void characteristics of hardened concrete can be determined by ASTM C 457 methods. This test is used to determine void spacing factor, specific surface of entrained air, and number of voids per inch of traverse.

See Chapter 14 for more information on tests for determining air content.

RECOMMENDED AIR CONTENTS

The amount of air to be used in air-entrained concrete depends on (1) type of structure, (2) climatic conditions, (3) number of freeze-thaw cycles, (4) extent of exposure to deicers, and (5) extent of exposure to sulfates or other aggressive chemicals in soil or waters. Concrete mixtures with low water-cement ratios may

___
*Reference 5-35.

not require as much entrained air for durability as do concretes of lower quality.

Building Code Requirements for Reinforced Concrete (ACI 318) states that a concrete that will be exposed to moist freezing and thawing or deicer chemicals shall be air entrained within the limits of Table 5-4 for severe and moderate exposures. Fig. 5-16 illustrates the effect of increased air on reducing expansion due to saturated freezing and thawing and illustrates the need to follow the requirements of Table 5-4 for severe exposure. This subject is also discussed in Chapters 1, 7, and 12.

When entrained air is not required for protection against freeze-thaw or deicers, the air contents for mild exposure given in Table 5-4 can be used. Higher air

contents can also be used as long as the design strength is achieved. The entrained air helps to reduce bleeding and segregation and will improve the workability and finishability of concrete.

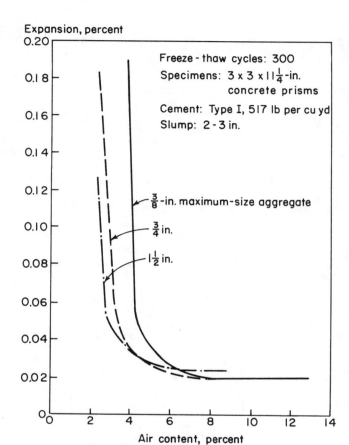

Fig. 5-16. Relationship between air content and expansion of concrete test specimens during 300 cycles of freezing and thawing for various maximum aggregate sizes. Reference 5-11.

Table 5-4. Total Target Air Content for Concrete

Nominal maximum aggregate size, in.	Air content, percent*		
	Severe exposure**	Moderate exposure**	Mild exposure**
3/8	7½	6	4½
½	7	5½	4
¾	6	5	3½
1	6	4½	3
1½	5½	4½	2½
2†	5	4	2
3†	4½	3½	1½

*Project specifications often allow the air content of the delivered concrete to be within −1 to +2 percentage points of the table target values.

**Severe exposure is an environment in which concrete is exposed to wet freeze-thaw conditions, deicers, or other aggressive agents. Moderate exposure is an environment in which concrete is exposed to freezing but will not be continually moist, not exposed to water for long periods before freezing, and will not be in contact with deicers or aggressive chemicals. Mild exposure is an environment in which concrete is not exposed to freezing conditions, deicers, or aggressive agents. Adapted from References 5-32 and 5-34.

†These air contents apply to total mix, as for the preceding aggregate sizes. When testing these concretes, however, aggregate larger than 1½ in. is removed by handpicking or sieving and air content is determined on the minus 1½-in. fraction of mix. (Tolerance on air content as delivered applies to this value.) Air content of total mix is computed from value determined on the minus 1½-in. fraction.

REFERENCES

5-1. Gonnerman, H. F., *Tests of Concretes Containing Air-Entraining Portland Cements or Air-Entraining Materials Added to Batch at Mixer,* Research Department Bulletin RX013, Portland Cement Association, 1944.

5-2. Cordon, W. A., *Entrained Air—A Factor in the Design of Concrete Mixes,* Materials Laboratories Report No. C-310, Research and Geology Division, Bureau of Reclamation, Denver, March 15, 1946.

5-3. Elfert, R. J., *Investigation of the Effect of Vibration Time on the Bleeding Property of Concrete With and Without Entrained Air,* Materials Laboratories Report No. C-375, Research and Geol-

ogy Division, Bureau of Reclamation, Denver, January 26, 1948.

5-4. Stanton, Thomas E., "Durability of Concrete Exposed to Sea Water and Alkali Soils—California Experience," *Journal of the American Concrete Institute,* May 1948.

5-5. Taylor, Thomas G., *Effect of Carbon Black and Black Iron Oxide on Air Content and Durability of Concrete,* Research Department Bulletin RX023, Portland Cement Association, 1948.

5-6. Brewster, R. S., *Effect of Vibration Time upon Loss of Entrained Air from Concrete Mixes,* Materials Laboratories Report No. C-461, Research and Geology Division, Bureau of Reclamation, Denver, November 25, 1949.

5-7. Kretsinger, D. G., *Effect of Entrained Air on Expansion of Mortar Due to Alkali-Aggregate*

Reaction, Materials Laboratories Report No. C-425, Research and Geology Division, U. S. Bureau of Reclamation, Denver, February 16, 1949.

5-8. Powers, T. C., *The Air Requirements of Frost-Resistant Concrete,* Research Department Bulletin RX033, Portland Cement Association, 1949.

5-9. Bloem, D. L., *Air-Entrainment in Concrete,* National Sand and Gravel Association and National Ready Mixed Concrete Association, Silver Spring, Maryland, 1950.

5-10. Gonnerman, H. F., "Durability of Concrete in Engineering Structures," *Building Research Congress 1951,* collected papers, Division No. 2, Section D, Building Research Congress, London, England, 1951, pages 92-104.

5-11. Klieger, Paul, *Studies of the Effect of Entrained Air on the Strength and Durability of Concretes Made with Various Maximum Sizes of Aggregates,* Research Department Bulletin RX040, Portland Cement Association, 1952.

5-12. Bates, A. A.; Woods, H.; Tyler, I. L.; Verbeck, G.; and Powers, T. C., *Rigid-Type Pavement,* Association of Highway Officials of the North Atlantic States, 28th Annual Convention, Proceedings pages 164-200, March 1952.

5-13. Menzel, Carl A., and Woods, William M., *An Investigation of Bond, Anchorage and Related Factors in Reinforced Concrete Beams,* Research Department Bulletin RX042, Portland Cement Association, 1952.

5-14. Powers, T. C., and Helmuth, R. A., *Theory of Volume Changes in Hardened Portland Cement Paste During Freezing,* Research Department Bulletin RX046, Portland Cement Association, 1953.

5-15. Woods, Hubert, *Observations on the Resistance of Concrete to Freezing and Thawing,* Research Department Bulletin RX067, Portland Cement Association, 1954.

5-16. Walker, S., and Bloem, D. L., *Design and Control of Air-Entrained Concrete,* Publication No. 60, National Ready Mixed Concrete Association, Silver Spring, Maryland, 1955.

5-17. Powers, T. C., *Basic Considerations Pertaining to Freezing and Thawing Tests,* Research Department Bulletin RX058, Portland Cement Association, 1955.

5-18. Verbeck, George, and Klieger, Paul, *Studies of "Salt" Scaling of Concrete,* Research Department Bulletin RX083, Portland Cement Association, 1956.

5-19. Klieger, Paul, *Further Studies on the Effect of Entrained Air on Strength and Durability of Concrete with Various Sizes of Aggregate,* Research Department Bulletin RX077, Portland Cement Association, 1956.

5-20. Gilkey, H. J., "Re-Proportioning of Concrete Mixtures for Air Entrainment", *Journal of the American Concrete Institute,* vol. 29, no. 8, Proceedings, 54, February 1958, pages 633-645.

5-21. Mielenz, R. C.; Wokodoff, V. E.; Backstrom, J. E.; and Flack, H. L., "Origin, Evolution, and Effects of the Air-Void System in Concrete. Part 1—Entrained Air in Unhardened Concrete," July 1958, "Part 2—Influence of Type and Amount of Air-Entraining Agent," August 1958, "Part 3—Influence of Water-Cement Ratio and Compaction," September 1958, and "Part 4—The Air-Void System in Job Concrete," October 1958, *Journal of the American Concrete Institute.*

5-22. Lerch, William, *Basic Principles of Air-Entrained Concrete,* T-101, Portland Cement Association, 1960.

5-23. Klieger, Paul, *Extensions to the Long-Time Study of Cement Performance in Concrete,* Research Department Bulletin RX157, Portland Cement Association, 1963.

5-24. *Scale-Resistant Concrete Pavements,* IS117P, Portland Cement Association, 1964.

5-25. Powers, T. C., *Topics in Concrete Technology: . . . (3) Mixtures Containing Intentionally Entrained Air; (4) Characteristics of Air-Void Systems,* Research Department Bulletin RX174, Portland Cement Association, 1965.

5-26. Powers, T. C., "The Mechanism of Frost Action in Concrete," *Stanton Walker Lecture Series on the Materials Sciences,* Lecture No. 3, National Sand and Gravel Association and National Ready Mixed Concrete Association, Silver Spring, Maryland, 1965.

5-27. Klieger, Paul, *Air-Entraining Admixtures,* Research Department Bulletin RX199, Portland Cement Association, 1966.

5-28. Greening, Nathan R., *Some Causes for Variation in Required Amount of Air-Entraining Agent in Portland Cement Mortars,* Research Department Bulletin RX213, Portland Cement Association, 1967.

5-29. Verbeck, G. J., *Field and Laboratory Studies of the Sulphate Resistance of Concrete,* Research Department Bulletin RX227, Portland Cement Association, 1967.

5-30. Brown, F. P., and Cady, P. D., "Deicer Scaling Mechanisms in Concrete," *Durability of Concrete,* ACI SP-47, American Concrete Institute, 1975, pages 101-119.

5-31. "Entrained Air Voids in Concrete Help Prevent Salt Damage," *Civil Engineering,* American Society of Civil Engineers, New York, May 1982.

5-32. *Guide to Durable Concrete,* ACI 201.2R, reaffirmed 1982, ACI Committee 201 Report, American Concrete Institute.

5-33. Whiting, D., and Stark, D., *Control of Air Content in Concrete,* National Cooperative Highway Research Program Report No. 258 and Addendum, Transportation Research Board and National Research Council, Washington, D.C., May 1983.

5-34. *Building Code Requirements for Reinforced Concrete,* ACI 318-83, ACI Committee 318 Report, American Concrete Institute, 1983.

5-35. Gebler, S. H., and Klieger, P., *Effect of Fly Ash on the Air-Void Stability of Concrete,* Research and Development Bulletin RD085T, Portland Cement Association, 1983.

5-36. Stark, David, *Longtime Study of Concrete Durability in Sulfate Soils,* Research and Development Bulletin RD086T, Portland Cement Association, 1984.

5-37. Sayward, John M., *Salt Action on Concrete,* Special Report 84-25, U.S. Army Cold Regions Research and Engineering Laboratory, Hanover, New Hampshire, August 1984.

5-38. Stark, David C., *Effect of Vibration on the Air-Void System and Freeze-Thaw Durability of Concrete,* Research and Development Bulletin RD092T, Portland Cement Association, 1986.

5-39. *Standard Practice for Curing Concrete,* ACI 308-81, revised 1986, ACI Committee 308 Report, American Concrete Institute.

CHAPTER 6
Admixtures for Concrete

Admixtures are those ingredients in concrete other than portland cement, water, and aggregates that are added to the mixture immediately before or during mixing. Admixtures can be classified by function as follows:

1. Air-entraining admixtures
2. Water-reducing admixtures
3. Retarding admixtures
4. Accelerating admixtures
5. Superplasticizers
6. Finely divided mineral admixtures
7. Miscellaneous admixtures such as workability, bonding, dampproofing, permeability-reducing, grouting, gas-forming, coloring, corrosion inhibiting, and pumping admixtures

Table 6-1 provides a more extensive admixture classification.

Concrete should be workable, finishable, strong, durable, watertight, and wear resistant. These qualities can often be obtained easily and economically by the selection of suitable materials rather than by resorting to admixtures (except air-entraining admixtures when needed).

The major reasons for using admixtures are

1. To reduce the cost of concrete construction
2. To achieve certain properties in concrete more effectively than by other means
3. To ensure the quality of concrete during the stages of mixing, transporting, placing, and curing in adverse weather conditions
4. To overcome certain emergencies during concreting operations

Despite these considerations, it should be borne in mind that no admixture of any type or amount can be considered a substitute for good concreting practice.

The effectiveness of an admixture depends upon such factors as type, brand, and amount of cement; water content; aggregate shape, gradation, and proportions; mixing time; slump; and temperatures of concrete and air.

Admixtures being considered for use in concrete should meet applicable specifications as presented in Table 6-1. Trial mixtures should be made with the admixture and the job materials at temperatures and humidities anticipated on the job. In this way the

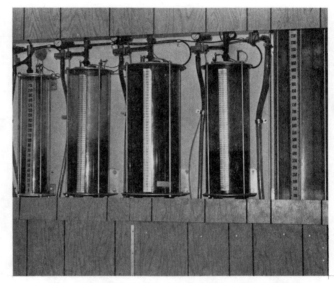

Fig. 6-1. Liquid admixtures in automatic dispenser tanks. Most air-entraining, water-reducing, accelerating, and retarding admixtures are liquids that are generally dark and syruplike. The admixtures are added to the concrete mixture in specific quantities through automatic dispenser tanks. Liquid admixtures are usually dispensed individually into the batch water; however, they should not be intermixed prior to batching.

compatibility of the admixture with other admixtures and job materials, as well as the effects of the admixture on the properties of the fresh and hardened concrete, can be observed. The amount of admixture recommended by the manufacturer or the optimum amount determined by laboratory test should be used.

Even though an admixture may produce concrete with the desired properties, the same results can often be obtained just as economically by changing the mix proportions or by selecting other concrete ingredients. Whenever possible, a comparison should be made between the cost of changing the basic concrete mixture and the additional cost of using an admixture. The latter should include, in addition to the cost of the admixture, any effect the use of the admixture will have on the cost of transporting, placing, finishing, curing, and protecting the concrete.

Table 6-1. Concrete Admixtures by Classification

Type of admixture	Desired effect	Material
Accelerators (ASTM C 494, Type C)	Accelerate setting and early-strength development	Calcium chloride (ASTM D 98) Triethanolamine, sodium thiocyanate, calcium formate, calcium nitrite, calcium nitrate
Air detrainers	Decrease air content	Tributyl phosphate, dibutyl phthalate, octyl alcohol, water-insoluble esters of carbonic and boric acid, silicones
Air-entraining admixtures (ASTM C 260)	Improve durability in environments of freeze-thaw, deicers, sulfate, and alkali reactivity Improve workability	Salts of wood resins (Vinsol resin) Some synthetic detergents Salts of sulfonated lignin Salts of petroleum acids Salts of proteinaceous material Fatty and resinous acids and their salts Alkylbenzene sulfonates Salts of sulfonated hydrocarbons
Alkali-reactivity reducers	Reduce alkali-reactivity expansion	Pozzolans (fly ash, silica fume), blast-furnace slag, salts of lithium and barium, air-entraining agents
Bonding admixtures	Increase bond strength	Rubber, polyvinyl chloride, polyvinyl acetate, acrylics, butadiene-styrene copolymers.
Coloring agents	Colored concrete	Modified carbon black, iron oxide, phthalocyanine, umber, chromium oxide, titanium oxide, cobalt blue (ASTM C 979)
Corrosion inhibitors	Reduce steel corrosion activity in a chloride environment	Calcium nitrite, sodium nitrite, sodium benzoate, certain phosphates or fluosilicates, fluoaluminates
Dampproofing admixtures	Retard moisture penetration into dry concrete	Soaps of calcium or ammonium stearate or oleate Butyl stearate Petroleum products
Finely divided mineral admixtures		
Cementitious	Hydraulic properties Partial cement replacement	Ground granulated blast-furnace slag (ASTM C 989) Natural cement Hydraulic hydrated lime (ASTM C 141)
Pozzolans	Pozzolanic activity Improve workability, plasticity, sulfate resistance; reduce alkali reactivity, permeability, heat of hydration Partial cement replacement Filler	Diatomaceous earth, opaline cherts, clays, shales, volcanic tuffs, pumicites (ASTM C 618, Class N); fly ash (ASTM C 618, Classes F and C), silica fume
Pozzolanic and cementitious	Same as cementitious and pozzolan categories	High calcium fly ash (ASTM C 618, Class C) Ground granulated blast-furnace slag (ASTM C 989)
Nominally inert	Improve workability Filler	Marble, dolomite, quartz, granite
Fungicides, germicides, and insecticides	Inhibit or control bacterial and fungal growth	Polyhalogenated phenols Dieldrin emulsions Copper compounds
Gas formers	Cause expansion before setting	Aluminum powder Resin soap and vegetable or animal glue Saponin Hydrolized protein
Grouting agents	Adjust grout properties for specific applications	See Air-entraining admixtures, Accelerators, Retarders, Workability agents
Permeability reducers	Decrease permeability	Silica fume Fly ash (ASTM C 618) Ground slag (ASTM C 989) Natural pozzolans Water reducers Latex

Table 6-1. Concrete Admixtures by Classification (continued)

Type of admixture	Desired effect	Material
Pumping aids	Improve pumpability	Organic and synthetic polymers Organic flocculents Organic emulsions of paraffin, coal tar, asphalt, acrylics Bentonite and pyrogenic silicas Natural pozzolans (ASTM C 618, Class N) Fly ash (ASTM C 618, Classes F and C) Hydrated lime (ASTM C 141)
Retarders (ASTM C 494, Type B)	Retard setting time	Lignin Borax Sugars Tartaric acid and salts
Superplasticizers* (ASTM C 1017, Type 1)	Flowing concrete Reduce water-cement ratio	Sulfonated melamine formaldehyde condensates Sulfonated naphthalene formaldehyde condensates Lignosulfonates
Superplasticizer* and retarder (ASTM C 1017, Type 2)	Flowing concrete with retarded set Reduce water	See Superplasticizers and also Water reducers
Water reducer (ASTM C 494, Type A)	Reduce water demand at least 5%	Lignosulfonates Hydroxylated carboxylic acids Carbohydrates (Also tend to retard set so accelerator is often added)
Water reducer and accelerator (ASTM C 494, Type E)	Reduce water (minimum 5%) and accelerate set	See Water reducer, Type A (Accelerator is added)
Water reducer and retarder (ASTM C 494, Type D)	Reduce water (minimum 5%) and retard set	See Water reducer, Type A
Water reducer—high range (ASTM C 494, Type F)	Reduce water demand (minimum 12%)	See Superplasticizers
Water reducer—high range—and retarder (ASTM C 494, Type G)	Reduce water demand (minimum 12%) and retard set	See Superplasticizers and also Water reducers
Workability agents	Improve workability	Air-entraining admixtures Finely divided admixtures, except silica fume Water reducers

*Superplasticizers are also referred to as high-range water reducers or plasticizers. These admixtures often meet both ASTM C 494 and C 1017 specifications simultaneously.

AIR-ENTRAINING ADMIXTURES

Air-entraining admixtures are used to purposely entrain microscopic air bubbles in concrete. Air-entrainment will dramatically improve the durability of concrete exposed to moisture during cycles of freezing and thawing. Entrained air greatly improves concrete's resistance to surface scaling caused by chemical deicers. The workability of fresh concrete is also improved significantly, and segregation and bleeding are reduced or eliminated.

Air-entrained concrete contains minute air bubbles that are distributed uniformly throughout the cement paste. Entrained air can be produced in concrete by use of an air-entraining cement, by introduction of an air-entraining admixture, or by a combination of both methods. An air-entraining cement is a portland cement with an air-entraining addition interground with the clinker during manufacture. An air-entraining admixture, on the other hand is added directly to the concrete materials either before or during mixing. The primary ingredients used in air-entraining admixtures are listed in Table 6-1. Specifications and methods of testing air-entraining admixtures are given in ASTM C 260 and C 233. Air-entraining additions for use in the manufacture of air-entraining cements must meet requirements of ASTM C 226. Applicable requirements for air-entraining cements are given in ASTM C 150. Refer to Chapter 5 for more information.

WATER-REDUCING ADMIXTURES

Water-reducing admixtures are used to reduce the quantity of mixing water required to produce concrete of a certain slump, reduce water-cement ratio, or increase slump. Typical water reducers reduce the water content by approximately 5% to 10%. High-range water reducers reduce water content by 12% to 30% (see

"Superplasticizers"). Adding a water-reducing admixture to a mix without reducing the water content can produce a mixture with a much higher slump. The rate of slump loss, however, is not reduced and in most cases is increased. Rapid slump loss results in reduced workability and less time to place concrete.

An increase in strength is generally obtained with water-reducing admixtures as the water-cement ratio is reduced. Despite reduction in water content, water-reducing admixtures can cause significant increases in drying shrinkage. Using a water reducer to reduce the cement and water content of a concrete mixture while maintaining a constant water-cement ratio, can result in equal or reduced compressive strength, and can increase slump loss by a factor of two or more.

Water reducers decrease, increase, or have no effect on bleeding, depending on the chemical composition. Many water-reducing admixtures can also retard the setting time of concrete. Some are modified to give varying degrees of retardation while others do not significantly affect the setting time. Some water-reducing admixtures such as lignosulfonates may also entrain some air in concrete.

The effectiveness of water reducers on concrete is a function of their chemical composition, concrete temperature, cement composition and fineness, cement content, and the presence of other admixtures. Some water reducers are more effective in lean mixtures and with cements of low alkali or low tricalcium aluminate contents. The classifications and components of water reducers are listed in Table 6-1.

RETARDING ADMIXTURES

Retarding admixtures are used to retard the rate of setting of concrete. High temperatures of fresh concrete ($85°F$ to $90°F$ and higher) are often the cause of an increased rate of hardening that makes placing and finishing difficult. One of the most practical methods of counteracting this effect is to reduce the temperature of the concrete by cooling the mixing water or the aggregates. Retarders do not decrease the initial temperature of concrete.

Retarders are sometimes used to (1) offset the accelerating effect of hot weather on the setting of concrete, (2) delay the initial set of concrete or grout when difficult or unusual conditions of placement occur, such as placing concrete in large piers and foundations, cementing oil wells, or pumping grout or concrete over considerable distances, or (3) delay the set for special finishing processes such as an exposed aggregate surface.

Because most retarders also act as water reducers, they are frequently called water-reducing retarders. Retarders may also entrain some air in concrete.

In general, some reduction in strength at early ages (one to three days) accompanies the use of retarders. The effects of these materials on the other properties of concrete, such as shrinkage, may not be predictable. Therefore, acceptance tests of retarders should be made with job materials under anticipated job conditions. The classifications and components of retarders are listed in Table 6-1.

ACCELERATING ADMIXTURES

An accelerating admixture is used to accelerate strength development of concrete at an early age. The strength development of concrete can also be accelerated by (1) using Type III high-early-strength portland cement, (2) lowering the water-cement ratio by adding 100 to 200 pounds of additional cement per cubic yard of concrete, or (3) curing at higher temperatures.

Calcium chloride ($CaCl_2$) is the material most commonly used in accelerating admixtures. It should conform to the requirements of ASTM D 98 and should be sampled and tested in accordance with ASTM D 345. The widespread use of calcium chloride accelerating admixtures has provided much data and experience on their effect on the properties of concrete. Besides accelerating strength gain, calcium chloride causes an increase in drying shrinkage, potential reinforcement corrosion, and discoloration (darkens concrete).*

Calcium chloride is not an antifreeze agent. When used in allowable amounts, it will not reduce the freezing point of concrete by more than a few degrees. Attempts to protect concrete from freezing by this method are foolhardy. Instead, proven reliable precautions should be taken during cold weather.**

Calcium chloride should be added to the concrete mix in solution form as part of the mixing water. If added to the concrete in dry form, all of the dry particles may not be completely dissolved during mixing. Undissolved lumps in the mix can cause popouts or dark spots in hardened concrete.

The amount of calcium chloride added should be no more than is necessary to produce the desired results and in no case exceed 2% by weight of cement. When calculating the chloride content of commercially available calcium chloride, it can be assumed that

1. Regular flake contains a minimum of 77% $CaCl_2$
2. Concentrated flake, pellet, or granular forms contain a minimum of 94% $CaCl_2$

An overdose can result in placement problems and can be detrimental to concrete, since it may cause rapid stiffening, cause a large increase in drying shrinkage, corrode reinforcement, and cause loss of strength at later ages.

Applications where calcium chloride should be used with caution are in

1. Concrete subjected to steam curing
2. Concrete containing embedded dissimilar metals, especially if electrically connected to steel reinforcement
3. Concrete slabs supported on permanent galvanized-steel forms
4. Colored concrete

The use of calcium chloride or admixtures containing soluble chlorides is *not recommended* under the following conditions:

---

*Reference 6-20.
**See Chapter 12, "Cold-Weather Concreting."

1. In prestressed concrete because of possible corrosion hazards
2. In concrete containing embedded aluminum (for example, conduit) since serious corrosion of the aluminum can result, especially if the aluminum is in contact with embedded steel and the concrete is in a humid environment
3. In concrete subjected to alkali-aggregate reaction or exposed to soil or water containing sulfates
4. In floor slabs intended to receive dry-shake metallic finishes
5. In hot weather generally
6. In massive concrete placements

The maximum chloride-ion content for corrosion protection of reinforced concrete as recommended by ACI 318 is presented in Table 6-2. Resistance to the corrosion of embedded steel is further improved with an increase in the depth of concrete cover over reinforcing steel, and a lower water-cement ratio.

Several nonchloride, noncorrosive accelerators are available for use in concrete where chlorides are not recommended (see Table 6-1). However, many nonchloride accelerators are not as effective as calcium chloride and are more expensive.

Table 6-2. Maximum Chloride-Ion Content for Corrosion Protection

Type of member	Maximum water-soluble chloride ion (Cl$^-$) in concrete, percent by weight of cement
Prestressed concrete	0.06
Reinforced concrete exposed to chloride in service	0.15
Reinforced concrete that will be dry or protected from moisture in service	1.00
Other reinforced concrete construction	0.30

Reference 6-19.

SUPERPLASTICIZERS (HIGH-RANGE WATER REDUCERS)

Superplasticizers* are high-range water reducers meeting ASTM C 1017 and C 494 Types F and G specifications that are added to concrete with a low-to-normal slump and water-cement ratio to make high-slump flowing concrete. Flowing concrete is a highly fluid but workable concrete that can be placed with little or no vibration or compaction and can still be free of excessive bleeding or segregation. Flowing concrete is used (1) in thin section placements, (2) in areas of closely spaced and congested reinforcing steel, (3) in tremie pipe (underwater) placements, (4) in pumped concrete to reduce pump pressure, thereby increasing lift and distance capacity, (5) in areas where conventional consolidation methods are impractical or can not be used, and (6) for reducing handling costs. The addition of a superplasticizer to a 3-in. slump concrete can easily produce a concrete with a 9-in. slump. Flowing concrete is defined by ASTM C 1017 as having a slump greater than 7½ in., yet maintaining cohesive properties. Excessively high slumps, 10 in. or more, may cause the concrete to segregate.

High-range water reducers (ASTM C 1017 and C 494 Types F and G) can also be used to make low water-cement ratio, high-strength concrete with workability in the ranges generally specified for consolidation by internal vibration. A water reduction of 12% to 30% can be obtained through the use of these admixtures. The reduced water content and water-cement ratio can produce concretes with (1) ultimate compressive strengths in excess of 10,000 psi, (2) increased early strength gain, and (3) reduced chloride-ion penetration as well as other beneficial properties associated with low water-cement ratio concrete.

High-range water reducers are generally more effective, but more expensive, than regular water-reducing admixtures in producing workable concrete. The effect of most superplasticizers in increasing workability or making flowing concrete is short-lived, 30 to 60 minutes, and is followed by a rapid loss in workability (slump loss). Due to this slump loss, these admixtures are often added to the concrete at the jobsite. Extended-slump-life high-range water reducers added at the batch plant help reduce slump-loss problems. Setting time may be accelerated or retarded based on the individual admixture chemistry, dosage rate, and interaction with other admixtures present in the concrete mix.

Tests have shown that some superplasticized concretes bleed more than do control concretes of equal water-cement ratio, but bleed significantly less than do control concretes of equal high slump.** High-slump, low-water-content, superplasticized concrete has less drying shrinkage than a high-slump, high-water-content conventional concrete but has similar to or higher drying shrinkage than conventional low-slump, low-water-content concrete.

The effectiveness of the superplasticizer is increased with an increasing amount of cement and fines in the concrete. It is also affected by the initial slump of the concrete.

Superplasticized concrete has larger entrained air voids and higher void-spacing factors than normal air-entrained concrete. This would normally indicate a reduced resistance to freezing and thawing; however, laboratory tests have shown that superplasticized concrete has very good freeze-thaw durability, even with its higher void-spacing factors. This may be the result of lower water-cement ratios often associated with superplasticized concrete. Table 6-1 lists the primary components and specifications of superplasticizing admixtures. The earlier information on water reducers is also applicable to superplasticizers.

---

*The terms *high-range water reducer* and *superplasticizer* are often used synonymously.
**The control concretes used a high water content to achieve the high slump. See References 6-3 and 6-6.

FINELY DIVIDED MINERAL ADMIXTURES

Finely divided mineral admixtures are powdered or pulverized materials added to concrete before or during mixing to improve or change some of the plastic or hardened properties of portland cement concrete. These admixtures are generally natural or byproduct materials (see Table 6-3). Based on their chemical or physical properties, they are classified as (1) cementitious materials, (2) pozzolans, (3) pozzolanic and cementitious materials, and (4) nominally inert materials.

Cementitious Materials

Cementitious materials are substances that alone have hydraulic cementing properties (set and harden in the presence of water). Cementitious materials include ground granulated blast-furnace slag, natural cement, hydraulic hydrated lime, and combinations of these and other materials.

Ground granulated blast-furnace slag made from iron blast-furnace slag is a nonmetallic product consisting essentially of silicates and aluminosilicates of calcium and other bases developed in a molten condition simultaneously with iron in a blast furnace. The molten slag at a temperature of about 2730°F is rapidly chilled by quenching in water to form a glassy sandlike granulated material. The granulated material, which is ground to less than 45 microns, has a surface area fineness of about 400 to 600 m²/kg Blaine. The rough and angular-shaped ground slag in the presence of water and an activator, NaOH or CaOH, both supplied by portland cement, hydrates and sets in a manner similar to portland cement. Air-cooled slag does not have the hydraulic properties of water-cooled slag. ASTM C 989 classifies slag by its reactivity as Grade 80, 100, or 120.

Table 6-3. Specifications and Classes of Finely Divided Mineral Admixtures

Hydraulic hydrated lime—ASTM C 141

Ground granulated iron blast-furnace slags—ASTM C 989

    Grade 80
        Slags with a low activity index

    Grade 100
        Slags with a moderate activity index

    Grade 120
        Slags with a high activity index

Fly ash and natural pozzolans—ASTM C 618

    Class N
        Raw or calcined natural pozzolans including
            Diatomaceous earths
            Opaline cherts and shales
            Tuffs and volcanic ashes or pumicites
            Some calcined clays and shales

    Class F
        Fly ash with pozzolanic properties

    Class C
        Fly ash with pozzolanic and cementitious properties

Natural cement is formed by calcining argillaceous limestone just below the melting point and then grinding the material to a fine powder.

Hydraulic hydrated lime, ASTM C 141, is formed by calcining limestone containing silica and alumina to a point where sufficient free calcium oxide and unhydrated calcium silicates are present for the hydration and hydraulic properties of the material.

Pozzolanic Materials

A pozzolan is a siliceous or aluminosiliceous material that in itself possesses little or no cementitious value but will, in finely divided form and in the presence of water, chemically react with the calcium hydroxide released by the hydration of portland cement to form compounds possessing cementitious properties.

A number of natural materials such as diatomaceous earth, opaline cherts, clays, shales, volcanic tuffs, and pumicites are used as pozzolans. Most natural pozzolans must be ground before use and many must be calcined at 1200°F to 1800°F to activate their clay constituents. These materials are classified by ASTM C 618 as Class N pozzolans.

Pozzolans also include fly ash and silica fume. Fly ash, the most widely used mineral admixture in concrete, is a finely divided residue (powder resembling cement) that results from the combustion of pulverized coal in electric power generating plants. Upon ignition in the furnace, most of the volatile matter and carbon in the coal are burned off. During combustion, the coal's mineral impurities (such as clay, feldspar, quartz, and shale) fuse in suspension and are carried away from the combustion chamber by the exhaust gas. In the process, the fused material cools and solidifies into spherical particles called fly ash. The fly ash is then collected from the exhaust gases by electrostatic precipitators or bag filters. Generally no further processing is needed for using fly ash in blended cement or concrete.

Most of the fly ash particles are solid spheres and some are hollow cenospheres. Also present are plerospheres, which are spheres containing smaller spheres. Ground materials, such as portland cement, have solid angular particles. The spherical nature of fly ash is shown in Fig. 6-2. The particle sizes in fly ash vary from less than 1 μm (micron) to more than 100 μm with a typical particle size under 20 μm. Only 10% to 30% of the particles by weight are larger than 45 μm. The surface area is typically 300 to 500 m²/kg, although some fly ashes can have surface areas as low as 200 m²/kg and as high as 700 m²/kg.

Fly ash is primarily silicate glass containing silica, alumina, iron, and calcium. Minor constituents are magnesium, sulfur, sodium, potassium, and carbon. A small amount of crystalline compounds are present. The specific gravity of fly ash generally ranges between 2.2 and 2.8 and the color is generally tan or gray.

ASTM C 618 Class F and Class C fly ashes are commonly used as pozzolanic admixtures for concrete. Class F materials are generally low-calcium (less than 10% CaO) fly ashes with carbon contents usually less than 5%, but some may be as high as 10%. Class C

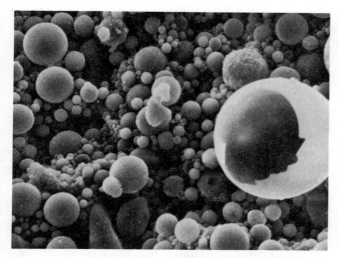

Fig. 6-2. Scanning electron microscope micrograph of fly ash particles at 1000X. Although most fly ash spheres are solid, some spheres are hollow (as shown in the micrograph) and some encapsulate additional smaller spheres.

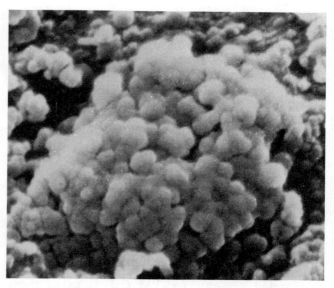

Fig. 6-3. Scanning electron microscope micrograph of silica-fume particles, 20,000X.

materials are often high-calcium (10% to 30% CaO) fly ashes with carbon contents usually less than 2%. Some fly ashes meet both Class F and Class C classifications.

Silica fume, also referred to as microsilica or condensed silica fume, is another material that is used as a pozzolanic admixture. This light to dark gray or sometimes bluish-green-gray powdery product is a result of the reduction of high-purity quartz with coal in an electric arc furnace in the manufacture of silicon or ferrosilicon alloy. Silica fume rises as an oxidized vapor from the 3630°F furnaces. It cools, condenses and is collected in huge cloth bags. The condensed silica fume is then processed to remove impurities and to control particle size.

Condensed silica fume is essentially silicon dioxide (more than 90%) in noncrystalline form. Since it is an airborne material like fly ash, it has a spherical shape (Fig. 6-3). It is extremely fine with particles less than 1 μm in diameter and with an average diameter of about 0.1 μm, about 100 times smaller than average cement particles. Condensed silica fume has a surface area of about 20,000 m^2/kg (nitrogen adsorption method). Tobacco smoke's surface area is about 10,000 m^2/kg. Type I and Type III cements have surface areas of about 300 to 400 m^2/kg and 500 to 600 m^2/kg (Blaine), respectively. The specific gravity of silica fume is generally in the range of 2.10 to 2.25 but can be as high as 2.55. Portland cement has a specific gravity of about 3.15. The bulk density (uncompacted unit weight) of silica fume is about 16 to 19 pcf. Silica fume is sold in powder form but is more commonly available in a liquid. ASTM is presently working on a standard for silica fume.*

Pozzolanic and Cementitious Materials

Certain ground granulated blast-furnace slags and fly ashes exhibit both pozzolanic and cementitious properties. ASTM C 618 Class C fly ashes with a calcium

oxide content of approximately 15% to 30% by weight are predominant in this class. Many of these ashes when exposed to water will hydrate and harden in less than 45 minutes.

The practice of using fly ash and ground granulated blast-furnace slag in portland cement concrete mixes has been growing in the United States in recent years. A major reason for the increase is concern over energy conservation as well as the reduction in the cost of concrete realized when ash or slag are used to partially replace cement.

Nominally Inert Materials

Nominally inert materials have little to no cementitious or pozzolanic properties. Some of the inert materials are finely divided raw quartz, dolomites, many limestones, marble, granite, and other materials. Inert materials are often used in addition to the cement and as a partial replacement of sand in concrete to improve poor workability often caused by a lack of fines in the sand. Pulverized limestone is sometimes added to sand-gravel aggregate concretes to reduce alkali-silica reactivity.

EFFECT OF MINERAL ADMIXTURES ON CONCRETE

The objective of this section is to provide a brief understanding of the properties mineral admixtures affect and the degree of influence of various mineral admixtures.

Finely divided mineral admixtures vary considerably in their effect on concrete mixes. Before such an

---

*Reference 6-31.

admixture is accepted for use, it should be tested in combination with the specific cement and aggregates being used to ascertain its suitability with regard to water requirements, strength development, shrinkage, heat of hydration, durability, or special properties such as prevention of alkali-aggregate reaction or reduction of sulfate attack. Finely divided materials are added in addition to or as a partial replacement of cement in concrete depending on the properties of the materials and the desired effect on concrete.

Fly ash, ground granulated blast-furnace slag, and condensed silica fume are the primary finely divided materials used in concrete today. Extensive studies, both long term and short term, have been performed on these materials. For these reasons, this section primarily refers to the use of fly ash, ground granulated blast-furnace slag, and silica fume in concrete, except where noted. In addition, the effects of fly ash and the ground slag on properties discussed below also apply to blended cements (ASTM C 595) using these materials.

Effect on Freshly Mixed Concrete

Water requirements. Concrete mixes containing fly ash or ground granulated blast-furnace slag will generally require less water (about 1% to 10%) for a given slump than concrete containing only portland cement. There are, however, some fly ashes, ground slags, and natural pozzolans in which the reverse is true; that is, concrete made with them will require more water than concrete without them. Fly ash reduces water demand in the same manner as do liquid chemical water reducers.[*]

Silica-fume concrete requires more water for a given slump, unless a water reducer or superplasticizer is used. Some lean mixes may not experience a significant increase in water demand when only a small amount of silica fume is present.

Air content. The amount of air-entraining admixture required to obtain a specified air content is normally greater when fly ash or silica fume is used. Class C ash generally requires less air-entraining admixture than Class F ash and tends to lose less air with mixing time. Ground slags have variable effects on the required dosage rate of air-entraining admixtures.

The inclusion of ground slags, fly ash, and silica fume in non-air-entrained concrete will generally reduce the amount of entrapped air. Fly ash and silica fume generally have a greater effect in this reduction than ground slag.

The amount of air-entraining admixture required for a certain air content is a function of the fineness, carbon content, alkali content, organic material content, loss on ignition, and presence of impurities in fly ash. Increases in alkalies decrease air-entraining agent dosages, while increases in the others increase dosage requirements. The Foam Index test (see Chapter 14) provides a good indication of the required dosages of air-entraining admixture for various fly ashes.[**]

Workability. Fly ash, ground slag, and several inert materials will generally improve the workability of

concretes of equal slump and strength. Silica fume may reduce workability; thus high-range water reducers are usually added to concrete containing silica fume to maintain workability.

Segregation and bleeding. Concretes using fly ash or silica fume generally show less segregation and bleeding than plain concretes. This effect makes fly ash particularly valuable for use in concretes made with aggregates deficient in fines. Concretes using some ground granulated blast-furnace slags tend to have slightly higher bleeding than plain concretes. Slags have no adverse effect on segregation.

Heat of hydration. The use of fly ash and ground slag will reduce the amount of heat built up in a concrete structure because of lower heat of hydration. Some pozzolans have a heat of hydration of only 40% of that of cement. This reduction in temperature rise is especially beneficial in concrete used for massive structures.[†] Silica fume may or may not reduce the heat of hydration; however, the heat of hydration is increased by the presence of superplasticizing admixtures to an amount greater than would be obtained with ordinary portland cement.[††]

Setting time. The use of fly ash, natural pozzolans, and ground granulated blast-furnace slag will generally retard the setting time of concrete. For example, in one study, fly ash caused the initial set to be retarded by 10 to 55 minutes and the final set to be retarded by 5 to 130 minutes; the mix contained 388 lb of Type I cement and 129 lb of fly ash compared to a control with 517 lb of cement per cu yd and no ash. The degree of set retardation depends on factors such as the amount of portland cement, water requirement, the type of finely divided material, and the temperature of the concrete. With fly ashes that increase water requirements, setting time increases. Significant delays in setting usually result in increased lateral pressure on the forms, which should be taken into account when designing forms for fluid pressure. Accelerating admixtures can be used to decrease the setting time.[‡]

Finishability. Finely divided admixtures will generally improve finishability of concrete or have little effect when compared to similar concrete mixtures without them.

Pumping. The use of finely divided mineral admixtures generally aids the pumpability of concrete.

Proportioning. Finely divided mineral admixtures are added to concrete as an addition to or as a partial replacement of the cement in concrete or as a combination of addition and replacement. The use of these admixtures as cement replacements can substantially reduce the early and 28-day strengths of concrete if

---

*Reference 6-28.
**Reference 6-8.
†Reference 6-33.
††Reference 6-10, pages 665 to 676, and Reference 6-23, pages 923 to 940 and 1231 to 1260.
‡Reference 6-21.

proportioned strictly as a cement replacement rather than as a combination.

Fly ash, when used in concrete in the United States, generally consists of 20% of the cement plus pozzolan by weight.* However, some high-calcium Class C ashes have been successfully used in concrete with the ash constituent being 30% to 50% and even up to a rare 80% of the cement-plus-pozzolan material.**

Ground granulated blast-furnace slag, when used in concrete in the United States, commonly constitutes an average of about 40% of the cementing material in the mix. Some slag concretes have a slag component of 70% or more of the cementitious material.

Silica fume has been used as a partial cement replacement or cement addition in amounts between 5% and 10% and up to 30% by weight of the total cementitious material.

When attempting to improve a particular property, such as resistance to sulfate attack or resistance to alkali-aggregate reactivity, the optimum mineral admixture content should be established by testing to (1) determine whether the admixture is indeed improving the property and (2) determine the correct dosage rate, as an overdose or underdose can be harmful or have no effect. Mineral admixtures also react differently with different cements.

Curing. The effects of temperature and moisture conditions on setting properties and strength development of concretes containing finely divided admixtures are similar to the effects on concrete made with only portland cement; however, the effective curing time may need to be longer. A concrete using portland cement and a finely divided admixture may develop strength more slowly than a concrete with only portland cement. Consideration of this should be taken into account when proportioning the concrete (see following discussion on "Strength").

Relatively high dosages of silica fume can make highly cohesive concrete with very little aggregate segregation or bleeding. With little or no bleed water available at the concrete surface for evaporation, plastic cracking can readily develop on hot, windy days if special precautions are not taken.

Proper curing of all concrete, especially those containing mineral admixtures, should commence immediately after finishing. Low curing temperatures, 40°F, can reduce early strength gain and freeze-thaw and deicer durability of concretes containing certain fly ashes.†

Effect on Hardened Concrete

Strength. Fly ash, ground granulated blast-furnace slag, silica fume and other finely divided admixtures contribute to the strength gain of concrete. However, the rate of strength gain of a concrete containing these admixtures will often vary from the strength gain of concrete using portland cement as the only cementitious material. Tensile, flexural, and torsional strength are affected in the same manner as compressive strength. Due to the lower rate of hydration when using some of these admixtures, the early strength gain can be lower than that of comparable concrete without the admixture, especially at low curing temperatures.

Because of the slow pozzolanic reaction of some mineral admixtures, continuous wet curing and favorable curing temperatures may need to be provided for longer periods than normally required. However, concrete containing silica fume is less affected by this and generally equals or exceeds the one-day strength of a cement-only control mix.†† Silica fume contributes to strength development primarily between 3 and 28 days, during which time a silica-fume concrete exceeds the strength of a control concrete. Silica fume also aids the early strength gain of fly ash-cement concretes.‡

The strength development of concrete with fly ash or ground slag is similar to normal concrete when cured around 70°F. Concretes made with certain highly reactive fly ashes (especially high-calcium Class C ashes) or ground slags, with a low to moderate cement replacement, can equal or exceed the control strength in 1 to 28 days. Some fly ashes and natural pozzolans often require 28 to 90 days or more to equal or exceed a 28-day control strength, depending on the mix proportions. Concretes made with other fly ashes or natural pozzolans may never meet the strength of a control concrete. Concretes containing Class C ashes generally develop higher early-age strength than concretes with Class F ashes.‡‡

Strength gain can be increased by increasing the amount of cement in the concrete, decreasing the water-cement plus pozzolan ratio, improving curing conditions, or using an accelerating admixture. Mass concrete design often takes advantage of the delayed strength gain of pozzolans as these structures are often not put into full service immediately.

Mineral admixtures are often essential to the production of high-strength concrete. Fly ash especially has been used in production of high-strength concrete of between 6000 and 14,000 psi. With silica fume, ready mix producers now have the ability to make concrete with strengths up to 20,000 psi or more if proper aggregates and a high-range water reducer are used.

The bond strength of concrete to concrete or steel, and the impact and abrasion resistance of concrete are related to compressive strength. Mineral admixtures generally do not affect these properties beyond their influence on strength.

Drying shrinkage and creep. When used in low to moderate amounts, the effect of fly ash, ground granulated blast-furnace slag, and silica fume on the drying shrinkage and creep of concrete is generally small and of little practical significance. Concrete containing 40%

---

*Reference 6-17.
**Reference 6-23, page 353.
†Reference 6-22.
††The cement content of the control mix in this discussion is assumed equal to the cement-plus-pozzolan contents of the test mix by weight.
‡Reference 6-10, pages 765 to 784.
‡‡Reference 6-21 discusses strength gain of air-cured and moist-cured concrete with fly ash.

to 65% ground slag by weight of total cementitious material may exhibit somewhat greater drying shrinkage than plain concrete.* With high replacement levels, creep may increase with increases in ash content.** Little information is available on the effects of silica fume on creep; however, some studies indicate that silica fume may reduce creep.†

Permeability and absorption. With adequate curing, fly ash and ground slag generally reduce the permeability of concrete even when the cement content is relatively low; silica fume is especially effective in this regard. Tests show that the permeability of concrete decreases as the quantity of hydrated cementitious material increases and the water-cement ratio decreases. The absorption of fly-ash concrete is about the same as concrete without ash, although some ashes can reduce absorption by 20% or more.

Concrete color. Some finely divided materials may slightly color hardened concrete. Color effects are related to the color and amount of the admixture used in concrete. Many mineral admixtures resemble cement and therefore have little effect on color. Some silica fumes may give concrete a slightly bluish or dark gray tint and tan fly ash may impart a tan color to concrete when used in large quantities.

Alkali-aggregate reactivity. Alkali-silica reactivity between cement alkalies and reactive silica in aggregate can be controlled with certain mineral admixtures. Silica fume, fly ash, and ground granulated blast-furnace slag have significantly reduced alkali-silica reactivity. Class F ashes have reduced reactivity expansion up to 70% or more in some cases. Most Class C ashes also reduce reactivity but to a lesser degree than most Class F ashes, plus a higher dosage is required. Mineral admixtures containing large amounts of water-soluble alkali, such as certain Class C ashes, should be avoided as they may increase rather than decrease reactivity.†† Mineral admixtures provide additional calcium silicate hydrate to chemically tie up alkalies in the concrete. Determination of effective and optimum mineral admixture dosage rates is important to maximize the reduction in reactivity. Also see Chapter 4.

A mineral admixture that reduces alkali-silica reactions may not necessarily reduce alkali-carbonate reactions, a type of reaction involving cement alkalies and certain dolomitic limestones.

Resistance to sulfate attack. With proper proportioning, silica fume, fly ash, and ground slag generally improve the resistance of concrete to sulfate or seawater attack, primarily by reducing the amount of reactive elements (such as calcium) needed for expansive sulfate reactions.

The relative improvement in sulfate resistance through use of fly ash is greater for low-cement-content concrete than for high-cement-content concrete. Also, the effectiveness of fly ash is increased with increased sulfate concentrations. For improved sulfate resistance of lean concrete, one study showed that for a particular Class F ash, an adequate amount was approximately 20%‡ of the cement plus fly ash; this illustrates the

need to determine optimum ash contents, as high ash contents were detrimental with certain cements. The sulfate resistance of high-cement-content concrete made with a low C_3A cement is so great that fly ash has little opportunity to improve resistance. Class F ashes generally improve sulfate resistance more efficiently than Class C ashes.

In some reports, ground granulated blast-furnace slag appears to have less effect on the resistance of concrete to sulfate attack than fly ash. One long-term study shows a slightly reduced sulfate resistance in concrete containing ground slag compared to concrete containing only portland cement as the cementing material.‡‡ Other studies indicate that concrete with ground slag has a sulfate resistance equal to or greater than concrete made with Type V sulfate-resisting cement.§

Silica fume provides excellent sulfate resistance to concrete, better than fly ash or ground slag in some studies.

Corrosion of embedded steel. Some finely divided mineral admixtures reduce steel corrosion by reducing the permeability of properly cured concrete to water, air, and chloride ions. Fly ash can significantly reduce chloride-ion ingress. Silica fume greatly decreases permeability and chloride-ion ingress and also significantly increases electrical resistivity, thereby reducing the electrochemical reaction of corrosion. Concrete containing silica fume is often used in overlays and full-depth slab placements on bridges and parking garages, structures particularly vulnerable to corrosion due to chloride-ion ingress.

Carbonation. Carbonation of concrete is a process by which carbon dioxide from the air penetrates the concrete and reacts with the hydroxides, such as calcium hydroxide, to form carbonates. In the reaction with calcium hydroxide, calcium carbonate is formed. Carbonation increases shrinkage on drying (promoting crack development) and lowers the alkalinity of concrete. High alkalinity is needed to protect embedded steel from corrosion; consequently, concrete should be resistant to carbonation to help prevent steel corrosion. The amount of carbonation is significantly increased in concretes with a high water-cement ratio, low cement content, short curing period, low strength, and highly permeable or porous paste. The depth of carbonation of good quality, well-cured concrete is generally of little practical significance.

At normal dosages, silica fume, ground slag, and fly ash are reported to increase carbonation, but usually not to a concernable amount, in concrete with short moist-curing periods and have little effect on carbona-

---

*Reference 6-5.
**Reference 6-10, pages 87 to 102, 157, and 307 to 319.
†Reference 6-10, page 29.
††References 6-28, 6-29, 6-30, 6-31, and 6-32.
‡Reference 6-7.
‡‡Reference 6-13.
§Reference 6-5.

tion with long moist-curing periods.*

Freeze-thaw resistance. Air-entrained concretes containing fly ash, ground granulated blast-furnace slag, or silica fume are generally reported to have good freeze-thaw durability. For concrete containing finely divided admixtures to provide the same resistance to freezing and thawing cycles as a concrete made using only portland cement as a binder, three conditions must be met:

1. Both concretes must have approximately the same compressive strength
2. Both concretes must have an equivalent adequate entrained air void system
3. Both concretes must be properly cured

The requirements presented in Chapter 5 for frost resistance should be followed.

Deicer scaling. The deicer-scaling resistance of concrete made without mineral admixtures is usually better than concrete made with mineral admixtures. Scaling resistance decreases as the mineral admixture content increases.

Deicer-scaling resistance of all concrete is significantly improved with the use of a low water-cement ratio, moderate to high cement content, air-entrainment, proper finishing and curing, and a drying period prior to exposure of the concrete to salts and freezing temperatures. Lean concrete with about 400 lb or less of cementitious material per cubic yard of concrete can be especially vulnerable to deicer scaling. A minimum of 564 lb of cement and a maximum water to cement plus pozzolan ratio of 0.45 is recommended for deicer-scaling resistance. Refer to Chapter 5 for more information.**

Chemical resistance. The resistance of concrete to strong chemicals, such as acids, is usually unaffected or slightly improved (especially where permeability and absorption is reduced) by the addition of mineral admixtures.

Soundness. Soundness here refers to the volume stability of cementitious paste. Free CaO and MgO (periclase) in excessive amounts can cause unsoundness (deleterious expansion). Most mineral admixtures do not contribute to unsoundness and some may reduce unsoundness when used at normal dosage rates. The autoclave test required in ASTM C 618 helps to prevent unsound material from being used.†

WORKABILITY AGENTS

Fresh concrete is sometimes harsh due to faulty mixture proportions or certain aggregate characteristics such as particle shape and improper grading. Under these conditions, improved workability may be needed, especially if the concrete requires a troweled finish. Good workability is also important for concrete placed in heavily reinforced members or placed by pumping or tremie methods. Frequently, increasing the cement content or the amount of fine aggregate will give the desired workability.

One of the best workability agents is entrained air. It acts like a lubricant and is especially effective in improving the workability of lean, harsh mixtures. Regular and high-range water reducers (superplasticizers) also improve workability.

Some organic materials, such as alginates and cellulose derivatives, when added to concrete with a given water content, will increase the slump. The composition of these workability agents may be similar to that of some water-reducing and retarding admixtures. They may also entrain air.

Finely divided mineral admixtures are used to improve the workability of mixes deficient in material passing the No. 50 and No. 100 sieves.

CORROSION INHIBITORS

Concrete protects embedded steel from corrosion through its highly alkaline nature. The high pH environment (usually greater than 12.5) causes a passive and noncorroding protective oxide film to form on steel. However, carbonation or the presence of chloride ions from deicers or seawater can destroy or penetrate the film. Once this happens, an electric cell is formed along the steel or between steel bars and the electrochemical process of corrosion begins. Some steel areas along the bar become the anode discharging current in the electric cell and there the iron goes into solution. Steel areas that receive current are the cathodes where hydroxide ions are formed. The iron and hydroxide ions form iron hydroxide, FeOH, which further oxidizes to form rust, iron oxide. Rusting is an expansive process—rust expands up to four times its original volume—which induces internal stress and eventual spalling in concrete over reinforcing steel. The cross-sectional area of the steel can also be significantly reduced. Once it starts, the rate of steel corrosion is influenced by the concrete's electrical resistivity, moisture content, and the rate at which oxygen migrates through the concrete to the steel. Chloride ions alone can also penetrate the passive film on the reinforcement and combine with iron ions to form a soluble iron chloride complex that carries the iron into the concrete to be later oxidized and form rust.

Corrosion-inhibiting admixtures chemically arrest the corrosion reaction. Calcium nitrite, the most commonly used liquid corrosion inhibitor, blocks the corrosion reaction of the chloride ions by chemically reinforcing and stabilizing the passive film. The nitrite ion causes the ferric oxides to become insoluble. In effect, the chloride ions are prevented from penetrating the passive film and making contact with the steel. A certain amount of calcium nitrite can stop corrosion up to a certain threshold level of chloride ion. There-

---

*References 6-10 and 6-30.
**References 6-1, 6-5, 6-10, 6-22, 6-25, 6-30, and 6-32.
†Reference 6-28.

fore, increased threshold chloride levels require increased levels of calcium nitrite to stop corrosion. The threshold level at which corrosion starts in normal concrete with no inhibiting admixture is about 0.15% water-soluble chloride ion by weight of cement.* The onset of chloride ingress and corrosion is significantly decreased with lower water-cement ratios. Calcium nitrite is also an accelerator.**

Other methods of reducing corrosion include the use of epoxy-coated reinforcing steel (ASTM D 3963), surface treatments, concrete overlays, and cathodic protection.

Epoxy-coated reinforcing steel works by preventing chloride ions from reaching the steel. Surface treatments and concrete overlays attempt to stop or reduce chloride-ion penetration at the concrete surface. Silanes, siloxanes, methacrylates, epoxies, and other materials are used as surface treatments.

Impermeable materials, such as most epoxies, should not be used on slabs on ground or other concrete where moisture can freeze under the coating. The freezing water can cause surface delamination under the impermeable coating. Latex-modified portland cement concrete, low-slump concrete, and concrete with silica fume are used in overlays to reduce chloride-ion ingress.

Cathodic protection methods reverse the corrosion current flow through the concrete and reinforcing steel. This is done by inserting a nonstructural anode in the concrete and forcing the steel to be the cathode by electrically charging the system. The anode is connected to the positive pole of a rectifier. Since corrosion occurs where the current leaves the steel, the steel cannot corrode if it is receiving the induced current.†

DAMPPROOFING AGENTS

The passage of water through concrete can usually be traced to the existence of cracks or areas of incomplete consolidation. Sound, dense concrete made with a water-cement ratio of less than 0.50 by weight will be watertight if it is properly placed and cured.

Admixtures known as dampproofing agents include certain soaps, stearates, and petroleum products. They may, but generally do not, reduce the permeability of concretes that have low cement contents, high water-cement ratios, or a deficiency of fines in the aggregate. Their use in well-proportioned mixes, however, may increase the mixing water required and actually result in increased rather than reduced permeability.

Dampproofing admixtures are sometimes used to reduce the transmission of moisture through concrete that is in contact with water or damp earth. Many so-called dampproofers are not effective, especially when used in concretes that are in contact with water under pressure.

PERMEABILITY-REDUCING AGENTS

Permeability-reducing agents reduce the rate at which water under pressure is transmitted through concrete.

One of the best methods of decreasing permeability in concrete is to increase the cement content and moist-curing period and reduce the water-cement ratio to less than 0.5. Most admixtures that reduce water-cement ratio consequently reduce permeability. Some mineral admixtures, especially silica fume, reduce permeability through the hydration and pozzolanic-reaction process.

COLORING ADMIXTURES

Natural and synthetic materials are used to color concrete for aesthetic and safety reasons. Red concrete is often used around buried electrical or gas lines as a warning to anyone of their presence. Pigments (ASTM C 979) should generally not exceed 10% by weight of the cement. Pigments used in amounts less than 6% generally do not affect concrete properties.

Unmodified carbon black substantially reduces air content. Most carbon black for coloring concrete contains an admixture to offset the effect on air. Before a coloring admixture is used at a project, it should be tested for color fastness in sunlight and autoclaving, chemical stability in cement, and effects on concrete properties. Calcium chloride should not be used with pigments so as to avoid color distortions.

PUMPING AIDS

Pumping aids are added to concrete mixes to improve pumpability. Pumping aids cannot cure all unpumpable concrete mixes; they are best used to make marginally pumpable concrete more pumpable. These admixtures thicken the fluid (increase viscosity) in concrete to reduce dewatering of the paste while under pressure from the pump.

Some pumping aids may increase water demand, reduce compressive strength, cause air entrainment, or retard setting time. These side effects can be corrected by adjusting the mix proportions or adding another admixture to offset the side effect.

A partial list of materials used in pumping aids is given in Table 6-1. Some admixtures that serve other primary purposes but also improve pumpability are air-entraining agents, fly ash, and some water-reducing and set-retarding admixtures.

CHEMICAL ADMIXTURES TO REDUCE ALKALI REACTIVITY

Some chemicals have shown success in reducing alkali-aggregate expansion. Of these, lithium and barium salts show outstanding reductions.

---

*Reference 6-18, page 12.
**References 6-10, pages 719-723; 6-14; and 6-16.
†Reference 6-14.

The most practical methods of reducing alkali-aggregate expansion, however, are through the use of pozzolans known to reduce alkali-aggregate expansion, the use of nonreactive aggregate, or the use of blended cements.

BONDING ADMIXTURES AND BONDING AGENTS

Bonding admixtures are usually water emulsions of organic materials including rubber, polyvinyl chloride, polyvinyl acetate, acrylics, styrene butadiene copolymers, and other polymers. They are added to portland cement mixtures to increase the bond strength between old and new concrete. Flexural strength and resistance to chloride-ion ingress are also improved. They are added in proportions equivalent to 5% to 20% by weight of the cement, the actual quantity depending on job conditions and type of admixture used. Some bonding admixtures may increase the air content of mixtures to which they are added.

Nonreemulsifiable types are resistant to water, better suited to exterior application, and used in places where moisture is present.

The ultimate result obtained with a bonding admixture will be only as good as the surface to which it is applied. The surface must be dry, clean, sound, free of dirt, dust, paint, and grease, and at the proper temperature. Organic or polymer modified concretes are acceptable for patching and thin bonded overlayment, particularly where near-feathered edges are desired.

Bonding agents should not be confused with bonding admixtures. Admixtures are an ingredient in the concrete; bonding agents are applied to existing concrete surfaces immediately before the new concrete is placed. Bonding agents help "glue" the existing and the new materials together. Bonding agents are often used in restoration and repair work and consist of portland cement or latex-modified portland cement grout or polymers such as epoxy (ASTM C881) or latex (ASTM C1059). See Chapter 15 on polymer portland cement concrete.

GROUTING AGENTS

Portland cement grouts are used for a variety of purposes: to stabilize foundations, set machine bases, fill cracks and joints in concrete work, cement oil wells, fill cores of masonry walls, grout prestressing tendons and anchor bolts, and fill the voids in preplaced aggregate concrete. To alter the properties of grout for specific applications, various air-entraining admixtures, accelerators, retarders, and nonshrink and workability agents are often used.

GAS-FORMING AGENTS

Aluminum powder and other gas-forming materials are sometimes added to concrete and grout in very small quantities to cause a slight expansion prior to hardening. This may be of benefit where the complete grouting of a confined area is essential, such as under machine bases or in post-tensioning ducts of prestressed concrete. These materials are also used in larger quantities to produce lightweight cellular concretes (see Chapter 15). The amount of expansion that occurs is dependent upon the amount of gas-forming material used, the temperature of the fresh mixture, the alkali content of the cement, and other variables. Where the amount of expansion is critical, careful control of mixtures and temperatures must be exercised. Gas-forming agents will not overcome shrinkage after hardening caused by drying or carbonation.

AIR DETRAINERS

Air-detraining admixtures reduce the air content in concrete. They are used when the air content cannot be reduced by adjusting the mix proportions or by changing the dosage of the air-entraining agent and other admixtures. However, air-detrainers are rarely used and their effectiveness and dosage rate should be established on trial mixes prior to use on actual job mixes. Materials used in air-detraining agents are listed in Table 6-1.

FUNGICIDAL, GERMICIDAL, AND INSECTICIDAL ADMIXTURES

Bacteria and fungal growth on or in hardened concrete may be partially controlled through the use of fungicidal, germicidal, and insecticidal admixtures. The most effective materials are polyhalogenated phenols, dieldrin emulsions, and copper compounds. The effectiveness of these materials is generally temporary, and in high dosages they may reduce the compressive strength of concrete.

REFERENCES

6-1. Klieger, Paul, and Isberner, Albert W., *Laboratory Studies of Blended Cements—Portland Blast-Furnace Slag Cements*. Research Department Bulletin RX218, Portland Cement Association, 1967.

6-2. Klieger, Paul, and Perenchio, William F., *Laboratory Studies of Blended Cement: Portland-Pozzolan Cements,* Research and Development Bulletin RD013T, Portland Cement Association, 1972.

6-3. Whiting, David, *Effects of High-Range Water Reducers on Some Properties of Fresh and Hardened Concretes,* Research and Development Bulletin RD061, Portland Cement Association, 1979.

6-4. Berry, E. E., and Malhotra, V. M., "Fly Ash for Use in Concrete—A Critical Review," *Journal of the American Concrete Institute,* Detroit, March-April 1980, pages 59-73.

6-5. Hogan, F. J., and Meusel, J. W., "Evaluation for Durability and Strength Development of a Ground Granulated Blast Furnace Slag., *Cement, Concrete, and Aggregates,* vol. 3, no. 1, American Society for Testing and Materials, Philadelphia, Summer 1981, pages 40-52.

6-6. Gebler, S. H., *The Effects of High-Range Water Reducers on the Properties of Freshly Mixed and Hardened Flowing Concrete,* Research and Development Bulletin RD081T, Portland Cement Association, 1982.

6-7. Stark, David, *Longtime Study of Concrete Durability in Sulfate Soils,* Research and Development Bulletin RD086T, Portland Cement Association, 1982.

6-8. Gebler, S. H., and Klieger, P., *Effect of Fly Ash on the Air-Void Stability of Concrete,* Research and Development Bulletin RD085T, Portland Cement Association, 1983.

6-9. Malhotra, V. M., and Carette, G. G., "Silica Fume Concrete—Properties, Applications, and Limitations," *Concrete International,* American Concrete Institute, May 1983, pages 40-46.

6-10. *Fly Ash, Silica Fume, Slag and Other Mineral By-Products in Concrete,* ACI Publication SP-79, American Concrete Institute, 1983.

6-11. Aitcin, P. C., *Condensed Silica Fume,* University of Sherbrooke, Sherbrooke, Quebec, Canada, 1983.

6-12. Carette, G. G., and Malhotra, V. M., "Mechanical Properties, Durability, and Drying Shrinkage of Portland Cement Concrete Incorporating Silica Fume," *Cement, Concrete, and Aggregates,* American Society for Testing and Materials, Summer 1983.

6-13. Mielenz, Richard C., "Mineral Admixtures—History and Background," *Concrete International,* American Concrete Institute, August 1983.

6-14. Chou, Gee Kin, "Cathodic Protection: An Emerging Solution to the Rebar Corrosion Problem," *Concrete Construction,* Addison, Illinois, June 1984.

6-15. Ramachandran, V. S., *Concrete Admixtures Handbook,* Noyes Publications, Park Ridge, New Jersey, 1984.

6-16. Walitt, Arthur C., "Calcium Nitrite Offers Long-term Corrosion Prevention," *Concrete Construction,* April 1985.

6-17. *1989 Survey of Fly Ash Use in Ready Mixed Concrete,* Survey by National Ready Mixed Concrete Association, Silver Spring, Maryland, Analysis by Mark A. Justman, Portland Cement Association, December 1991.

6-18. *Corrosion of Metals in Concrete,* ACI 222R-85, ACI Committee 222 Report, American Concrete Institute, 1985.

6-19. *Building Code Requirements for Reinforced Concrete,* ACI 318-83, revised 1986, ACI Committee 318 Report, American Concrete Institute.

6-20. Kosmatka, Steven H., "Discoloration of Concrete—Causes and Remedies," *Concrete Technology Today,* PL861B, Portland Cement Association, 1986.

6-21. Gebler, Steven H., and Klieger, Paul, *Effect of Fly Ash on Some of the Physical Properties of Concrete,* Research and Development Bulletin RD089T, Portland Cement Association, 1986.

6-22. Gebler, Steven H., and Klieger, Paul, *Effect of Fly Ash on the Durability of Air-Entrained Concrete,* Research and Development Bulletin RD090T, Portland Cement Association, 1986.

6-23. *Fly Ash, Silica Fume, Slag, and Natural Pozzolans in Concrete,* SP-91, American Concrete Institute, 1986.

6-24. Rixom, M. R., and Mailvaganam, N. P., *Chemical Admixtures for Concrete,* E. & F. N. Spon, New York, 1986.

6-25. Whiting, David, "Deicer-Scaling Resistance of Lean Concretes Containing Fly Ash," *Concrete Technology Today,* PL892B, Portland Cement Association, 1989.

6-26. *Admixtures for Concrete,* ACI 212.1R-81, revised 1986, ACI Committee 212 Report, American Concrete Institute.

6-27. *Guide for Use of Admixtures in Concrete,* ACI 212.2R-81, revised 1986, ACI Committee 212 Report, American Concrete Institute.

6-28. Helmuth, Richard A., *Fly Ash in Cement and Concrete,* SP040T, Portland Cement Association, 1987.

6-29. Buck, Alan D., and Mather, Katharine, *Methods for Controlling Effects of Alkali-Silica Reaction,* Technical Report SL-87-6, Waterways Experiment Station, U.S. Army Corp. of Engineers, Vicksburg, Mississippi, 1987.

6-30. *Concrete Durability, Katharine and Bryant Mather International Conference,* SP100, American Concrete Institute, 1987.

6-31. "Silica Fume in Concrete," ACI Committee 226 Report, *ACI Materials Journal,* American Concrete Institute, March-April 1987.

6-32. Kosmatka, Steven H., *Effect of Fly Ash on Concrete,* SS379, Portland Cement Association, 1987.

6-33. *Concrete for Massive Structures,* IS128T, Portland Cement Association, 1987.

6-34. *Ground Granulated Blast-Furnace Slag as a Cementitious Constituent in Concrete,* ACI 226.1R-87, ACI Committee 226 Report, American Concrete Institute, 1987.

CHAPTER 7
Proportioning Normal Concrete Mixtures

The objective in designing concrete mixtures is to determine the most economical and practical combination of readily available materials to produce a concrete that will satisfy the performance requirements under particular conditions of use. To fulfill this objective, a properly proportioned concrete mix should possess these qualities:

1. Acceptable workability of freshly mixed concrete
2. Durability, strength, and uniform appearance of hardened concrete
3. Economy

Understanding the basic principles of mixture design is as important as the actual calculations. Only with proper selection of materials and mixture characteristics and proper proportioning can the above qualities be obtained in concrete production.

SELECTING MIX CHARACTERISTICS

Before a concrete mixture can be proportioned, mixture characteristics are selected based on the intended use of the concrete, the exposure conditions, the size and shape of members, and the physical properties of the concrete (such as strength) required for the structure. Once the characteristics are selected, the mixture can be proportioned from field or laboratory data (discussed further under "Proportioning"). Since most of the desirable properties of hardened concrete depend primarily upon the quality of the cement paste, the first step in proportioning a concrete mixture is the selection of the appropriate water-cement ratio for the durability and strength needed. Concrete mixtures should be kept as simple as possible, as an excessive number of ingredients often make a concrete mixture difficult to control.

Water-Cement Ratio and Strength Relationship

Compressive strength, because it can easily be determined, is the most universally used measure for concrete quality. Although it is an important characteristic, other properties such as durability, permeability, and wear resistance may be equally or more important.

Within the normal range of strengths in concrete construction, the compressive strength is inversely related to the water-cement ratio:

> For fully compacted concrete made with sound and clean aggregates, the strength and other desirable properties of concrete under given job conditions are governed by the quantity of mixing water used per unit of cement.

While there is agreement among authorities that the ratio of water to cement has a major influence on the strength of concrete, there is less agreement on the form of the relationship. The significance of the amount of water upon strength is the parameter, as stated above, proposed by Abrams.* Many concrete technologists prefer to use the water-cement ratio, because the strength and other desirable properties of concrete are almost linearly related to this index. A more rational parameter is the relative density of the cement paste, which is also linearly related to strength.

The strength of the cement paste binder in concrete depends on the quality and quantity of the reacting components and on the degree to which the hydration reaction is completed. Concrete becomes stronger with time as long as there is moisture available and a favorable temperature. Therefore, the strength at any particular age is not so much a function of the original water-cement ratio as it is the degree to which cement has hydrated. The importance of prompt and thorough curing is easily recognized in this rationale.

Differences in strength for a given water-cement ratio may result from changes in the aggregate size, grading, surface texture, shape, strength, and stiffness; differences in cement types and sources; entrained-air content; the presence of admixtures; and the length of curing time.

Strength

The specified compressive strength at 28 days, f'_c, is the strength that is expected to be equaled or exceeded by the average of any set of three consecutive strength tests, with no individual test (average of two cylinders)

---

*D. A. Abrams, *Design of Concrete Mixtures,* Bulletin No. 1, Structural Materials Research Laboratory, Lewis Institute, Chicago, 1918.

more than 500 psi below the specified strength when specimens are cured under laboratory conditions for an individual class of concrete.*

The average strength should equal specified strength plus an allowance to account for variations in materials; variations in methods of mixing, transporting, and placing the concrete; and variations in making, curing, and testing concrete cylinder specimens. The average strength, which is greater than f'_c, is called f'_{cr}; it is the strength required in the mix design. Requirements for f'_{cr} are discussed in detail under "Proportioning" in this chapter.

Water-Cement Ratio

The water-cement ratio is simply the weight of water divided by the weight of cement. The water-cement ratio selected for mix design must be the lowest value required to meet the design exposure considerations. Tables 7-1 and 7-2 are guides for selecting the water-cement ratio for various exposure conditions.

When durability does not control, the water-cement ratio should be selected on the basis of concrete compressive strength. In such cases the water-cement ratio and mixture proportions for the required strength should be based on adequate field data or trial mixtures made with actual job materials to determine the relationship between water-cement ratio and strength (Fig. 7-1). Table 7-3 can be used to select a water-cement ratio with respect to the required average strength, f'_{cr}, for trial mixtures when no other data are

available. Table 7-4 can be used only with permission of the project engineer when past data is not available and trial mixes will not be made.

When a pozzolan is used in the concrete, a water to cement plus pozzolan ratio, $W/(C + P)$, may be used instead of the traditional water to cement only ratio (W/C) if allowed by the project engineer or specifications. In weight equivalency, $W/(C + P)$ is simply the weight of water divided by the summation of the weights of cement plus pozzolan.**

Aggregates

Two characteristics of aggregates have an important influence on proportioning concrete mixtures because they affect the workability of the fresh concrete. They are

1. Grading (particle size and distribution)
2. Nature of particles (shape, porosity, surface texture)

Grading is important for attaining an economical mixture because it affects the amount of concrete that can be made with a given amount of cement and water. Coarse aggregates should be graded up to the largest size practical under job conditions. The maximum size that can be used depends on the size and shape of the concrete member to be cast and on the amount and distribution of reinforcing steel in the member.

The maximum size of coarse aggregate should not exceed one-fifth the narrowest dimension between sides of forms nor three-fourths the clear space between individual reinforcing bars or wire, bundles of bars, or prestressing tendons or ducts. For unreinforced slabs on ground, the maximum size should not exceed one-third the slab thickness.† Smaller sizes can be used when availability or economic consideration require them. It is also good practice to limit aggregate size to not more than three-fourths the clear space between reinforcement and the forms.

The amount of mixing water required to produce a cubic yard of concrete of a given slump is dependent on the maximum size and shape and the amount of coarse aggregate. Larger sizes minimize the water requirement and thus allow the cement content to be reduced. Also, rounded aggregate requires less water than a crushed aggregate in concretes of equal slump (see "Water Content").

The maximum size of coarse aggregate that will produce concrete of maximum strength for a given cement content depends upon the aggregate source as well as its shape and grading. For high compressive strength concrete (6000 psi plus) with a cement content exceeding 600 lb per cubic yard, the optimum maximum size is about ¾ in. Higher strengths can also sometimes be achieved through the use of crushed-stone aggregate rather than rounded-gravel aggregate.

The most desirable fine-aggregate grading will depend upon the type of work, the richness of the mixture, and the size of the coarse aggregate. For leaner

Table 7-1. Maximum Water-Cement Ratios for Various Exposure Conditions*

Exposure condition	Maximum water-cement ratio by weight for normal-weight concrete
Concrete protected from exposure to freezing and thawing or application of deicer chemicals	Select water-cement ratio on basis of strength, workability, and finishing needs
Concrete intended to be watertight:	
a. Concrete exposed to fresh water	0.50
b. Concrete exposed to brackish water or seawater	0.45
Concrete exposed to freezing and thawing in a moist condition:**	
a. Curbs, gutters, guardrails, or thin sections	0.45
b. Other elements	0.50
c. In presence of deicing chemicals	0.45
For corrosion protection for reinforced concrete exposed to deicing salts, brackish water, seawater, or spray from these sources	0.40†

*Adapted from Reference 7-9.
**Air-entrained concrete.
†If minimum concrete cover required by ACI 318 Section 7.7 is increased by 0.5 in., water-cement ratio may be increased to 0.45 for normal-weight concrete.

*Reference 7-9.
**Reference 7-6 discusses an equivalent absolute volume method.
†References 7-4 and 7-9.

Table 7-2. Requirements for Concrete Exposed to Sulfate-Containing Solutions*

Sulfate exposure	Water-soluble sulfate (SO₄) in soil, percent by weight	Sulfate (SO₄) in water, ppm	Cement type**	Normal-weight aggregate concrete — Maximum water-cement ratio, by weight	Lightweight aggregate concrete — Minimum compressive strength, psi
Negligible	0.00–0.10	0–150	—	—	—
Moderate†	0.10–0.20	150–1500	II, IP(MS), IS(MS), P(MS), I(PM)(MS), I(SM)(MS)	0.50	3750
Severe	0.20–2.00	1500–10,000	V	0.45	4250
Very severe	Over 2.00	Over 10,000	V plus pozzolan††	0.45	4250

*Adapted from References 1-15, 7-9, and 7-10.
**Cement Types II and V are specified in ASTM C150 and the remaining types, blended cements, are specified in ASTM C595.
†Seawater.
††Pozzolan (ASTM C618 or silica fume) that has been determined by test or service record to improve sulfate resistance when used in concrete containing Type V cement.

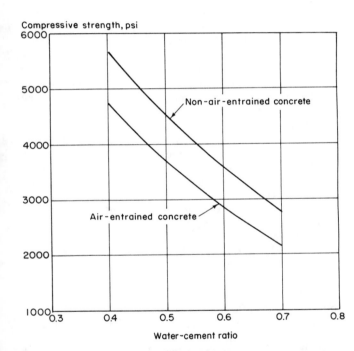

Fig. 7-1. Typical trial mixture or field data strength curves.

Table 7-4. Maximum Permissible Water-Cement Ratios for Concrete When Strength Data from Field Experience or Trial Mixtures Are Not Available

Specified 28-day compressive strength, f_c', psi	Water-cement ratio by weight — Non-air-entrained concrete	Air-entrained concrete
2500	0.67	0.54
3000	0.58	0.46
3500	0.51	0.40
4000	0.44	0.35
4500	0.38	*
5000	*	*

With most materials, water-cement ratios shown will provide average strengths greater than required. This table should be used only with special permission from the project engineer. It is not intended for use in designing trial batches. Use Table 7-3 for trial batch design.

*For strength above 4500 psi (non-air-entrained concrete) and 4000 psi (air-entrained concrete), concrete proportions shall be established from field data or trial mixtures.

Adapted from Reference 7-9.

Table 7-3. Typical Relationship Between Water-Cement Ratio and Compressive Strength of Concrete

Compressive strength at 28 days, psi*	Water-cement ratio by weight — Non-air-entrained concrete	Air-entrained concrete
6000	0.41	—
5000	0.48	0.40
4000	0.57	0.48
3000	0.68	0.59
2000	0.82	0.74

*Values are estimated average strengths for concrete containing not more than the percentage of air shown in Table 7-6. For a constant water-cement ratio, the strength of concrete is reduced as the air content is increased.

Strength is based on 6x12-in. cylinders moist-cured 28 days at 73.4°F ± 3°F in accordance with Section 9b of ASTM C31.

Relationship assumes maximum size of aggregate about ¾ in. to 1 in.

Adapted from Reference 7-6.

Table 7-5. Volume of Coarse Aggregate Per Unit of Volume of Concrete

Maximum size of aggregate, in.	Volume of dry-rodded coarse aggregate* per unit volume of concrete for different fineness moduli of fine aggregate — 2.40	2.60	2.80	3.00
⅜	0.50	0.48	0.46	0.44
½	0.59	0.57	0.55	0.53
¾	0.66	0.64	0.62	0.60
1	0.71	0.69	0.67	0.65
1½	0.75	0.73	0.71	0.69
2	0.78	0.76	0.74	0.72
3	0.82	0.80	0.78	0.76
6	0.87	0.85	0.83	0.81

*Bulk volumes are based on aggregates in dry-rodded condition as described in ASTM C29. These volumes are selected from empirical relationships to produce concrete with a degree of workability suitable for usual reinforced construction. For less workable concrete such as required for concrete pavement construction, they may be increased about 10%. For more workable concrete, such as may sometimes be required when placement is to be by pumping, they may be reduced up to 10%.

Adapted from Reference 7-6.

mixtures a fine grading (lower fineness modulus) is desirable for workability. For richer mixtures a coarse grading (higher fineness modulus) is used for greater economy.

The volume of coarse aggregate can be determined from Table 7-5.

Entrained Air

Entrained air must be used in all concrete that will be exposed to freezing and thawing and deicing chemicals and can be used to improve workability even where not required.

Air entrainment is accomplished by using an air-entraining portland cement or by adding an air-entraining admixture at the mixer. The amount of admixture should be adjusted to meet variations in concrete ingredients and job conditions. The amount recommended by the admixture manufacturer will, in most cases, produce the desired air content.

Recommended target air contents for air-entrained concrete are shown in Table 7-6. Note that the amount of air required to provide adequate freeze-thaw resistance is dependent upon the maximum size of aggregate and the level of exposure. Air is entrained in the mortar fraction of the concrete; in properly proportioned mixes, the mortar content decreases as maximum aggregate size increases, thus decreasing the required concrete air content. The levels of exposure are defined by ACI 211.1* as follows:

Mild exposure. This exposure includes indoor or outdoor service in a climate where concrete will not be exposed to freezing or deicing agents. When air entrainment is desired for a beneficial effect other than durability, such as to improve workability or cohesion or in low cement content concrete to improve strength, air contents lower than those needed for durability can be used.

Moderate exposure. Service in a climate where freezing is expected but where the concrete will not be continually exposed to moisture or free water for long periods prior to freezing and will not be exposed to deicing or other aggressive chemicals. Examples include exterior beams, columns, walls, girders, or slabs that are not in contact with wet soil and are so located that they will not receive direct applications of deicing chemicals.

Severe exposure. Concrete that is exposed to deicing or other aggressive chemicals or where the concrete may become highly saturated by continual contact with moisture or free water prior to freezing. Examples include pavements, bridge decks, curbs, gutters, sidewalks, canal linings, or exterior water tanks or sumps.

When mixing water is held constant, the entrainment of air will increase slump. When cement content and slump are held constant, the entrainment of air results in the need for less mixing water, particularly in leaner concrete mixtures. In batch adjustments, in order to maintain a constant slump while changing the air content, the water should be decreased by about 5 lb for each percentage point increase in air content or increased 5 lb for each percentage point decrease. Refer to Chapter 5 for more information.

A specific air content cannot be readily or repeatedly achieved because of the many variables affecting air content; therefore, a permissible range of air contents around a target value must be provided. Although a range of ±1% of the Table 7-6 values is often used in project specification, it is sometimes an impractically tight limit. The solution is to use a wider range such as −1 to +2 percentage points of the target values. For a target value of 6%, the specified range for the concrete delivered to the job could be 5% to 8%.

Slump

Concrete must always be made with a workability, consistency, and plasticity suitable for job conditions. Workability is a measure of how easy or difficult it is to place, consolidate, and finish concrete. Consistency is the ability of freshly mixed concrete to flow. Plasticity determines concrete's ease of molding. If more aggregate is used in a concrete mixture or if less water is added, the mixture becomes more stiff (less plastic and less workable) and difficult to mold. Neither very dry, crumbly mixtures nor very watery, fluid mixtures can be regarded as having plasticity.

The slump test is a measure of concrete consistency. For given proportions of cement and aggregate without admixtures, the higher the slump, the wetter the mixture. Slump is indicative of workability when assessing similar mixtures. However, it should not be used to compare mixtures of totally different proportions. When used with different batches of the same mixture, a change in slump indicates a change in consistency and in the characteristics of materials, mixture proportions, or water content.

Different slumps are needed for various types of concrete construction. Slump is usually indicated in the job specifications as a range, such as 2 to 4 in., or as a maximum value not to be exceeded.** When it is not specified, an approximate value can be selected from Table 7-7 for concrete consolidated by mechanical vibration. For batch adjustments, the slump can be increased by about 1 in. by adding 10 lb of water per cubic yard of concrete.

Water Content

The water content of concrete is influenced by a number of factors: aggregate size and shape, slump, water-cement ratio, air content, cement content, admixtures, and environmental conditions. Increased air content and aggregate size, reduction in water-cement ratio and slump, rounded aggregates, and the use of water-reducing admixtures or fly ash reduce water demand. On the other hand, increased temperatures, cement contents, slump, water-cement ratio, aggregate angularity, and a decreased proportion of coarse aggregate to fine aggregate increase water demand.

---

*Reference 7-6.
**ASTM C 94 discusses slump tolerances in detail.

Table 7-6. Approximate Mixing Water and Target Air Content Requirements for Different Slumps and Maximum Sizes of Aggregate

Slump, in.	Water, pounds per cubic yard of concrete, for indicated maximum sizes of aggregate*							
	3⁄8 in.	1⁄2 in.	3⁄4 in.	1 in.	1 1⁄2 in.	2 in.**	3 in.**	6 in.**
Non-air-entrained concrete								
1 to 2	350	335	315	300	275	260	220	190
3 to 4	385	365	340	325	300	285	245	210
6 to 7	410	385	360	340	315	300	270	—
Approximate amount of entrapped air in non-air-entrained concrete, percent	3	2.5	2	1.5	1	0.5	0.3	0.2
Air-entrained concrete								
1 to 2	305	295	280	270	250	240	205	180
3 to 4	340	325	305	295	275	265	225	200
6 to 7	365	345	325	310	290	280	260	—
Recommended average total air content, percent, for level of exposure:†								
Mild exposure	4.5	4.0	3.5	3.0	2.5	2.0	1.5	1.0
Moderate exposure	6.0	5.5	5.0	4.5	4.5	4.0	3.5	3.0
Severe exposure	7.5	7.0	6.0	6.0	5.5	5.0	4.5	4.0

*These quantities of mixing water are for use in computing cement factors for trial batches. They are maximums for reasonably well shaped angular coarse aggregates graded within limits of accepted specifications.

**The slump values for concrete containing aggregate larger than 1 1⁄2 in. are based on slump tests made after removal of particles larger than 1 1⁄2 in. by wet screening.

†The air content in job specifications should be specified to be delivered within −1 to +2 percentage points of the table target value for moderate and severe exposures.

Adapted from References 7-6 and 7-9.

Table 7-7. Recommended Slumps for Various Types of Construction

Concrete construction	Slump, in.	
	Maximum*	Minimum
Reinforced foundation walls and footings	3	1
Plain footings, caissons, and substructure walls	3	1
Beams and reinforced walls	4	1
Building columns	4	1
Pavements and slabs	3	1
Mass concrete	2	1

*May be increased 1 in. for consolidation by hand methods such as rodding and spading.

Adapted from Reference 7-6.

The approximate water contents in Table 7-6, used in proportioning, are for angular coarse aggregates (crushed stone). For some concretes and aggregates, the water estimates in Table 7-6 can be reduced by approximately 20 lb for subangular aggregate, 35 lb for gravel with some crushed particles, and 45 lb for a rounded gravel to produce the slumps shown. This illustrates the need for trial batch testing of local materials, as each aggregate source is different and can influence concrete properties differently.

It should be kept in mind that changing the amount of any single ingredient in a concrete mixture can have significant effects on the proportions of other ingredients as well as alter the properties of the mixture. For example, the addition of 10 lb of water per cubic yard will increase the slump by approximately 1 in. and will also increase the air content. In mixture adjustments, a decrease in air content by 1 percentage point will increase the water demand by about 5 lb per cu yd of concrete for the same slump.

Cement Content and Cement Type

The cement content is usually determined from the selected water-cement ratio and water content, although a minimum cement content frequently is included in specifications in addition to a maximum water-cement ratio. Minimum cement requirements serve to ensure satisfactory durability and finishability, improved wear resistance of slabs, and suitable appearance of vertical surfaces. This is important even though strength requirements may be met at lower cement contents.

For severe freeze-thaw, deicer, and sulfate exposures, it is desirable to specify a minimum cement content of 564 lb per cubic yard of concrete and only enough mixing water to achieve the desired consistency without exceeding the maximum water-cement ratios shown in Tables 7-1 and 7-2. For placing concrete underwater, usually not less than 650 lb of cement per cubic yard of concrete should be used. For placeability, finishability, abrasion resistance, and durability in flatwork, the quantity of cement to be used should be not less than shown in Table 7-8.

To obtain economy, proportioning should minimize the amount of cement required without sacrificing concrete quality. Since quality depends primarily on water-cement ratio, the water content should be held to a minimum to reduce the cement requirement. Steps to minimize water and cement requirements include use of (1) the stiffest practical mixture, (2) the largest practical maximum size of aggregate, and (3) the optimum ratio of fine-to-coarse aggregate.

Concrete that will be exposed to sulfate conditions should be made with the type of cement shown in Table 7-2. Refer to Chapter 2 for more information.

Seawater contains significant amounts of sulfates and chlorides. Although sulfates in seawater are capable of attacking concrete, the presence of chlorides inhibits the expansive reaction that is characteristic of sulfate attack. Calcium sulfoaluminate, the reaction product of sulfate attack, is more soluble in a chloride solution and can be more readily leached out of the concrete, thus resulting in less destructive expansion. This is the major factor explaining observations from a number of sources that the performance of concretes in seawater with portland cements having tricalcium aluminate (C_3A) contents as high as 10%, and sometimes greater, have shown satisfactory durability, providing the permeability of the concrete is low and the reinforcing steel has adequate cover. Cements meeting the requirements of ASTM C150 and C595 and meeting a C_3A requirement of not more than 10% or less than 4% (to ensure durability of reinforcement) are acceptable.*

Admixtures

Water-reducing admixtures are added to concrete to reduce the water-cement ratio or to improve the workability of a concrete without changing the water-cement ratio. Water reducers will usually decrease water contents by 5% to 10% and several will also increase air contents by $\frac{1}{2}$ to 1 percentage point. Retarders may also increase the air content.

High-range water reducers (superplasticizers) reduce water contents between 12% and 30% and some can simultaneously increase the air content up to 1 percentage point; others can reduce or not affect the air content.

Calcium chloride-based admixtures reduce water contents by about 3% and increase the air content by about $\frac{1}{2}$ percentage point.

Finely divided admixtures can have varied effects on water demand and air contents. The addition of fly ash will generally reduce water demand and decrease the air content if no adjustment in the amount of air-entraining admixture is made. Silica fume increases water demand and decreases air content.

When using a chloride-based admixture, the risks of reinforcing steel corrosion should be considered. Table 7-9 provides recommended limits on the water-soluble chloride-ion content in reinforced and prestressed concrete for different conditions.

Table 7-8. Minimum Cement Requirements for Normal-Weight Concrete Used in Flatwork

Maximum size of aggregate, in.	Cement, lb per cubic yard*
1½	470
1	520
¾	540
½	590
⅜	610

*Cement quantities may need to be greater for severe exposure.
Adapted from Reference 7-2.

Table 7-9. Maximum Chloride-Ion Content for Corrosion Protection

Type of member	Maximum water-soluble chloride ion (CL⁻) in concrete, percent by weight of cement*
Prestressed concrete	0.06
Reinforced concrete exposed to chloride in service	0.15
Reinforced concrete that will be dry or protected from moisture in service	1.00
Other reinforced concrete construction	0.30

*In hardened concrete at an age of 28 to 42 days. The test procedure should conform to that described in Federal Highway Administration Report No. FHWA-RD-77-85, "Sampling and Testing for Chloride Ion in Concrete." (Also see Chapter 14.)
Adapted from Reference 7-9.

When using more than one admixture in a concrete, the compatibility of intermixing admixtures should be assured by the admixture manufacturer or the combination of admixtures should be tested in trial batches. The water contained in many admixtures should be considered part of the mixing water if the admixture's water content is sufficient to affect the water-cement ratio by 0.01 or more (refer to Chapter 6 for more details).

PROPORTIONING

Proportioning methods have evolved from the arbitrary volumetric method (1:2:3—cement:sand:coarse aggregate) of the early 1900's to the present-day weight and absolute-volume methods described in the American Concrete Institute's Committee 211 standard practice for proportioning concrete mixes.** Weight-proportioning methods are fairly simple and quick for estimating mix proportions using an assumed or known weight of the concrete per unit volume. A more accurate method, absolute volume, involves use of specific gravity values for all the ingredients to calculate the absolute volume each will occupy in a unit volume of

---

*See References 7-5 and 7-9 for additional information.
**Reference 7-6.

concrete. The absolute volume method will be illustrated. A concrete mixture can be proportioned from field experience (statistical data) or from trial mixtures.*

Proportioning from Field Data

A presently or previously used concrete mixture design can be used for a new project if strength-test data and standard deviations** show that the mixture is acceptable. Durability aspects previously presented must also be met. The statistical data should essentially represent the same materials, proportions, and concreting conditions to be used in the new project. The data used for proportioning should also be from a concrete with an f_c' within 1000 psi of the strength required for the proposed work. Also, the data should represent at least 30 consecutive tests or two groups of consecutive tests totaling at least 30 tests (one test is the average strength of two cylinders from the same sample). If only 15 to 29 consecutive tests are available, an adjusted standard deviation can be obtained by multiplying the standard deviation (S) for the 15 to 29 tests and a modification factor from Table 7-10. The data must represent 45 or more days of tests.

The standard or modified deviation is then used in Equations 7-1 and 7-2. The average compressive strength from the test record must equal or exceed the ACI 318 required average compressive strength, f_{cr}', in order for the concrete proportions to be acceptable. The f_{cr}' for the selected mixture proportions is equal to the larger of Equations 7-1 and 7-2.

$$f_{cr}' = f_c' + 1.34S \qquad (7\text{-}1)$$
$$f_{cr}' = f_c' + 2.33S - 500 \qquad (7\text{-}2)$$

where

f_{cr}' = required average compressive strength of concrete used as the basis for selection of concrete proportions, psi

f_c' = specified compressive strength of concrete, psi

S = standard deviation, psi

When field strength test records do not meet the previously discussed requirements, f_{cr}' can be obtained from Table 7-11. A field strength record, several strength test records, or tests from trial mixtures must be used for documentation showing that the average strength of the mixture is equal to or greater than f_{cr}'.

If less than 30 but not less than 10 tests are available, the tests may be used for average strength documentation if the time period is not less than 45 days. Mixture proportions may also be established by interpolating between two or more test records if each meets the above and project requirements. If a significant difference exists between the mixtures that are used in the interpolation, a trial mixture should be considered to check strength gain. If the test records meet the above requirements and limitations of ACI 318,† the proportions for the mixture may then be considered acceptable for the proposed work.

If the average strength of the mixtures with the statistical data is less than f_{cr}', or statistical data or test

Table 7-10. Modification Factor for Standard Deviation When Less Than 30 Tests Are Available

Number of tests*	Modification factor for standard deviation**
Less than 15	Use Table 7-11
15	1.16
20	1.08
25	1.03
30 or more	1.00

*Interpolate for intermediate numbers of tests.
**Modified standard deviation to be used to determine required average strength, f_{cr}'.
Adapted from Reference 7-9.

Table 7-11. Required Average Compressive Strength When Data Are Not Available to Establish a Standard Deviation

Specified compressive strength, f_c', psi	Required average compressive strength, f_{cr}', psi
Less than 3000	$f_c' + 1000$
3000 to 5000	$f_c' + 1200$
Over 5000	$f_c' + 1400$

Adapted from Reference 7-9.

records are insufficient or are not available, the mixture should be proportioned by the trial-mixture method. The approved mixture must have a compressive strength that meets or exceeds f_{cr}'. Three trial mixtures using three different water-cement ratios or cement contents should be tested. A water-cement ratio/strength curve (Fig. 7-1) can then be plotted and the proportions interpolated from the data. It is also good practice to test the properties of the newly proportioned mixture in a trial batch.

---

*References 7-4, 7-6, 7-8, and 7-9.
**The standard deviation of strength tests of a concrete mixture with at least 30 consecutive tests can be determined as follows:

$$S = [\Sigma(X_i - \overline{X})^2 \div (n-1)]^{1/2}$$

where

S = standard deviation, psi
X_i = individual strength test (average strength of two cylinders at 28 days)
$\overline{X}$ = average of n strength-test results
n = number of consecutive strength tests

If two records are used to obtain at least 30 tests, the standard deviation used should be the statistical average of the values calculated from each test record in accordance with the following formula:

$$\overline{S} = \left[\frac{(n_1 - 1)(S_1)^2 + (n_2 - 1)(S_2)^2}{(n_1 + n_2 - 2)} \right]^{1/2}$$

where

$\overline{S}$ = statistical average deviation where two test records are used to estimate the standard deviation.
S_1, S_2 = standard deviations calculated from two test records, 1 and 2, respectively.
n_1, n_2 = number of tests in test records 1 and 2, respectively.

If less than 30, but at least 15, tests are available, the calculated standard deviation is increased by the factor given in Table 7-10.

See References 7-3, 7-8, and 7-9 for additional information on determining standard deviation.

The coefficient of variation, $V = S/\overline{X}$
†Reference 7-9.

If test data are not available and it is impractical to make trial batches, ACI 318* provides special provisions for mix proportioning using Table 7-4. However, special permission must be obtained to use these provisions which produce conservative mixtures that are generally used only in emergencies or for small placements.

ACI 214** provides statistical analysis methods for monitoring the strength of the concrete in the field to ensure that the mix properly meets or exceeds the design strength, f'_c.

Proportioning by Trial Mixtures

When field test records are not available or are insufficient for proportioning by field experience methods, the concrete proportions selected should be based on trial mixtures. The trial mixtures should use the same materials proposed for the work. Three mixtures with three different water-cement ratios or cement contents should be made to produce a range of strengths that encompass f'_{cr}. The trial mixtures should have a slump and air content within ±0.75 in. and ±0.5%, respectively, of the maximum permitted. Three cylinders per water-cement ratio should be made and cured according to ASTM C192. At 28 days or the designated test age, the compressive strength of the concrete is determined by testing the cylinders in compression. The test results are plotted to produce a strength versus water-cement ratio curve (Fig. 7-1) that is used to proportion a mixture.

A number of different methods of proportioning concrete ingredients have been used at one time or another, including

> Arbitrary assignment (1:2:3), volumetric
> Void ratio
> Fineness modulus
> Surface area of aggregates
> Cement content

Any one of these methods can produce approximately the same final mixture after adjustments are made in the field. The best approach is to select proportions based on past experience and reliable test data with an established relationship between strength and water-cement ratio for the materials to be used in the concrete. The mixtures can be relatively small batches made with laboratory precision or job-size batches made during the course of normal concrete production. Use of both is often necessary to reach a satisfactory job mixture.

The following parameters must be selected first: required strength, minimum cement content or maximum water-cement ratio, maximum size of aggregate, air content, and desired slump. Trial batches are then made varying the relative amounts of fine and coarse aggregates as well as other ingredients. Based on considerations of workability and economy, the proper mixture proportions are selected.

When the quality of the concrete mixture is specified by water-cement ratio, the trial-batch procedure consists essentially of combining a paste (water, cement, and, generally, an air-entraining admixture) of the correct proportions with the necessary amounts of fine and coarse aggregates to produce the required slump and workability. Quantities per cubic yard are then calculated.

Representative samples of the cement, water, aggregates, and admixtures must be used. To simplify calculations and eliminate error caused by variations in aggregate moisture content, the aggregates should be prewetted then dried to a saturated surface-dry condition and placed in covered containers to keep them in this condition until they are used. The moisture content of the aggregates should be determined and the batch weights corrected accordingly.

The size of the trial batch is dependent on the equipment available and on the number and size of test specimens to be made. If the batches will be mixed by hand and no test specimens are required, a batch made with 10 lb of cement may be adequate. However, larger batches will produce more accurate data. Machine mixing is recommended since it more nearly represents job conditions; it is mandatory if the concrete is to contain entrained air. The mixing procedures of ASTM C192 should be used.

Measurements and Calculations

Tests for slump, air content, and temperature should be made on the trial mixture, and the following measurements and calculations should also be performed.

Unit weight and yield. The unit weight of freshly mixed concrete is expressed in pounds per cubic foot. The yield is the volume of fresh concrete produced in a batch, usually expressed in cubic feet. The yield is calculated by dividing the total weight of the materials batched by the unit weight of the freshly mixed concrete. Unit weight and yield are determined in accordance with ASTM C138.

Absolute volume. The volume (yield) of freshly mixed concrete is equal to the sum of the absolute volumes of the cement, water (exclusive of that in the aggregate), aggregates, admixtures when applicable, and air. The absolute volume† is computed from a material's weight and specific gravity as follows:

Absolute volume

$$= \frac{\text{Weight of material}}{\text{Specific gravity of material} \times \text{unit weight of water}}$$

A value of 3.15 can be used for the specific gravity of portland cement. The specific gravity of water is 1 and the unit weight of water is 62.4 pcf. Specific gravity of normal-weight aggregate usually ranges between 2.4 and 2.9. The specific gravity of aggregate as used in mixture design calculations is the bulk specific gravity of either saturated surface-dry material or ovendry

---

*Reference 7-9.
**Reference 7-3.
†The absolute volume of a granular material (cement, aggregate) is the volume of the solid matter in the particles; it does not include the volume of spaces between particles.

material. Specific gravities of admixtures, such as water reducers or finely divided materials, must also be considered. Absolute volume is usually expressed in cubic feet.

The absolute volume of air in concrete, expressed as cubic feet per cubic yard, is equal to the air content percentage divided by 100 (e.g., 7% ÷ 100) and then multiplied by the volume of the concrete batch.

The volume of concrete in the batch can be determined by either of two methods: (1) If the specific gravities of the aggregates and cement are known, these can be used to calculate concrete volume. (2) If specific gravities are unknown or varying, the volume can be computed by dividing the total weight of materials in the mixer by the unit weight of concrete (see "Unit Weight and Yield"). In some cases, both determinations are made, one serving as a check on the other.

EXAMPLES OF MIXTURE PROPORTIONING

Example 1. Absolute Volume Method

ACI 211.1* illustrates both a weight method and a volume method for concrete mixture proportioning. The volumetric method is more accurate and is the method illustrated below.**

Conditions and specifications. Concrete is required for a loading-dock slab that will be exposed to moisture in a severe freeze-thaw climate but not subjected to deicers. A specified compressive strength, f_c', of 3500 psi is required at 28 days using a Type I cement. The member thickness is 12 in. and the design calls for a minimum of 3 in. of concrete cover over the reinforcing steel. The minimum distance between reinforcing bars is 4 in. The only admixture allowed is for air entrainment. No statistical data on previous mixes are available. The materials available are as follows:

Cement:	Type I, ASTM C 150.
Coarse aggregate:	¾-in. maximum-size gravel containing some crushed particles (ASTM C 33) with an ovendry specific gravity of 2.68, absorption of 0.5% (moisture content at SSD condition) and ovendry rodded unit weight of 100 lb per cu ft. The laboratory sample for trial batching has a moisture content of 2%.
Fine aggregate:	Natural sand (ASTM C 33) with an ovendry specific gravity of 2.64 and absorption of 0.7%. The laboratory sample moisture content is 6%. The fineness modulus is 2.80.
Air-entraining admixture:	Wood-resin type, ASTM C 260.

From this information, the task is to proportion a trial mixture that will meet the above specifications and be appropriate for the exposure conditions.

Strength. Since no statistical data is available, f_{cr}' (required compressive strength for proportioning) from Table 7-11 is equal to $f_c' + 1200$. Therefore $f_{cr}' = 3500 + 1200 = 4700$ psi.

Water-cement ratio. For an environment with moist freezing and thawing, Table 7-1 requires a maximum water-cement ratio of 0.50. The recommended water-cement ratio for an f_{cr}' of 4700 psi is 0.42 interpolated from Table 7-3.† Since the lower water-cement ratio governs, the mix must be designed for 0.42. If a curve from trial batches or field tests had been available, the water-cement ratio could have been taken from that data (Fig. 7-1).

Coarse-aggregate size. From the specified information, a ¾-in. maximum-size aggregate should be adequate as it is less than ⅓ the slab thickness and less than ¾ of the distance between reinforcing bars.

Air content. For severe freeze-thaw exposure, Table 7-6 recommends a target air content of 6.0%. Therefore, design the mix for 6% ±1.0% air and use 7% (or the maximum allowable) for batch proportions. The trial batch air content must be within ±0.5% of the maximum allowable air content.

Slump. As no slump was specified, a slump of 1 to 3 in. would be adequate as indicated by Table 7-7. Use 3 in. for proportioning purposes, the maximum recommended for slabs.

Water content. Table 7-6 recommends that a 3-in. slump, air-entrained concrete made with ¾-in. maximum-size aggregate should have a water content of about 305 lb per cu yd. However, gravel with some crushed particles should reduce the water content of the table value by about 35 lb. Therefore, the water content can be estimated to be about 305 lb minus 35 lb, which is 270 lb.

Cement content. The cement content is based on the maximum water-cement ratio and the water content. Therefore, 270 lb of water divided by a water-cement ratio of 0.42 requires a cement content of 643 lb, which is greater than the 564 lb minimum commonly specified for severe freeze-thaw climates. The 643 lb also meets the minimum cement requirements of Table 7-8 for concrete used in flatwork.

Coarse-aggregate content. The quantity of ¾-in. maximum-size coarse aggregate can be estimated from Table 7-5. The bulk volume of coarse aggregate recommended when using a sand with a fineness modulus of 2.80 is 0.62. Since it weighs 100 lb per cu ft, the ovendry weight of coarse aggregate for a cubic yard of concrete (27 cu ft) is

$100 \times 27 \times 0.62 = 1674$ lb per cubic yard of concrete

Admixture content. For a 7% air content, the air-entraining admixture manufacturer recommends a dosage rate of 0.9 fl oz per 100 lb of cement. From this

---

*Reference 7-6.
**Other examples of mix design are in PCA's *Concrete for Massive Structures*, IS128T. SI examples are in EB101T, the Canadian metric edition of *Design and Control of Concrete Mixtures*.
†$W/C = [(5000 - 4700)(0.48 - 0.40)/(5000 - 4000)] + 0.40 = 0.42$

information, the amount of air-entraining admixture is

$$0.9 \times \frac{643}{100} = 5.8 \text{ fl oz per cu yd}$$

Fine-aggregate content. At this point, the amount of all ingredients except the fine aggregate are known. In the absolute volume method, the volume of fine aggregate is determined by subtracting the absolute volume of the known ingredients from 27 cu ft (1 cu yd). The absolute volume of the water, cement, and coarse aggregate is calculated by dividing the known weight of each by the product of their specific gravity and the unit weight of water. Volume computations are as follows:

Water $= \dfrac{270}{1 \times 62.4} = 4.33$ cu ft

Cement $= \dfrac{643}{3.15 \times 62.4} = 3.27$ cu ft

Air $= \dfrac{7.0}{100} \times 27 = 1.89$ cu ft

Coarse aggregate $= \dfrac{1674}{2.68 \times 62.4} = \underline{10.01 \text{ cu ft}}$

Total volume of known
ingredients   19.50 cu ft

The liquid admixture volume is generally too insignificant to include in these calculations. However, certain admixtures such as some accelerators, superplasticizers, and corrosion inhibitors are exceptions due to their large dosage rates, and their volumes should be included.

The calculated absolute volume of fine aggregate is then

$$27 - 19.50 = 7.50 \text{ cu ft}$$

The weight of dry fine aggregate is

$$7.50 \times 2.64 \times 62.4 = 1236 \text{ lb}$$

The mixture then has the following proportions before trial mixing for one cubic yard of concrete:*

Water	270 lb
Cement	643 lb
Coarse aggregate (dry)	1674 lb
Fine aggregate (dry)	1236 lb
Total weight	3823 lb
Air-entraining admixture**	5.8 fl oz

Slump   3 in. (±¾ in. for trial batch)

Air content   7% (±0.5% for trial batch)

Estimated unit $= [270 + 643 + (1674 \times 1.005†)$
weight (using SSD $+ (1236 \times 1.007†)] \div 27$
aggregate) $= 142.22$ lb per cubic foot

Moisture. Corrections are needed to compensate for moisture in the aggregates. In practice, aggregates will contain some measurable amount of moisture. The dry-batch weights of aggregates, therefore, have to be increased to compensate for the moisture that is absorbed in and contained on the surface of each particle

and between particles. The mixing water added to the batch must be reduced by the amount of free moisture contributed by the aggregate. Tests indicate that for this example, coarse-aggregate moisture content is 2% and fine-aggregate moisture content is 6%.

With the aggregate moisture contents (MC) indicated, the trial batch aggregate proportions become

Coarse aggregate (2% MC) $= 1674 \times 1.02 = 1707$ lb
Fine aggregate (6% MC) $= 1236 \times 1.06 = 1310$ lb

Absorbed water does not become part of the mixing water and must be excluded from the water adjustment. Surface moisture contributed by the coarse aggregate amounts to $2\% - 0.5\% = 1.5\%$, that contributed by the fine aggregate, $6\% - 0.7\% = 5.3\%$. The estimated requirement for added water becomes

$$270 - (1674 \times 0.015) - (1236 \times 0.053) = 179 \text{ lb}$$

The estimated batch weights for one cubic yard of concrete are revised to include aggregate moisture as follows:

Water (to be added)	179 lb
Cement	643 lb
Coarse aggregate (2% MC, wet)	1707 lb
Fine aggregate (6% MC, wet)	1310 lb
Total	3839 lb
Air-entraining admixture	5.8 fl oz

Trial batch. At this stage, the estimated batch weights should be checked by means of trial batches or by full-size field batches. Enough concrete must be mixed for appropriate air and slump tests and for the three 6x12-in. cylinders required for compressive-strength tests at 28 days. For a laboratory trial batch it is convenient, in this case, to scale down the weights to produce 2.0 cu ft of concrete or 2/27 cu yd.

Laboratory Trial Batch

Water	$179 \times 2/27 =$	13.26 lb
Cement	$643 \times 2/27 =$	47.63 lb
Coarse aggregate (wet)	$1707 \times 2/27 =$	126.44 lb
Fine aggregate (wet)	$1310 \times 2/27 =$	97.04 lb
Total		284.37 lb
Air-entraining admixture	$5.8 \times 2/27 =$	0.43 fl oz††

The above concrete when mixed had a measured slump of 4 in., air content of 8%, and unit weight of 141.49 lb per cubic foot. During mixing, some of the

---

*If proper concrete background information had been available using these materials and the water-cement ratio selection had been based on acceptable data, these proportions could have been submitted as the mixture design to the project engineer without trial batching. A submitted mixture would show the slump as 1 to 3 in. and air content as 5% to 7%.

**The amount of air-entraining admixture can be adjusted to obtain proper air content in the field.

†(0.5% absorption ÷ 100) + 1 = 1.005
(0.7% absorption ÷ 100) + 1 = 1.007

††Due to the small amount of admixture, laboratories often convert fluid ounces to milliliters by multiplying fluid ounces by 29.57353 to improve measurement accuracy. Also, most laboratory pipets used for measuring fluids are in milliliter units.

premeasured water may remain unused or additional water may be added to approach the required slump. In this example, although 13.26 lb of water was calculated to be added, the trial batch actually used only 13.12 lb. The mixture excluding admixture therefore becomes

Water	13.12 lb
Cement	47.63 lb
Coarse aggregate (2% MC)	126.44 lb
Fine aggregate (6% MC)	97.04 lb
Total	284.23 lb

The yield of the trial batch is

$$\frac{284.23}{141.49} = 2.009 \text{ cu ft}$$

The mixture water content is determined from the added water plus the free water on the aggregates and is calculated as follows:

Water added		$= 13.12$ lb
Free water on coarse aggregate	$= \dfrac{126.44}{1.02*} \times 0.015** =$	1.86 lb
Free water on fine aggregate	$= \dfrac{97.04}{1.06*} \times 0.053** =$	4.85 lb
Total		19.83 lb

The mixing water required for a cubic yard of the same slump concrete as the trial batch is

$$\frac{19.83 \times 27}{2.009} = 267 \text{ lb}$$

Batch adjustments. The measured 4-in. slump is unacceptable (>0.75 in. above 3-in. max.), the yield was slightly high, and the 8.0% air content as measured in this example is also too high (>0.5% above 7% max.). Adjust the yield and reestimate the amount of air-entraining admixture required for a 7% air content and adjust the water to obtain a 3-in. slump. Increase the mixing water content by 5 lb for each 1% by which the air content is decreased from that of the trial batch and reduce the water content by 10 lb for each 1-in. reduction in slump.

The adjusted mixture water for the reduced slump and air content is

$$(5 \times 1) - (10 \times 1) + 267 = 262 \text{ lb per cubic yard}$$

With decreased mixing water needed in the trial batch, less cement is needed to maintain the desired water-cement ratio of 0.42. The new cement content is

$$\frac{262}{0.42} = 624 \text{ lb per cubic yard}$$

The amount of coarse aggregate remains unchanged because workability is satisfactory.

The new batch weights based on the new cement and water contents are calculated as follows:

Water	$\dfrac{262}{1 \times 62.4} =$	4.20 cu ft
Cement	$\dfrac{624}{3.15 \times 62.4} =$	3.17 cu ft

Coarse aggregate	$\dfrac{1674}{2.68 \times 62.4} =$	10.01 cu ft
Air	$\dfrac{7.0 \times 27}{100} =$	1.89 cu ft
Total		19.27 cu ft

Fine aggregate volume $= 27 - 19.27 = 7.73$ cu ft

The weight of dry fine aggregate required is

$$7.73 \times 2.64 \times 62.4 = 1273 \text{ lb}$$

Air-entraining admixture† $= \dfrac{0.8 \times 624}{100} = 5.0$ fl oz

Adjusted batch weights per cubic yard of concrete are

Water	262 lb
Cement	624 lb
Coarse aggregate (dry)	1674 lb
Fine aggregate (dry)	1273 lb
Total	3833 lb
Air-entraining admixture	5.0 fl oz

Estimated concrete unit weight (aggregates at SSD)

$$= \frac{262 + 624 + (1674 \times 1.005) + (1273 \times 1.007)}{27}$$
$$= 142.60 \text{ lb per cubic foot}$$

Upon completion of checking these proportions in a trial batch, it was found that the proportions were adequate for the desired slump, air content, and yield. The 28-day test cylinders had an average compressive strength of 4900 psi, which exceeds the f'_{cr} of 4700 psi. Due to fluctuations in moisture content, absorption rates, and specific gravity of the aggregate, the unit weight determined by volume calculations may not always equal the unit weight determined by ASTM C 138. Occasionally the proportion of fine to coarse aggregate is kept constant in adjusting the batch weights to maintain workability or other properties obtained in the first trial batch. After adjustments to the cement, water, and air content have been made, the volume remaining for aggregate is appropriately proportioned between the fine and coarse aggregates.

Concrete mixtures with water-cement ratios above and below 0.42 should also be tested to develop a strength curve (Fig. 7-1). From the curve, a new mixture with a compressive strength closer to f'_{cr}, yet with a water-cement ratio less than 0.50 for exposure conditions, can be proportioned and tested. Fig. 7-4 illustrates the general mix design concept. The final mixture would probably look similar to the above mixture with a slump range of 1 in. to 3 in. and an air content of 5% to 7%. The amount of air-entraining admixture must be adjusted to field conditions to maintain the specified air content.

---

*1 + (2% MC/100) = 1.02
1 + (6% MC/100) = 1.06
**(2% MC − 0.5% absorption)/100 = 0.015
(6% MC − 0.7% absorption)/100 = 0.053
†An air-entraining agent dosage of 0.8 fl oz per 100 lb of cement is expected to achieve the 7% air content.

Water reducers. Water reducers are used to increase workability without the addition of water or to reduce the water-cement ratio of a concrete mix to improve permeability or other properties.

Using the final mixture developed in the last example, assume that the project engineer approves the use of a water reducer to increase the slump to 5 in. to improve workability for a difficult placement area. Assuming that the water reducer has a manufacturer's recommended dosage rate of 4 oz per 100 lb of cement to increase slump 2 in., the admixture amount becomes $\frac{624}{100} \times 4 = 25.0$ oz per cu yd. The amount of air-entraining agent may also need to be reduced (up to 50%), as many water reducers entrain air. If a water reducer was used to reduce the water-cement ratio, the water and sand content would also need adjustment.

Pozzolans. Pozzolans are sometimes added in addition to or as a partial replacement of cement to aid in workability and resistance to sulfate attack and alkali reactivity. If a pozzolan or other finely divided mineral admixture (Chapter 6) were required for this example mixture, it would have been entered in the first volume calculation used in determining fine aggregate content.

Assuming that 75 lb of fly ash with a specific gravity of 2.5 would be used in addition to the originally derived cement content, the ash volume would be

$$\frac{75}{2.5 \times 62.4} = 0.48 \text{ cu ft}$$

The water-cement plus pozzolan ratio would be

$$\frac{W}{C + P} = \frac{270}{643 + 75} = 0.38 \text{ by weight}$$

The water to cement only ratio would still be

$$\frac{W}{C} = \frac{270}{643} = 0.42 \text{ by weight}$$

The fine aggregate volume would have to be reduced by 0.48 cu ft to allow for the volume of ash.

The pozzolan amount and volume computation could also have been derived in conjunction with the first cement content calculation using a $\frac{W}{C+P}$ ratio of 0.42 (or equivalent). For example, assume 15% of the cementitious material is specified to be a pozzolan and $W/(C + P) = 0.42$. Then with $W = 270$ lb and $C + P = 643$ lb, $P = 643 \times 15/100 = 96$ lb and $C = 643 - 96 = 547$ lb. Appropriate proportioning computations for these and other mix ingredients would follow.

Example 2. Laboratory Trial Mixture Using the PCA Water-Cement Ratio Method

Although the following specifications could be met using the methods of Example 1, the following method develops the mixture proportions directly from laboratory trial proportions, rather than through the development of a cubic yard design first as in Example 1.

Specifications. Air-entrained concrete is required for a reinforced foundation wall that will be exposed to moderate sulfate soils. A compressive strength, f'_c, of 3000 psi at 28 days using Type II cement is specified.

Minimum thickness of the wall is 10 in. and concrete cover over ½-in.-diameter reinforcement is 3 in. The clear distance between reinforcing bars is 3 in. Assume the water-cement ratio versus compressive strength relationship based on field data is illustrated by Fig. 7-1. Based on the test records of the materials to be used, the standard deviation is 300 psi. Proportion and check test by trial batch a mixture meeting the above specifications. All data are entered in the appropriate blanks on a trial-mixture data sheet (Fig. 7-2).

Data and Calculations for Trial Batch (saturated surface-dry aggregates)

Cement weight per batch: 10 lb _____ 20 lb ✓ 40 lb _____

Note: Complete Columns 1 through 4, fill in items below, then complete 5 and 6.

1 Material	2 Initial weight, lb	3 Final weight, lb	4 Weight used, Col. 2 minus Col. 3	5 Weight per cubic yard, (C) × Col. 4	6 Remarks
Cement	20.0	0	20.0	539	
Water	10.0	0	10.0	269	
Fine aggregate	66.2	27.9	38.3	1032 (a)	Percent fine aggregate = $\frac{a}{a+b} \times 100$ = 33.5%
Coarse aggregate	89.8	13.8	76.0	2048 (b)	
Air-entraining admixture	0.3oz	Total = 144.3 (T)		3888	
		T × C = 144.3 × 26.943 = 3888			Math check

Measured slump: **3** in. Measured air content: **5.4** %

Appearance: Sandy _____ Good ✓ Rocky _____

Workability: Good ✓ Fair _____ Poor _____

Weight of container + concrete = **93.4** lb

Weight of container = **21.4** lb

Weight of concrete (A) = **72.0** lb

Volume of container (B) = **0.50** cu ft

Unit weight of concrete (W) = $\frac{A}{B}$ = $\frac{72.0}{0.50}$ = **144.0** lb/cu ft

Yield (volume of concrete produced) = $\frac{\text{Total weight of material per batch}}{\text{Unit weight of concrete}}$

= $\frac{144.3}{144.0}$ = **1.0021** cu ft

Number of **144.3** lb batches per cu yd (C) = $\frac{27 \text{ cu ft*}}{\text{Yield}}$ = $\frac{27}{1.0021}$ = **26.943** batches

*One cubic yard has 27 cu ft.

Fig. 7-2. Trial mixture data sheet.

Water-cement ratio. For these exposure conditions, Table 7-2 indicates that concrete with a maximum water-cement ratio of 0.50 should be used.

The water-cement ratio for strength is selected from a curve plotted to show the relationship between the water-cement ratio and compressive strength on the basis of data from laboratory trial batches or, as in this case, field experience (Fig. 7-1).

For a standard deviation of 300 psi, f'_{cr} must be the larger of

$$f'_{cr} = f'_c + 1.34S = 3000 + 1.34(300) = 3400 \text{ psi}$$

or

$$f'_{cr} = f'_c + 2.33S - 500 = 3000 + 2.33(300) - 500 = 3200 \text{ psi}$$

Therefore, $f'_{cr} = 3400$ psi

From Fig. 7-1, the water-cement ratio for air-entrained concrete is 0.53 for an f'_{cr} of 3400. This is greater than the 0.50 permitted for exposure conditions; therefore, the exposure requirements govern. A water-cement ratio of 0.50 should be used, even though this may produce strengths higher than needed to satisfy structural requirements.

Aggregate size. Assuming it is economically available, 1½-in. maximum-size aggregate is satisfactory; it is less than ⅕ the wall thickness and less than ¾ the clear distance between reinforcing bars and between reinforcing bars and the form. If this size were not available, the next smaller available size would be used. Aggregates are to be in a saturated surface-dry condition for these trial mixtures.

Air content. Because of the exposure conditions and to improve workability, a moderate level of entrained air was specified. From Table 7-6, the target air content for concrete with 1½-in. aggregate in moderate exposure is 4.5%. Therefore, proportion the mixture with an air-content range of 4.5% ±1% and aim for 5.5% ±0.5% in the trial batch.

Slump. The recommended slump range for placing a reinforced concrete foundation wall is 1 to 3 in., assuming that the concrete will be consolidated by vibration (Table 7-7). Batch for 3 in. ±0.75 in.

Batch quantities. For convenience, a batch containing 20 lb of cement is to be made. The quantity of mixing water required is 20 × 0.50 = 10 lb. Representative samples of fine and coarse aggregates are weighed into suitable containers. The values are indicated as initial weights in Column 2 of the trial-batch data sheet (Fig. 7-2).

All of the measured quantities of cement, water, and air-entraining admixture are used and added to the mixer. Fine and coarse aggregates, previously brought to a saturated surface-dry condition, are added, in proportions similar to the mixes from which Fig. 7-1 was developed, until a workable concrete mixture with a slump deemed adequate for placement is produced.

The relative proportions for workability can readily be judged by an experienced concrete technician or engineer.

Workability. Results of tests for slump, air content, unit weight, and a description of the appearance and workability ("Good" for this example) are noted on the data sheet.

Weights of the fine and coarse aggregates not used are recorded on the data sheet in Column 3, and weights of aggregates used (Column 2 minus Column 3) are noted in Column 4. If the slump when tested had been greater than that required, additional fine or coarse aggregates (or both) would have been added. Had the slump been less than required, water and cement in the appropriate ratio (0.50) would have been added to produce the desired slump. It is important that any additional quantities be measured accurately and recorded on the data sheet.

Mixture proportions. Mixture proportions for a cubic yard of concrete are calculated in Column 5 of Fig. 7-2 by using the batch yield and unit weight. For example, the number of pounds of cement per cubic yard is determined by dividing 27 cu ft (1 cu yd) by the volume of concrete in the batch and multiplying the result by the number of pounds of cement in the batch. The percentage of fine aggregate by weight of total aggregate is also calculated. In this trial batch, the cement content was 539 lb per cubic yard, and the fine aggregate made up 33.5% of the total aggregate by weight. The air content and slump were acceptable. The 28-day strength was 3850 psi ($>f'_{cr}$). The mixture in Column 5, along with slump and air content limits of 1 to 3 in. and 3.5% to 5.5%, respectively, is now ready for submission to the project engineer.

Mixture adjustments. To determine the most workable and economical proportions, additional trial batches could be made varying the percentage of fine aggregate. In each batch the water-cement ratio, aggregate gradation, air content, and slump should be maintained approximately the same. Results of four such trial batches are summarized in Table 7-12.

From the data summarized in Table 7-12, the percentage of fine aggregate is plotted against the cement content as shown in Fig. 7-3. Note that for the combination of materials used, the minimum cement content (539 lb per cubic yard) occurs at a fine-aggregate content of about 32% of total aggregate. Since the water-cement ratio is 0.50 and the unit weight of the concrete for an air content of 5% is about 144 lb per cubic foot,

Table 7-12. Example of Results of Laboratory Trial Mixtures*

Batch no.	Slump, in.	Air content, percent	Unit weight, pcf	Cement content, lb per cubic yard	Fine aggregate, percent of total aggregate	Workability
1	3	5.4	144	539	33.5	Good
2	2¾	4.9	144	555	27.4	Harsh
3	2½	5.1	144	549	35.5	Excellent
4	3	4.7	145	540	30.5	Good

*Water-cement ratio was 0.50.

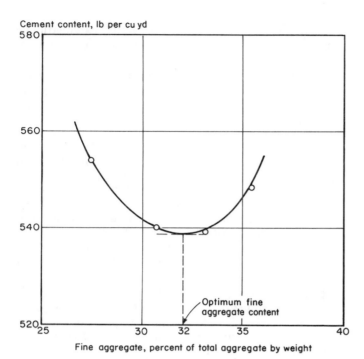

Cement content, lb per cu yd

Fine aggregate, percent of total aggregate by weight

Optimum fine aggregate content

Fig. 7-3. Example of relationship between percentage of fine aggregate and cement content for a given water-cement ratio and slump.

the quantities of materials required per cubic yard can be calculated:

Cement	539 lb
Water	269 lb
Total	808 lb
Concrete	$144 \times 27 = 3888$ lb
Aggregates	$3888 - 808 = 3080$ lb
Fine aggregate (SSD)	$\frac{32}{100} \times 3080 = 986$ lb
Coarse aggregate (SSD)	$3080 - 986 = 2094$ lb

Unless care is taken to control the slump and air content and unless weights are accurately determined, it may be difficult to obtain the data necessary to plot a curve such as Fig. 7-3. For well-graded, rounded aggregates the curve may be nearly flat.

If there is a wide difference in the cost of fine and coarse aggregates, the optimum percentage of fine aggregate for economy may be different from that at which the minimum cement content occurs. For example, if coarse aggregate costs more than fine aggregate, it may be more economical to use more fine aggregate with the consequent increase in cement content. Tables 7-13 and 7-14 are provided to illustrate the change in mix proportions for various types of concrete mixes using particular aggregate sources.

Table 7-13. Example Trial Mixtures for Non-Air-Entrained Concrete of Medium Consistency, 3- to 4-in. slump

Water-cement ratio, lb per lb	Maximum size of aggregate, in.	Air content (entrapped air), percent	Water, lb per cu yd of concrete	Cement, lb per cu yd of concrete	With fine sand, fineness modulus = 2.50			With coarse sand, fineness modulus = 2.90		
					Fine aggregate, percent of total aggregate	Fine aggregate, lb per cu yd of concrete	Coarse aggregate, lb per cu yd of concrete	Fine aggregate, percent of total aggregate	Fine aggregate, lb per cu yd of concrete	Coarse aggregate, lb per cu yd of concrete
0.40	3/8	3	385	965	50	1240	1260	54	1350	1150
	1/2	2.5	365	915	42	1100	1520	47	1220	1400
	3/4	2	340	850	35	960	1800	39	1080	1680
	1	1.5	325	815	32	910	1940	36	1020	1830
	1 1/2	1	300	750	29	880	2110	33	1000	1990
0.45	3/8	3	385	855	51	1330	1260	56	1440	1150
	1/2	2.5	365	810	44	1180	1520	48	1300	1400
	3/4	2	340	755	37	1040	1800	41	1160	1680
	1	1.5	325	720	34	990	1940	38	1100	1830
	1 1/2	1	300	665	31	960	2110	35	1080	1990
0.50	3/8	3	385	770	53	1400	1260	57	1510	1150
	1/2	2.5	365	730	45	1250	1520	49	1370	1400
	3/4	2	340	680	38	1100	1800	42	1220	1680
	1	1.5	325	650	35	1050	1940	39	1160	1830
	1 1/2	1	300	600	32	1010	2110	36	1130	1990
0.55	3/8	3	385	700	54	1460	1260	58	1570	1150
	1/2	2.5	365	665	46	1310	1520	51	1430	1400
	3/4	2	340	620	39	1150	1800	43	1270	1680
	1	1.5	325	590	36	1100	1940	40	1210	1830
	1 1/2	1	300	545	33	1060	2110	37	1180	1990
0.60	3/8	3	385	640	55	1510	1260	58	1620	1150
	1/2	2.5	365	610	47	1350	1520	51	1470	1400
	3/4	2	340	565	40	1200	1800	44	1320	1680
	1	1.5	325	540	37	1140	1940	41	1250	1830
	1 1/2	1	300	500	34	1090	2110	38	1210	1990

Table 7-13. Example Trial Mixtures for Non-Air-Entrained Concrete of Medium Consistency, 3- to 4-in. slump (continued)

Water-cement ratio, lb per lb	Maximum size of aggregate, in.	Air content (entrapped air), percent	Water, lb per cu yd of concrete	Cement, lb per cu yd of concrete	With fine sand, fineness modulus = 2.50			With coarse sand, fineness modulus = 2.90		
					Fine aggregate, percent of total aggregate	Fine aggregate, lb per cu yd of concrete	Coarse aggregate, lb per cu yd of concrete	Fine aggregate, percent of total aggregate	Fine aggregate, lb per cu yd of concrete	Coarse aggregate, lb per cu yd of concrete
0.65	⅜	3	385	590	55	1550	1260	59	1660	1150
	½	2.5	365	560	48	1390	1520	52	1510	1400
	¾	2	340	525	41	1230	1800	45	1350	1680
	1	1.5	325	500	38	1180	1940	41	1290	1830
	1½	1	300	460	35	1130	2110	39	1250	1990
0.70	⅜	3	385	550	56	1590	1260	60	1700	1150
	½	2.5	365	520	48	1430	1520	53	1550	1400
	¾	2	340	485	41	1270	1800	45	1390	1680
	1	1.5	325	465	38	1210	1940	42	1320	1830
	1½	1	300	430	35	1150	2110	39	1270	1990

Table 7-14. Example Trial Mixtures for Air-Entrained Concrete of Medium Consistency, 3- to 4-in. slump

Water-cement ratio, lb per lb	Maximum size of aggregate, in.	Air content, percent	Water, lb per cu yd of concrete	Cement, lb per cu yd of concrete	With fine sand, fineness modulus = 2.50			With coarse sand, fineness modulus = 2.90		
					Fine aggregate, percent of total aggregate	Fine aggregate, lb per cu yd of concrete	Coarse aggregate, lb per cu yd of concrete	Fine aggregate, percent of total aggregate	Fine aggregate, lb per cu yd of concrete	Coarse aggregate, lb per cu yd of concrete
0.40	⅜	7.5	340	850	50	1250	1260	54	1360	1150
	½	7.5	325	815	41	1060	1520	46	1180	1400
	¾	6	300	750	35	970	1800	39	1090	1680
	1	6	285	715	32	900	1940	36	1010	1830
	1½	5	265	665	29	870	2110	33	990	1990
0.45	⅜	7.5	340	755	51	1330	1260	56	1440	1150
	½	7.5	325	720	43	1140	1520	47	1260	1400
	¾	6	300	665	37	1040	1800	41	1160	1680
	1	6	285	635	33	970	1940	37	1080	1830
	1½	5	265	590	31	930	2110	35	1050	1990
0.50	⅜	7.5	340	680	53	1400	1260	57	1510	1150
	½	7.5	325	650	44	1200	1520	49	1320	1400
	¾	6	300	600	38	1100	1800	42	1220	1680
	1	6	285	570	34	1020	1940	38	1130	1830
	1½	5	265	530	32	980	2110	36	1100	1990
0.55	⅜	7.5	340	620	54	1450	1260	58	1560	1150
	½	7.5	325	590	45	1250	1520	49	1370	1400
	¾	6	300	545	39	1140	1800	43	1260	1680
	1	6	285	520	35	1060	1940	39	1170	1830
	1½	5	265	480	33	1030	2110	37	1150	1990
0.60	⅜	7.5	340	565	54	1490	1260	58	1600	1150
	½	7.5	325	540	46	1290	1520	50	1410	1400
	¾	6	300	500	40	1180	1800	44	1300	1680
	1	6	285	475	36	1100	1940	40	1210	1830
	1½	5	265	440	33	1060	2110	37	1180	1990
0.65	⅜	7.5	340	525	55	1530	1260	59	1640	1150
	½	7.5	325	500	47	1330	1520	51	1450	1400
	¾	6	300	460	40	1210	1800	44	1330	1680
	1	6	285	440	37	1130	1940	40	1240	1830
	1½	5	265	410	34	1090	2110	38	1210	1990
0.70	⅜	7.5	340	485	55	1560	1260	59	1670	1150
	½	7.5	325	465	47	1360	1520	51	1480	1400
	¾	6	300	430	41	1240	1800	45	1360	1680
	1	6	285	405	37	1160	1940	41	1270	1830
	1½	5	265	380	34	1110	2110	38	1230	1990

CONCRETE FOR SMALL JOBS

Although well-established ready mixed concrete mixtures are generally used for most construction, ready mix is not always practical for small jobs, especially those requiring one cubic yard or less. Small batches of concrete mixed at the site are required for such jobs.

If mixture proportions or mixture specifications are not available, Tables 7-15 and 7-16 can be used to select proportions for such concrete. Recommendations with respect to exposure conditions discussed earlier should be followed.

The proportions in Tables 7-15 and 7-16 are only a guide and may need adjustments to obtain a workable mix with locally available aggregates. Packaged, combined, dry concrete ingredients (ASTM C 387) are also available. For more information, see Reference 7-1.

DESIGN REVIEW

In practice, concrete mixture proportions will be governed by the limits of data available on the properties of materials, the degree of control exercised over the production of concrete at the plant, and the amount of supervision at the jobsite. It should not be expected that field results will be an exact duplicate of laboratory trial batches. An adjustment of the selected trial mixture is usually necessary on the job.

The mixture proportioning procedures presented here are applicable to normal-weight concrete. For concrete requiring some special property, using special admixtures or materials—lightweight aggregates, for example—different proportioning principles may be involved. A number of special concretes are discussed in Chapter 15.

The Fig. 7-4 flowchart summarizes mix design procedures.

REFERENCES

7-1. *Concrete for Small Jobs,* IS174T, Portland Cement Association, 1980.

7-2. *Guide for Concrete Floor and Slab Construction,* ACI 302.1R-80, ACI Committee 302 Report, American Concrete Institute, Detroit, 1980.

7-3. *ACI Standard Recommended Practice for Evaluation of Strength Test Results of Concrete,* ACI 214-77, reaffirmed 1983, ACI Committee 214 Report, American Concrete Institute.

7-4. *Commentary on Building Code Requirements for Reinforced Concrete (ACI 318),* ACI 318R-83, ACI Committee 318 Report, American Concrete Institute, 1983.

7-5. *Guide for the Design and Construction of Fixed Offshore Concrete Structures,* ACI 357R-84, ACI Committee 357 Report, American Concrete Institute, 1984.

7-6. *Standard Practice for Selecting Proportions for Normal, Heavyweight and Mass Concrete,* ACI 211.1-81, revised 1985, ACI Committee 211 Report, American Concrete Institute.

Table 7-15. Proportions by Weight to Make One Cubic Foot of Concrete for Small Jobs

Maximum-size coarse aggregate, in.	Air-entrained concrete				Non-air-entrained concrete			
	Cement, lb	Wet fine aggregate, lb	Wet coarse aggregate, lb*	Water, lb	Cement, lb	Wet fine aggregate, lb	Wet coarse aggregate, lb*	Water, lb
3/8	29	53	46	10	29	59	46	11
1/2	27	46	55	10	27	53	55	11
3/4	25	42	65	10	25	47	65	10
1	24	39	70	9	24	45	70	10
1 1/2	23	38	75	9	23	43	75	9

*If crushed stone is used, decrease coarse aggregate by 3 lb and increase fine aggregate by 3 lb.
Reference 7-1.

Table 7-16. Proportions by Volume* of Concrete for Small Jobs

Maximum-size coarse aggregate, in.	Air-entrained concrete				Non-air-entrained concrete			
	Cement	Wet fine aggregate	Wet coarse aggregate	Water	Cement	Wet fine aggregate	Wet coarse aggregate	Water
3/8	1	2 1/4	1 1/2	1/2	1	2 1/2	1 1/2	1/2
1/2	1	2 1/4	2	1/2	1	2 1/2	2	1/2
3/4	1	2 1/4	2 1/2	1/2	1	2 1/2	2 1/2	1/2
1	1	2 1/4	2 3/4	1/2	1	2 1/2	2 3/4	1/2
1 1/2	1	2 1/4	3	1/2	1	2 1/2	3	1/2

*The combined volume is approximately 2/3 of the sum of the original bulk volumes.
Reference 7-1.

7-7. Seghal, J. Paul, "Concrete Mix Design and Development," *Concrete,* Pit and Quarry Publications, Inc., Cleveland, February 1985.

7-8. *Specifications for Structural Concrete for Buildings,* ACI 301-84, revised 1985, ACI Committee 301 Report, American Concrete Institute.

7-9. *Building Code Requirements for Reinforced Concrete,* ACI 318-83, revised 1986, ACI Committee Report, American Concrete Institute.

7-10. Kosmatka, Steven H., "Sulfate-Resistant Concrete," *Concrete Technology Today,* PL883B, Portland Cement Association, October 1988.

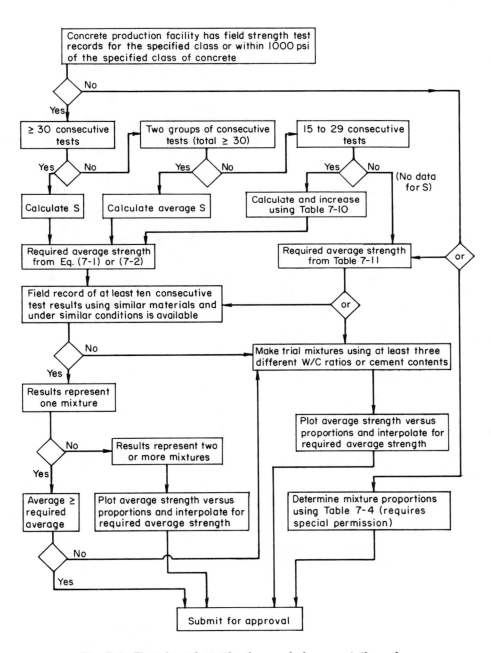

Fig. 7-4. Flowchart for selection and documentation of concrete proportions. Reference 7-4.

CHAPTER 8
Batching, Mixing, Transporting, and Handling Concrete

Production and delivery of concrete is achieved in different ways. The basic steps and common techniques are explained here.

BATCHING

Batching is the process of weighing or volumetrically measuring and introducing into the mixer the ingredients for a batch of concrete. To produce concrete of uniform quality, the ingredients must be measured accurately for each batch. Most specifications require that batching be done by weight rather than by volume because of inaccuracies in measuring aggregate (especially damp sand) by volume. Use of the weight system for batching provides greater accuracy and simplicity and avoids the problem created by bulking of damp sand. Water and liquid admixtures can be measured accurately by either volume or weight. Volumetric batching is used for concrete mixed in a continuous mixer and for certain work around the farm and home where weighing facilities are not at hand.*

Specifications generally require that materials be measured in individual batches within the following percentages of accuracy: cement 1%, aggregates 2%, water 1%, admixtures 3%.

Equipment should be capable of measuring quantities within these tolerances for the smallest batch regularly used as well as large batches (Fig. 8-1). The accuracy of batching equipment should be checked periodically and adjusted when necessary.

Air-entraining admixtures, calcium chloride, and other chemical admixtures should be charged into the mix as solutions and the liquid should be considered part of the mixing water. Admixtures that cannot be added in solution can be either weighed or measured by volume as directed by the manufacturer. Admixture dispensers should be checked daily since errors in dispensing admixtures, particularly overdoses, can lead to serious problems in both freshly mixed and hardened concrete.

Fig. 8-1. Control panel for weight-batching equipment in a typical ready mixed concrete plant.

MIXING CONCRETE

All concrete should be mixed thoroughly until it is uniform in appearance, with all ingredients evenly distributed. Mixers should not be loaded above their rated capacities and should be operated at approximately the speeds for which they were designed. Increased output should be obtained by using a larger mixer or additional mixers, rather than by speeding up or overloading the equipment on hand. If the blades of the mixer become worn or coated with hardened concrete, the mixing action will be less efficient. Badly worn blades should be replaced and hardened concrete should be removed periodically, preferably after each day's production of concrete.

If the concrete has been adequately mixed, samples taken from different portions of a batch will have essentially the same unit weight, air content, slump, and coarse-aggregate content. Maximum allowable differences in test results within a batch of ready mixed concrete are given in ASTM C 94.

*The procedure is described in Reference 8-2.

Structural lightweight concrete can be mixed the same way as normal-weight concrete when the aggregates have less than 10% total absorption by weight or when the absorption is less than 2% by weight during the first hour after immersion in water.*

Stationary Mixing

Concrete is sometimes mixed at the jobsite in a stationary mixer or a paving mixer (Fig. 8-2). Stationary mixers include both onsite mixers and central mixers in ready mix plants. They are available in sizes from 2 cu ft to 12 cu yd and can be of the tilting or nontilting type or the open-top revolving blade or paddle type. All types may be equipped with loading skips and some are equipped with a swinging discharge chute. Many stationary mixers have timing devices, some of which can be set for a given mixing time and locked so that the batch cannot be discharged until the designated time has elapsed.

Fig. 8-2. Concrete can be mixed at the jobsite in a stationary mixer.

Specifications usually require a minimum of one-minute mixing for stationary mixers of up to 1-cu yd capacity with an increase of 15 seconds for each additional 1 cu yd, or fraction thereof, of capacity. Highway specifications usually permit a shorter mixing period for pavers or central mixers.

The mixing period should be measured from the time all cement and aggregate are in the mixer drum, provided that all the water is added before one-fourth of the mixing time has elapsed. Some specifications, including ASTM C 94, permit mixing time to be re-

duced if mixer performance tests show that the concrete as discharged is adequately uniform.

Under usual conditions, up to about 10% of the mixing water should be placed in the drum before the solid materials are added. Water then should be added uniformly with the solid materials, leaving about 10% to be added after all other materials are in the drum. When heated water is used in cold weather, this order of charging may require some modification to prevent possible rapid stiffening. In this case, addition of the cement should be delayed until most of the aggregate and water have intermingled in the drum. Where the mixer is charged directly from batchers, the materials should be added simultaneously at such rates that the charging time is about the same for all materials.

If retarding or water-reducing admixtures are used, they should be added in the same sequence in the charging cycle. If not, significant variations in the time of initial set and percentage of entrained air may result. Addition of the admixture should be completed not later than one minute after addition of water to the cement has been completed or prior to the start of the last three-fourths of the mixing cycle, whichever occurs first. If two or more admixtures are used in the same batch of concrete, they should be added separately to avoid any interaction that might interfere with the efficiency of any of the admixtures and adversely affect the concrete.

Ready Mixed Concrete

Ready mixed concrete is proportioned and mixed off the project site and is delivered to the construction area in a freshly mixed and unhardened state. It can be manufactured by any of the following methods:

1. Central-mixed concrete is mixed completely in a stationary mixer (Fig. 8-3) and is delivered either in a truck agitator, a truck mixer operating at agitating speed, or a special nonagitating truck.

---

*For aggregates not meeting these limits, mixing procedures are described in Reference 8-1.

Fig. 8-3. Central mixing in stationary mixers of the tilting drum type.

2. Shrink-mixed concrete is mixed partially in a stationary mixer and completed in a truck mixer.

3. Truck-mixed concrete is mixed completely in a truck mixer (Fig. 8-4).

Fig. 8-4. Truck-mixed concrete is mixed completely in a truck mixer, which can usually hold 7 to 10 cu yd of concrete.

Fig. 8-5. Mobile batcher measures materials by volume and continuously mixes concrete as the dry ingredients, water, and admixtures are fed into the mixer.

ASTM C 94 notes that when a truck mixer is used for complete mixing, 70 to 100 revolutions of the drum or blades at the rate of rotation designated by the manufacturer as *mixing speed* are usually required to produce the specified uniformity of concrete. No more than 100 revolutions at mixing speed should be used. All revolutions after 100 should be at a rate of rotation designated by the manufacturer as *agitating speed*. Agitating speed is usually about 2 to 6 rpm, and mixing speed is generally about 6 to 18 rpm. Mixing at high speeds for long periods of time, about 1 or more hours, can result in concrete strength loss, temperature rise, excessive loss of entrained air, and accelerated slump loss.

ASTM C 94 also requires that concrete be delivered and discharged within 1½ hours or before the drum has revolved 300 times after introduction of water to the cement and aggregates or the cement to the aggregates. Mixers and agitators should always be operated within the limits of the volume and speed of rotation designated by the equipment manufacturer.

Mobile Batcher Mixed Concrete (Continuous Mixer)

Mobile batcher mixers are special trucks (Fig. 8-5) that batch by volume and continuously mix concrete as the dry materials, water, and admixtures are continuously fed into the mixer. The concrete must conform to ASTM C 685 specifications and is proportioned and mixed at the jobsite only as it is needed. The concrete mixture is also easily adjusted for project placement and weather conditions.

High-Energy Mixers

High-energy mixers, unlike conventional concrete mixers, first blend cement and water into a slurry with high-speed rotating blades. The slurry is then added to aggregates and mixed with conventional mixing equipment to produce a uniform concrete mixture. High-energy mixing causes water to intermix more completely with cement particles, resulting in more complete cement hydration. This results in more efficient use of cement, higher strength, and improvements in several other concrete properties over those developed by conventional mixing alone. High-energy mixers were first used in ready mixed concrete production in the United States in 1987.

Remixing Concrete

Fresh concrete that is left to agitate in the mixer drum tends to stiffen before initial set develops. Such concrete may be used if upon remixing it becomes sufficiently plastic to be compacted in the forms. Under careful supervision a small amount of water may be added to remix the concrete providing the following conditions are met: (1) maximum allowable water-cement ratio is not exceeded; (2) maximum allowable slump is not exceeded; (3) maximum allowable mixing and agitating time (or drum revolutions) are not exceeded; and (4) concrete is remixed for at least half the minimum required mixing time or number of revolutions.

Indiscriminate addition of water to make the concrete more fluid should not be allowed because this lowers the quality of the concrete. Remixed concrete can be expected to harden rapidly. Subsequently, a cold joint may develop when concrete is placed adjacent to or above remixed concrete.

TRANSPORTING AND HANDLING CONCRETE

Although there is no best way to transport and handle concrete, good advance planning can help choose the proper method before a crisis occurs, thus avoiding problems. Advance planning must consider three occurrences that, should they occur during handling and placing, could seriously affect the quality of the finished work:

1. Delays. The objective in planning any work schedule is to produce the fastest work with the best labor force and the proper equipment for the work at hand. Machines for transporting and handling concrete are being improved all the time. The greatest productivity will be achieved if the work is planned to get the most out of personnel and equipment and if the equipment is selected to reduce the delay time during concrete placement.

2. Early stiffening and drying out. Concrete begins to stiffen as soon as the cement and water are mixed, but the degree of stiffening that occurs in the first 30 minutes is not usually a problem; concrete that is kept agitated generally can be placed and compacted within 1½ hours after mixing. Planning should eliminate or minimize any variables that would allow the concrete to stiffen to the extent that full consolidation is not achieved and finishing becomes difficult. Less time is available during conditions that hasten the stiffening process, such as hot, dry weather; use of accelerators; and use of heated concrete.

3. Segregation. Segregation is the tendency for coarse aggregate to separate from the sand-cement mortar. This results in part of the batch having too little coarse aggregate and the remainder having too much. The former is likely to shrink more and crack and have poor resistance to abrasion. The latter is too harsh for full consolidation and finishing and is a frequent cause of honeycomb. The method and equipment used to transport and handle the concrete must not result in segregation of the concrete materials.

Equipment for Transporting and Handling Concrete

Table 8-1 summarizes the most common methods and equipment for moving concrete to the point where it is needed.

There have been few, if any, major changes in the principles of conveying concrete in the last 50 years.

Table 8-1. Methods and Equipment for Transporting and Handling Concrete

Equipment	Type and range of work for which equipment is best suited	Advantages	Points to watch for
Belt conveyors	For conveying concrete horizontally or to a higher or lower level. Usually used between main discharge point and secondary discharge point.	Belt conveyors have adjustable reach, traveling diverter, and variable speed both forward and reverse. Can place large volumes of concrete quickly when access is limited.	End-discharge arrangements needed to prevent segregation, leave no mortar on return belt. In adverse weather (hot, windy) long reaches of belt need cover.
Belt conveyors mounted on truck mixers	For conveying concrete to a lower, horizontal, or higher level.	Conveying equipment arrives with the concrete. Adjustable reach and variable speed.	End-discharge arrangements needed to prevent segregation, leave no mortar on return belt.
Buckets	Used with cranes, cableways, and helicopters for construction of buildings and dams. Convey concrete directly from central discharge point to formwork or to secondary discharge point.	Enable full versatility of cranes, cableways, and helicopters to be exploited. Clean discharge. Wide range of capacities.	Select bucket capacity to conform to size of the concrete batch and capacity of placing equipment. Discharge should be controllable.
Chutes	For conveying concrete to lower level, usually below ground level, on all types of concrete construction.	Low cost and easy to maneuver. No power required, gravity does most of the work.	Slopes range between 1 to 2 and 1 to 3 and chutes must be adequately supported in all positions. Arrange for discharge at end (downpipe) to prevent segregation.
Cranes	The right tool for work above ground level.	Can handle concrete, reinforcing steel, formwork, and sundry items in high-rise, concrete-framed buildings.	Has only one hook. Careful scheduling between trades and operations are needed to keep it busy.
Dropchutes	Used for placing concrete in vertical forms of all kinds. Some chutes are one piece, others are assembled from loosely connected segments.	Dropchutes direct concrete into formwork and carry it to bottom of forms without segregation. Their use avoids spillage of grout and concrete on the form sides, which is harmful when off-the-form surfaces are specified. They also will prevent segregation of coarse particles.	Dropchutes should have sufficiently large, splayed-top openings into which concrete can be discharged without spillage. The cross section of dropchute should be chosen to permit inserting into the formwork without interfering with reinforcing steel.

Table 8-1. Methods and Equipment for Transporting and Handling Concrete (continued)

Equipment	Type and range of work for which equipment is best suited	Advantages	Points to watch for
Mobile batcher mixers	Used for intermittent production of concrete at jobsite.	A combined materials transporter and mobile batching and mixing system for quick, precise proportioning of specified concrete. One-man operation.	Trouble-free operation requires good preventive maintenance program on equipment. Materials must be identical to those in original mix design.
Nonagitating trucks	Used to transport concrete on short hauls over smooth roadways.	Capital cost of nonagitating equipment is lower than that of truck agitators or mixers.	Concrete slump should be limited. Possibility of segregation. Height is needed for high lift of truck body upon discharge.
Pneumatic guns (shotcrete)	Used where concrete is to be placed in difficult locations and where thin sections and large areas are needed.	Ideal for placing concrete in free-form shapes, for repairing and strengthening buildings, for protective coatings, and thin linings.	Quality of work depends on skill of those using equipment. Only experienced nozzlemen should be employed.
Pumps	Used to convey concrete directly from central discharge point at jobsite to formwork or to secondary discharge point.	Pipelines take up little space and can be readily extended. Delivers concrete in continuous stream. Pump can move concrete both vertically and horizontally. Mobile pumps can be delivered when necessary to small or large projects. Stationary pump booms provide continuous concrete for tall building construction.	Constant supply of freshly mixed concrete is needed with average consistency and without any tendency to segregate. Care must be taken in operating pipeline to ensure an even flow and to clean out at conclusion of each operation. Pumping vertically, around bends, and through flexible hose will considerably reduce the maximum pumping distance.
Screw spreaders	Used for spreading concrete over flat areas, as in pavements.	With a screw spreader a batch of concrete discharged from bucket or truck can be quickly spread over a wide area to a uniform depth. The spread concrete has good uniformity of compaction before vibration is used for final compaction.	Screw spreaders are usually used as part of a paving train. They should be used for spreading before vibration is applied.
Tremies	For placing concrete underwater.	Can be used to funnel concrete down through the water into the foundation or other part of the structure being cast.	Precautions are needed to ensure that the tremie discharge end is always buried in fresh concrete, so that a seal is preserved between water and concrete mass. Diameter should be 10 to 12 in. unless pressure is available. Concrete mixture needs more cement, 7 to 8 bags per cubic yard, and greater slump, 6 to 9 in., because concrete must flow and consolidate without any vibration.
Truck agitators	Used to transport concrete for all uses in pavements, structures, and buildings. Haul distances must allow discharge of concrete within 1½ hours, but limit may be waived under certain circumstances.	Truck agitators usually operate from central mixing plants where quality concrete is produced under controlled conditions. Discharge from agitators is well controlled. There is uniformity and homogeneity of concrete on discharge.	Timing of deliveries to suit job organization. Concrete crew and equipment must be ready onsite to handle concrete.
Truck mixers	Used to transport concrete for all uses in pavements, structures, and buildings. Haul distances must allow discharge of concrete within 1½ hours, but limit may be waived under certain circumstances.	No central mixing plant needed, only a batching plant, since concrete is completely mixed in truck mixer. Discharge is same as for truck agitator.	Timing of deliveries to suit job organization. Concrete crew and equipment must be ready onsite to handle concrete. Control of concrete quality is not as good as with central mixing.
Wheelbarrows and buggies	For short flat hauls on all types of onsite concrete construction, especially where accessibility to work area is restricted.	Very versatile and therefore ideal inside and on jobsites where placing conditions are constantly changing.	Slow and labor intensive.

What has changed is the technology that led to development of better machinery to do the work more efficiently. The wheelbarrow and Georgia buggy, although still used, have advanced to become the power buggy (Fig. 8-6); the bucket hauled over a pulley wheel has become the bucket and crane (Fig. 8-7); and the horse-drawn wagon is now the ready mixed concrete truck (Figs. 8-8 and 8-9).

Years ago concrete was placed in reinforced concrete buildings by means of a tower and long chutes. This was a guyed tower centrally placed on the site with a hopper at the top to which concrete was hauled by winch. A series of chutes suspended from the tower allowed the concrete to flow by gravity directly to the point required. As concrete-framed buildings became taller, the need to hoist reinforcement and formwork as well as concrete to higher levels led to the development of the tower crane—a familiar sight on the building skyline today (Fig. 8-10). It is fast and versatile, but the fact that it has only one hook must be considered when planning a job.

The conveyer belt is old in concept but little changed over the years (Fig. 8-11). Recently, truck mixer mounted conveyor belts have come into use (Fig. 8-12). The pneumatic process for shotcreting was patented in 1911 and is literally unchanged (see Chapter 15). The first mechanical concrete pump was developed and used in the 1930's and the hydraulic pump was developed in the 1950's. The advanced mobile pump with hydraulic placing boom (Fig. 8-13) is probably the single most important innovation in concrete handling equipment. It is economical to use in placing both large and small quantities of concrete, depending on jobsite conditions. The screw spreader (Fig. 8-14) has been very effective in placing and distributing concrete for pavements. Screw spreaders can place a uniform depth of concrete quickly and efficiently.*

---

*Reference 8-8 provides extensive information on methods to transport and handle concrete.

Fig. 8-7. Concrete is easily lifted to its final location by bucket and crane. Bucket sizes vary from ½ to 12 cu yd.

Fig. 8-8. Ready mixed concrete can often be placed in its final location by direct chute discharge from a truck mixer.

Fig. 8-6. Versatile power buggy can move all types of concrete over short distances.

Fig. 8-9. In comparison to conventional rear-discharge trucks, front-discharge truck mixers provide the driver with more mobility and control for direct discharge into forms.

Fig. 8-11. The conveyor belt is an efficient, portable method of handling concrete. A dropchute prevents concrete from segregating as it leaves the belt; a scraper prevents loss of mortar. Conveyor belts can be operated in series and on extendable booms of hydraulic cranes.

Fig. 8-10. The tower crane and bucket can easily handle concrete from a truck mixer for tall-building construction.

Fig. 8-12. A conveyor belt mounted on a truck mixer places concrete up to about 40 feet without the need for additional handling equipment.

Choosing the Best Method

The first thing to look at is the type of job: its physical size, the total amount of concrete to be placed, and the time schedule. Studying the job details further tells how much of the work is below, at, or above ground level. This aids in choosing the concrete handling equipment necessary for placing concrete at the required levels.

Concrete must be moved from the mixer to the point of placement as rapidly as possible without segregation or loss of ingredients. The transporting and handling equipment must have the capacity to move sufficient concrete so that cold joints are eliminated.

Work At and Below Ground Level

The largest volumes of concrete in any job usually are either below or at ground level and therefore can be placed by methods different from those employed on the superstructure. Concrete work belowground can vary enormously—from filling large-diameter bored piles or massive mat foundations to the intricate work involved in basement and subbasement walls. A crane can be used to handle formwork, reinforcing steel, and concrete. However, the crane may be fully employed erecting formwork and reinforcing steel in advance of

Fig. 8-13(a). A truck-mounted pump and boom can conveniently move concrete vertically or horizontally to the desired location.

Fig. 8-14. The screw spreader quickly spreads concrete over a wide area to a uniform depth. Screw spreaders are used primarily in pavement construction.

Fig. 8-13(b). View of concrete discharging from flexible hose connected to rigid pipeline leading from the pump. Rigid pipe is used in pump booms and by itself to move concrete over relatively long distances. Up to 25 ft of flexible hose is attached to increase placement mobility.

the concrete, and other methods of handling the concrete may have to be used to place the largest volume in the least time.

Possibly the concrete can be chuted directly from the truck mixer to the point needed. Chutes should be metal or metal lined. They must not slope greater than 1 vertical to 2 horizontal or less than 1 vertical to 3 horizontal. Long chutes, over 20 ft, or those not meeting slope standards must discharge into a hopper before distribution to point of need.

Alternatively, a concrete pump can move the concrete to its final position (Fig. 8-13). Pumps must be of adequate capacity and capable of moving concrete without segregation. The loss of slump from pump

hopper to discharge at the end of the pipeline must be minimal—not greater than 2 in. The air content generally should not be reduced by more than 2 percentage points. Pipelines must not be of aluminum or aluminum alloys to avoid excessive entrainment of air by reaction of aluminum with cement alkali hydroxides resulting in serious reduction in concrete strength.

Belt conveyors are very useful for work near ground level. Since placing concrete belowground is frequently a matter of horizontal movement assisted by gravity, lightweight portable conveyors can be used for high output at relatively low cost.

Work Above Ground Level

For work above ground level, concrete can be lifted by conveyor belt, crane and bucket, hoist, pump, or the ultimate sky hook, the helicopter (Fig. 8-15). The tower crane (Fig. 8-10) and pumping boom (Fig. 8-16) are the right tools for tall buildings; high-speed cranes operating at 800 ft per minute or more are available. Crane cycle time can be reduced by using a concrete hoist for elevation and the crane for lateral distribution. Use of a pump is affected by the volume of concrete needed per floor; large volumes minimize pipeline movement in relation to output.

The specifications and performance of transporting and handling equipment are being continuously improved. The best results and lowest costs will be realized if the work is planned to get the most out of the equipment and if the equipment is flexibly employed to reduce total job cost. Any method is expensive if it does not get the job done.*

---

*Reference 8-8 is very helpful in deciding which method to use as it provides capacity and range information on various methods and equipment.

Fig. 8-15. For work aboveground or at inaccessible sites, a concrete bucket can be lifted by helicopter.

Fig. 8-16. A pump boom mounted on a mast and located near the center of a structure can frequently reach all points of placement. It is especially applicable to tall buildings where tower cranes cannot be tied up with placing concrete. Concrete is supplied to the boom through a pipeline from a ground-level pump. Concrete can be pumped hundreds of feet vertically with these pumping methods.

REFERENCES

8-1. *Structural Lightweight Concrete,* IS032T, Portland Cement Association, 1972, revised 1986.

8-2. *Concrete for Small Jobs,* IS174T, Portland Cement Association, 1980.

8-3. *Placing Concrete by Pumping Methods,* ACI 304.2R-71, revised 1982, ACI Committee 304 Report, American Concrete Institute, Detroit.

8-4. *Placing Concrete with Belt Conveyors,* ACI 304.4R-75, revised 1985, ACI Committee 304 Report, American Concrete Institute.

8-5. *Specification for Structural Concrete for Buildings,* ACI 301-84, revised 1985, ACI Committee 301 Report, American Concrete Institute.

8-6. *Guide for Measuring, Mixing, Transporting, and Placing Concrete,* ACI 304R-85, ACI Committee 304 Report, American Concrete Institute, 1985.

8-7. Haney, James T., and Meyers, Rodney A., *Ready Mixed Concrete—Plant and Truck Mixer Operations and Quality Control,* NRMCA Publication No. 172, National Ready Mixed Concrete Association, Silver Spring, Maryland, May 1985.

8-8. Panarese, William C., *Transporting and Handling Concrete,* IS178T, Portland Cement Association, 1974, revised 1987.

CHAPTER 9
Placing and Finishing Concrete

PREPARATION BEFORE PLACING

Preparation prior to placing concrete includes compacting, trimming, and moistening the subgrade; erecting the forms; and setting the reinforcing steel and other embedded items securely in place. Moistening the subgrade is important, especially in hot, dry weather to keep the dry subgrade from drawing too much water from the concrete and to increase the immediate air-moisture level thereby decreasing the amount of evaporation from the concrete surface.

In cold weather, the subgrade must not be frozen. Snow, ice, and other debris must be removed from within the forms before concrete is placed. Where concrete is to be deposited on rock or hardened concrete, all loose material must be removed, and cut faces should be nearly vertical or horizontal rather than sloping.

Newly placed concrete is usually roughened shortly after hardening to produce a better bond with the next placement. As long as no laitance, dirt, or loose particles are present, newly hardened concrete requires little preparation prior to placing freshly mixed concrete on it. Old hardened concrete usually requires mechanical cleaning and roughening prior to placement of new concrete. The subject of placing freshly mixed concrete on hardened concrete is discussed in more detail under the sections entitled "Placing on Hardened Concrete" and "Construction Joints."

Forms should be accurately set, clean, tight, adequately braced, and constructed of or lined with materials that will impart the desired off-the-form finish to the hardened concrete.* Wood forms, unless oiled or otherwise treated with a form-release agent, should be moistened before placing concrete, otherwise they will absorb water from the concrete and swell. Forms should be made for removal with minimum damage to the concrete. With wood forms, use of too large or too many nails should be avoided to facilitate removal and reduce damage. For architectural concrete, the form release agent should be a nonstaining material.

Reinforcing steel should be clean and free of loose rust or mill scale when concrete is placed. Mortar splattered on reinforcing bars from previous placements need not be removed from steel and other embedded items if the next lift is to be completed within a few hours; loose, dried mortar, however, must be removed from items that will be encased by later lifts of concrete.

All equipment used to place concrete must be clean and in good working condition. Standby equipment should be available in the event of a breakdown.

DEPOSITING THE CONCRETE

Concrete should be deposited continuously as near as possible to its final position (Fig. 9-1). In slab construction, placing should be started along the perimeter at one end of the work with each batch discharged against previously placed concrete. The concrete should not be dumped in separate piles and then leveled and worked together; nor should the concrete be deposited in large piles and moved horizontally into final position. Such practices result in segregation because mortar tends to flow ahead of coarser material.

In general, concrete should be placed in horizontal layers of uniform thickness, each layer being thoroughly consolidated before the next is placed. The rate of placement should be rapid enough so that the layer of

---

*See References 9-9 and 9-26 for more information on formwork.

Fig. 9-1. Concrete should be placed as near as possible to its final position.

concrete has not yet set when a new layer is placed upon it. This will avoid flow lines, seams, and planes of weakness (cold joints) that result when freshly mixed concrete is placed on hardened concrete. Layers should be about 6 to 20 in. thick for reinforced members, 15 to 20 in. thick for mass work, the thickness depending on the width between forms and the amount of reinforcement.

To avoid segregation, concrete should not be moved horizontally over too long a distance as it is being placed in forms or slabs. In some work, such as placing concrete in sloping wingwalls or beneath openings in walls, it is necessary to move the concrete horizontally within the forms, but this horizontal distance should be kept to a minimum.

In walls, beams, and girders, the first batches in each lift should be placed at the ends, with subsequent placements progressing toward the center. In all cases, water should be prevented from collecting at the ends, in corners, and along faces of forms.

In sloping wingwalls, water tends to collect along the sloping top surface, an area most vulnerable to weathering. To control this, the top formboards of the sloping face can be omitted at the start and the concrete placed directly in this section of the wall. The boards can be placed, if necessary, as concreting progresses.

Dropchutes will prevent segregation and spattering of mortar on reinforcement and forms. If the placement can be completed before the mortar dries, dropchutes may not be needed. The height of free fall of concrete need not be limited unless a separation of coarse particles occurs (resulting in honeycomb), in which case a limit of 3 to 4 ft may be adequate. Properly designed concrete has been allowed to drop by free fall for several hundred feet into caissons. Results of a field test to determine if concrete could be dropped vertically 50 ft into a caisson without segregation proved that there was no significant difference in aggregate gradation between control samples as delivered and free-fall samples taken from the bottom of the caisson.*

Concrete is sometimes placed through openings, called windows, in the sides of tall, narrow forms. When a chute discharges directly through the opening there is danger of segregation. A collecting hopper should be used outside the opening to permit the concrete to flow more smoothly through the opening, and decrease the tendency to segregate.

When concrete is placed in tall forms at a fairly rapid rate, some bleed water may collect on the top surface, especially with non-air-entrained concrete. Bleeding can be reduced by placing more slowly and by using concrete of a stiffer consistency. When practical, concrete should be placed to a level about a foot below the top of tall forms and an hour or so allowed for the concrete to partially set. Placing should resume before the surface hardens to avoid formation of a cold joint. It is good practice to overfill the form by an inch or so and cut off the excess concrete after it has stiffened and bleeding has ceased.

To avoid cracking, concrete in columns and walls should be allowed to stand for at least two hours, and preferably overnight, before concreting is continued in any slabs, beams, or girders framing into them. Haunches and column capitals are considered part of the floor or roof slab and should be placed integrally with them.

CONSOLIDATING CONCRETE

Consolidation is the process of compacting fresh concrete to mold it within the forms and around embedded items and reinforcement and to eliminate stone pockets, honeycomb, and entrapped air (Fig. 9-2). It should not remove significant amounts of intentionally entrained air in air-entrained concrete. Consolidation

Fig. 9-2. Honeycomb and rock pockets are the results of inadequate consolidation.

is accomplished by hand or by mechanical methods. The method chosen depends on the consistency of the mixture and the placing conditions, such as complexity of the formwork and amount and spacing of reinforcement.

Workable, flowing mixtures can be consolidated by hand rodding, that is, thrusting a tamping rod or other suitable tool repeatedly into the concrete. The rod should be long enough to reach the bottom of the form or lift and thin enough to pass between the reinforcing steel and the forms. Low-slump concrete can be transformed into flowing concrete for easier consolidation through the use of superplasticizers (Chapter 6) without the addition of water to the concrete mixture.

Spading can be used to improve the appearance of formed surfaces. A flat, spadelike tool should be repeatedly inserted and withdrawn adjacent to the form. This forces the larger coarse aggregates away from the forms and assists entrapped air voids in their upward movement toward the top surface where they can escape. A mixture that can be consolidated readily by

*Reference 9-4.

104

hand tools should not be consolidated by mechanical methods because the concrete is likely to segregate under intense mechanical action.

Proper mechanical consolidation makes possible the placement of stiff mixtures with the low water-cement ratios and high coarse-aggregate contents associated with high-quality concrete, even in highly reinforced elements (Fig. 9-3). Some of the mechanical methods are centrifugation, used to consolidate moderate-to-high-slump concrete in making pipes, poles, and piles; shock or drop tables, used to compact very stiff low-slump concrete in the manufacture of architectural precast units; and vibration—internal and external.

Fig. 9-4. **Internal vibrators are commonly used to consolidate concrete in walls, columns, beams, and slabs.**

Fig. 9-3. **Proper vibration makes possible the placement of concrete even in heavily reinforced concrete members.**

Vibration

Vibration, either internal or external, is the most widely used method for consolidating concrete. When concrete is vibrated, the internal friction between the aggregate particles is temporarily destroyed and the concrete behaves like a liquid; it settles in the forms under the action of gravity and the large entrapped air voids rise more easily to the surface. Internal friction is reestablished as soon as vibration stops.

Vibrators, whether internal or external, are usually characterized by their frequency of vibration, expressed as the number of vibrations per minute, vpm, and by the amplitude of vibration, which is the deviation in inches from the point of rest.

When vibration is used to consolidate concrete, a standby vibrator should be at hand at all times in the event of a mechanical breakdown.

Internal vibration. Internal or immersion-type vibrators, often called spud or poker vibrators (Fig. 9-4), are commonly used to consolidate concrete in walls, columns, beams, and slabs. Flexible-shaft vibrators consist of a vibrating head connected to a driving motor by a flexible shaft. Inside the head an unbalanced

weight connected to the shaft rotates at high speed, causing the head to revolve in a circular orbit. The motor can be powered by electricity, gasoline, or air. The vibrating heads are usually cylindrical with diameters ranging from 3/4 to 7 in. Some vibrators have an electric motor built into the head, which is generally at least 2 in. in diameter. The performance of a vibrator is affected by the dimensions of the vibrator head as well as its frequency and amplitude.

Small-diameter vibrators have high frequencies, ranging from 10,000 to 15,000 vpm, and low amplitudes ranging between 0.015 and 0.03 in. As the diameter increases, the frequency decreases and the amplitude increases. The effective radius of action of the vibrator increases with increasing diameter. Vibrators with a diameter of 3/4 to 1 1/2 in. have a radius of action in freshly mixed concrete ranging between 3 and 6 in. whereas the radius of action for vibrators of 2- to 3 1/2-in. diameter ranges between 7 and 14 in.

Proper use of internal vibrators is important for best results. Vibrators should not be used to move concrete horizontally since this causes segregation. Whenever possible, the vibrator should be lowered vertically into the concrete at regularly spaced intervals and allowed to descend by gravity. It should penetrate quickly to the bottom of the layer being placed and at least 6 in. into any previously placed layer. Each layer or lift should be about the length of the vibrator head or generally a maximum of 12 in. in regular formwork.

In thin slabs, the vibrator should be inserted at an angle or horizontally in order to keep the vibrator head completely immersed. For slabs on grade, the vibrator should not make contact with the subgrade. The distance between insertions should be about 1 1/2 times the radius of action so that the area visibly affected by the vibrator overlaps the adjacent previously vibrated area by a few inches.

The vibrator should be held stationary until adequate consolidation is attained and then slowly withdrawn. An insertion time of 5 to 15 seconds will usually provide adequate consolidation. The concrete should move to fill the hole left by the vibrator on withdrawals. If the hole does not refill, reinsertion of the vibrator at a nearby point should solve the problem.

Adequacy of internal vibration is judged by experience and by changes in the surface appearance of the concrete. Changes to watch for are the embedment of large aggregate particles, the leveling of the concrete surface, the appearance of a thin film of glistening paste around the vibrator head, and the cessation of large bubbles of entrapped air escaping at the surface.

Allowing a vibrator to remain immersed in concrete after paste accumulates over the head is bad practice and can result in nonuniformity. The length of time that a vibrator should be left in the concrete will depend on the slump of the concrete, the power of the vibrator, and the nature of the section being compacted.

Revibration of previously compacted concrete can be done intentionally or it may occur unintentionally when the underlying layer has partially hardened. Revibration is used to improve bond between concrete and reinforcing steel, release water trapped under horizontal reinforcing bars, and remove additional entrapped air voids. In general, if concrete becomes workable under revibration, this is not harmful and may be beneficial.

External vibration. External vibrators can be form vibrators, vibrating tables, or surface vibrators such as vibratory screeds, plate vibrators, vibratory roller screeds, or vibratory hand floats or trowels. Form vibrators, designed to be securely attached to the outside of the forms, are especially useful (1) for consolidating concrete in members that are very thin or congested with reinforcement, (2) to supplement internal vibration, and (3) for stiff mixes where internal vibrators cannot be used.

Attaching a form vibrator directly to the form generally is unsatisfactory. Rather, the vibrator should be attached to a steel plate that in turn is attached to steel I-beams or channels passing through the form stiffeners themselves in a continuous run. Loose attachments can result in significant vibration energy losses and inadequate consolidation.

Form vibrators can be either electrically or pneumatically operated. They should be spaced to distribute the intensity of vibration uniformly over the form; optimum spacing is best found by experimentation. Sometimes it may be necessary to operate some of the form vibrators at a different frequency for better results; therefore, it is recommended that form vibrators be equipped with controls to regulate their frequency and amplitude. Duration of external vibration is considerably longer than for internal vibration—generally between 1 and 2 minutes.

Form vibrators should not be applied within the top 30 in. of vertical forms. Vibration of the top of the form, particularly if the form is thin or inadequately stiffened, causes an in-and-out movement that can create a gap between the concrete and the form. Internal vibrators are recommended for use in this area of vertical forms.

In heavily reinforced sections where an internal vibrator cannot be inserted, it is sometimes helpful to vibrate the reinforcing bars by attaching a form vibrator to their exposed portions. This practice eliminates air and water entrapped underneath the reinforcing bars and increases bond between the bars and surrounding concrete, provided the concrete is still workable under the action of vibration. Internal vibrators should not be attached to reinforcing bars for this purpose because the vibrators may be damaged.

Vibrating tables are used in precasting plants. They should be equipped with controls so that the frequency and amplitude can be varied according to the size of the element to be cast and the consistency of the concrete. Workable mixtures generally require higher frequencies than stiff mixtures. Increasing the frequency and decreasing the amplitude as vibration progresses will improve consolidation.

Surface vibrators such as vibratory screeds (Fig. 9-5) are used to consolidate concrete in floors and other flatwork. Vibratory screeds give positive control of the strikeoff operation and save a great deal of labor; however, this equipment should not be used on concrete

Fig. 9-5. Vibratory screeds such as this truss-type unit reduce the work of strikeoff while consolidating the concrete.

with slumps in excess of 3 in. Surface vibration of such concrete will result in an excessive accumulation of mortar and fine material on the surface and thus reduce wear resistance. For the same reason, surface vibrators should not be operated after the concrete has been adequately consolidated. Floors, driveways, and sidewalks are often placed with a 3 to 5 in. slump, a consistency at which normal finishing operations provide adequate consolidation. Stiffer mixtures that require vibration are used where more durable, abrasion-resistant surfaces are required.

Because surface vibration of concrete slabs is least effective along the edges, a spud or poker-type vibrator should be used along the edge forms immediately before the vibratory screed is applied.

Vibratory screeds are used for consolidating slabs up to 10 in. thick, provided such slabs are nonreinforced or only lightly reinforced (welded-wire fabric). Internal vibration or a combination of internal and surface vibration is recommended for reinforced slabs.

CONCRETE SLABS

Concrete slabs can be finished in many ways, depending on the intended service use.* Various colors and textures, such as exposed-aggregate or a pattern-stamped surface, may be called for. Some surfaces may require only strikeoff and screeding to proper contour and elevation, while for other surfaces a broomed, floated, or troweled finish may be specified.

The mixing, transporting, and handling of concrete for slabs should be carefully coordinated with the finishing operations. Concrete should not be placed on the subgrade or into forms more rapidly than it can be spread, struck off, consolidated, and bullfloated or darbied. In fact, concrete should not be spread over too large an area before strikeoff, nor should a large area be struck off and bleed water allowed to accumulate before bullfloating or darbying.

Finishing crews should be large enough to correctly place, finish, and cure concrete slabs with due regard to the effects of concrete temperature and atmospheric conditions on the setting time of the concrete and the size of the placement to be completed.

Subgrade Preparation

Cracks, slab settlement, and structural failure can often be traced to an inadequately prepared and poorly compacted subgrade. The subgrade on which a slab on ground is to be placed should be well drained, of uniform bearing capacity, level or properly sloped, and free of sod, organic matter, and frost. The three major causes of nonuniform support are (1) the presence of soft and hard spots, (2) backfilling without adequate compaction, and (3) expansive soils. Uniform support cannot be achieved by merely dumping granular material on a soft area. To prevent bridging and settlement cracking, soft or mucky areas and hard spots (rocks) should be dug out and filled with soil similar to the rest of the subgrade or with granular material such as sand, gravel, or crushed stone, if a similar soil isn't available. All fill materials must be compacted to provide the same uniform support as the rest of the subgrade.

During subgrade preparation, it should be remembered that undisturbed soil is generally superior to compacted material for supporting concrete slabs. Expansive, compressible, and potentially troublesome soils should be evaluated by a geotechnical engineer; a special slab design may be required.

The subgrade should be moistened with water in advance of placing concrete, but should not have puddles or wet, soft, muddy spots when the concrete is placed.

Subbase

A satisfactory slab on ground can be built without a subbase, but a subbase is frequently placed on the subgrade as a leveling course to equalize minor surface irregularities, enhance uniformity of support, bring the site to the desired grade, and serve as a capillary break between the slab and the subgrade.

Where a subbase is used, the concrete contractor should place and compact to near maximum density a 4-in.-thick layer of granular material such as sand, gravel, crushed stone, or slag. If a thicker subbase or subgrade is needed for achieving the desired grade, the material should be compacted in thin layers about 4 in. deep. Subgrades and subbases can be compacted with small portable vibrators, rollers, and hand tampers. Unless the subbase is well compacted, it is better to leave the subgrade uncovered and undisturbed.

Vapor Barriers and Moisture-Problem Prevention

Many of the moisture problems associated with enclosed slabs on ground (floors) can be minimized or eliminated by (1) sloping landscaping away from buildings, (2) using a 4-in. granular subbase to form a capillary break between the soil and the slab, (3) providing drainage for the granular subbase to prevent water from collecting under the slab, (4) installing foundation drain tile, and (5) installing a vapor barrier, often polyethylene sheeting.

A vapor barrier should be placed under all concrete floors on ground that are likely to receive an impermeable floor finish or be used for any purpose where the passage of water vapor through the floor is undesirable.

Good quality, well-consolidated concrete at least 4 in. thick is practically impermeable to the passage of liquid water unless the water is under considerable pressure; however, such concrete—even concrete several times as thick—is not impermeable to the passage of water vapor.

Water vapor that passes through concrete slabs evaporates at the surface if it is not sealed. Floor coverings such as linoleum, vinyl tile, carpeting, wood, and synthetic surfacing effectively seal the moisture within the slab where it eventually may loosen, buckle, or blister the floor covering.

In addition, to prevent problems with floor covering materials caused by moisture from the concrete itself, the following steps should be taken: (1) use a low water-cement ratio concrete, (2) moist-cure the slab for 5 to 7 days, (3) allow the slab a 2-or-more-month drying period, and (4) test the slab moisture condition before installing the floor covering. In one commonly used test, a four-foot square polyethylene sheet is taped to the floor. If after 24 hours (or before the flooring

---

*Details are given in References 9-10, 9-14, 9-22, and 9-32.

material or its adhesive cures) no moisture accumulates on the underside of the plastic film, the slab is considered dry enough for placing the flooring material. Flooring-material manufacturers often have a recommended test to use before installing their product.*

Insulation is sometimes installed over the vapor barrier to assist in keeping the temperature of a concrete floor above the dew point; this helps prevent moisture in the air from condensing on the slab surface. This practice also creates a warm floor for thermal comfort. Codes and specifications often require insulation at the perimeter of a floor slab. Placing insulation under the entire slab on ground for energy conservation alone usually cannot be justified.

Vapor barriers directly under concrete may increase the time delay before final finishing due to slow water evaporation, particularly in cold weather. They may also increase warping or curling and aggravate shrinkage crack development in the slab surface. To minimize these effects, a 3-in.-thick sand layer may be placed over the vapor barrier (or insulation if present).** If concrete is placed directly on a vapor barrier, the water-cement ratio should be low (0.45 or less).

Forms

Edge forms and intermediate screeds should be set accurately and firmly to the specified elevation and contour for the finished surface. Slab forms are usually metal or wood braced firmly with wood or steel stakes to keep them in horizontal and vertical alignment. The forms should be straight and free from warping and have sufficient strength to resist concrete pressure without bulging. They should also be strong enough to support any mechanical placing and finishing equipment used.

Placing and Spreading

Placement should start at the far point and proceed toward the concrete supply source. The concrete, which should be placed as close as possible to its final position, should slightly overfill the forms and be roughly leveled with square ended shovels or concrete rakes. Large voids trapped in the concrete during placing should be removed by consolidation.

Screeding (Strikeoff)

Screeding or strikeoff is the process of cutting off excess concrete to bring the top surface of a slab to proper grade. The templet used in the manual method is called a straightedge although the lower edge may be straight or slightly curved, depending on the surface specified. It should be moved across the concrete with a sawing motion and advanced forward a short distance with each movement. There should be a surplus (surcharge) of concrete against the front face of the straightedge to fill in low areas as the straightedge passes over the slab. A 6-in. slab needs a surcharge of about 1 in. Straightedges are sometimes equipped with vibrators that consolidate the concrete and assist in

reducing the strikeoff work. This combination of screed and vibrator is termed a vibratory screed (Fig. 9-5). Vibratory screeds are discussed under "Consolidating Concrete." Screeding and consolidation must be completed before excess bleed water collects on the surface.

Bullfloating or Darbying

To eliminate high and low spots and to embed large aggregate particles, a bullfloat or darby should be used immediately after strikeoff. The long-handle bullfloat (Fig. 9-6) is used on areas too large to reach with a short-handle darby. For non-air-entrained concrete, these tools should preferably be made of wood; for air-

Fig. 9-6. Bullfloating must be completed before any bleed water accumulates on the surface.

entrained concrete they should be of aluminum or magnesium alloy.

Bullfloating or darbying must be completed before bleed water accumulates on the surface. Care must be taken not to overwork the concrete as this could result in a less durable surface.

The preceeding operations should level, shape, and smooth the surface and work up a slight amount of cement paste. Although sometimes no further finishing is required, on most slabs, bullfloating or darbying is followed by one or more of the following finishing operations: edging, jointing, floating, troweling, and brooming. A slight hardening of the concrete is necessary before any of these operations can be begun. When the bleed-water sheen has evaporated and the concrete will sustain foot pressure with only about ¼ in. inden-

---

*Reference 9-29 discusses additional tests for moisture.
**Reference 9-14, pages 10-11.

tation, the surface is ready for continued finishing operations (Fig. 9-7).

Warning: One of the principal causes of surface defects in concrete slabs is finishing while bleed water is on the surface. Any finishing operation performed on the surface of a concrete slab while bleed water is present can cause serious crazing, dusting, or scaling (see Chapter 5). Floating and troweling the concrete (discussed later) before the bleeding process is completed may also trap bleed water under the finished surface producing a weakened zone or void under the finished surface, which occasionally results in delamination. The use of low-slump, air-entrained concrete with an adequate cement content and properly graded fine aggregate will minimize bleeding and help ensure maintenance-free slabs.*

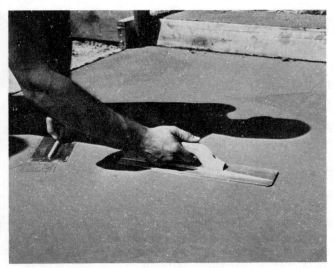

a. Hand floating the surface with a hand float held flat on the concrete surface and moved in a sweeping arc with a slight sawing motion.

b. Power floating. Footprints indicate proper timing.

Fig. 9-7. When the bleedwater sheen has evaporated and the concrete will sustain foot pressure with only slight indentation, the surface is ready for floating and final finishing operations.

Edging and Jointing

Edging is required along all edge forms and isolation and construction joints in floors and outdoor slabs such as walks, drives, and patios. Edging densifies and compacts the concrete next to the form where floating and troweling are less effective, making it more durable and less vulnerable to spalling and chipping.

In the edging operation, the concrete should be cut away from the forms to a depth of 1 in., using a pointed mason trowel or a margin trowel. Then an edger should be held almost flat on the surface and run with the front slightly raised to prevent the edger from leaving too deep an impression. Edging may be required after each subsequent finishing operation.

Proper jointing practices can eliminate unsightly random cracks. Contraction joints are made with a hand groover or by inserting strips of plastic, wood, metal, or preformed joint material into the unhardened concrete. When hand methods are used to form control joints, the slab should be jointed right after or during the edging operation. Contraction joints also can be made in hardened concrete by sawing. Jointing is discussed further under the heading "Making Joints in Floors and Walls."

Floating

After the concrete has been hand-edged and hand-jointed, it should be floated with a wood or metal hand float or with a finishing machine using float blades (Fig. 9-7).

The purpose of floating is threefold: (1) to embed aggregate particles just beneath the surface; (2) to remove slight imperfections, humps, and voids; and (3) to compact the mortar at the surface in preparation for additional finishing operations. The concrete should not be overworked as this may bring an excess of water and fine material to the surface and result in subsequent surface defects.

Hand floats usually are made of metal or wood. The metal float reduces the amount of work required because drag is reduced and the float slides more readily over the concrete surface. An aluminum or magnesium float is essential for hand floating air-entrained concrete because a wood float tends to stick to and tear the concrete surface. The light metal float also forms a smoother surface than the wood float.

The hand float should be held flat on the concrete surface and moved with a slight sawing motion in a sweeping arc to fill in holes, cut off lumps, and smooth ridges. When finishing large slabs, power floats can be used to reduce finishing time.

Floating produces a relatively even (but not smooth) texture that has good slip resistance and is often used as a final finish, especially for exterior slabs. Where a float finish is the desired final finish, it may be necessary to float the surface a second time after it has partially hardened.

---

*References 9-14 and 9-32 present the placing and finishing techniques in a detailed manner and Reference 9-31 discusses defects.

Marks left by hand edgers and groovers are normally removed during floating unless the marks are desired for decorative purposes, in which case these tools should be used again after final floating.

Troweling

Where a smooth, hard, dense surface is desired, floating should be followed by steel troweling. Troweling should not be done on a surface that has not been floated; troweling after only bullfloating or darbying is not an adequate finishing procedure.

It is customary when hand-finishing large slabs to float and immediately trowel an area before moving the kneeboards. These operations should be delayed until after the concrete has hardened enough so that water and fine material are not brought to the surface. Too long a delay, of course, will result in a surface that is too hard to float and trowel. The tendency, however, is to float and trowel the surface too soon. Premature floating and troweling can cause scaling, crazing, or dusting and will result in a surface with reduced wear resistance.

Spreading dry cement on a wet surface to take up excess water is bad practice and can cause crazing. Such wet spots should be avoided, if possible, by adjustments in aggregate gradation, mix proportions, and consistency. When wet spots do occur, finishing operations should be delayed until the water either evaporates or is removed with a rubber floor squeegee. If a squeegee is used, care must be taken so that cement is not removed with the water.

The first troweling may produce the desired surface free of defects. However, surface smoothness, density, and wear resistance can all be improved by additional trowelings. There should be a lapse of time between successive trowelings to permit the concrete to become harder. As the surface stiffens, each successive troweling should be made with smaller trowels, using progressively more tilt and pressure on the trowel blade. The final pass should make a ringing sound as the trowel moves over the hardening surface.

A power trowel is similar to a power float, except that it is fitted with smaller, individual steel trowel blades that are adjustable for tilt and pressure on the concrete surface. When the first troweling is done by machine, at least one additional troweling by hand should be done to remove small irregularities. If necessary, tooled edges and joints should be rerun after troweling to maintain uniformity and true lines.

Brooming

A slip-resistant surface can be produced by brooming before the concrete has thoroughly hardened, but it should be sufficiently hard to retain the scoring impression. Rough scoring can be achieved with a rake, a steel-wire broom, or a stiff, coarse, fiber broom; such coarse-texture brooming usually follows floating. If a finer texture is desired, the concrete should be floated and troweled to a smooth surface, and then brushed with a soft-bristled broom. Best results are obtained with brooms that are specially made for texturing concrete. Slabs are usually broomed transversely to the main direction of traffic.

Patterns and Textures

A variety of patterns and textures can be used to produce decorative finishes. Patterns can be formed with divider strips or by scoring or stamping the surface just before the concrete hardens. Textures can be produced with little effort and expense with floats, trowels, and brooms; more elaborate textures can be achieved with special techniques.*

An exposed-aggregate finish provides a rugged, attractive surface. Select aggregates, usually of uniform size such as ⅜ to ½ in. or larger, should be evenly distributed on the surface immediately after the slab has been bullfloated or darbied. Flat or elongated aggregate particles should not be used since they are easily dislodged when the aggregate is exposed.

Aggregates to be exposed should be washed thoroughly before use to assure satisfactory bond. The aggregate particles must be completely embedded in the concrete. This can be done by lightly tapping them with a wooden hand float, a darby, or the broad side of a piece of 2x4 lumber; then, when the concrete can support a finisher on kneeboards, the surface should be hand-floated with a magnesium float or darby until the mortar completely surrounds and slightly covers all the aggregate particles. When the concrete has hardened sufficiently, the aggregate should be exposed by simultaneously brushing and flushing with water.

Since timing is important, test panels should be made to determine the correct time for exposing the aggregate without dislodging the particles. On large jobs a reliable retarder can be sprayed or brushed on the surface immediately after floating, but on small jobs this may not be necessary.

Curing and Protection

All newly placed and finished concrete slabs should be cured and protected from rapid drying, from extreme changes in temperature, and from damage by subsequent construction and traffic.

Curing should begin as soon as possible after finishing (Fig. 9-8). Curing is needed to ensure continued hydration of the cement and strength gain of the concrete.

Special precautions are necessary when concrete work continues during periods of adverse weather conditions. In cold weather, arrangements should be made in advance for heating, covering, insulating, or enclosing the concrete. Hot-weather work may require special precautions against rapid evaporation and drying and excessively high temperatures.**

---

*See Reference 9-10.
**See also "Curing Concrete," Chapter 10; "Hot-Weather Concreting," Chapter 11; and "Cold-Weather Concreting," Chapter 12.

Fig. 9-8. An excellent method of wet curing is to completely cover the surface with wet burlap and keep it continuously wet during the curing period.

PLACING ON HARDENED CONCRETE

Bonded Construction Joints in Structural Concrete

A bonded construction joint is needed between the two structural concrete placements when freshly mixed concrete is placed in contact with existing hardened concrete and a high-quality bond and watertightness are required. Poorly bonded construction joints are usually the result of (1) lack of bond between old and new concrete or (2) a weak porous layer in the hardened concrete at the joint. The quality of a bonded joint therefore depends on the quality of the hardened concrete and preparation of its surface.

In columns and walls, the concrete near the top surface of a lift is often of inferior quality to the concrete below. This may be due to poor consolidation or use of badly proportioned or high-slump mixtures that cause excessive laitance, bleeding, and segregation. Even in well-proportioned and carefully consolidated mixtures, some aggregate particle settlement and water gain (bleeding) at the top surface is unavoidable, particularly with high rates of placement. Also, the encasing formwork prevents the escape of moisture from the fresh concrete. While the formwork is considered to provide adequate curing as long as it remains in place, at the top surface where there is no encasing formwork, the concrete may dry out too rapidly; this may result in a weak porous layer unless protection and curing are provided.

Preparing Hardened Concrete

When freshly mixed concrete is placed on recently hardened concrete, certain precautions must be taken to secure a well-bonded, watertight joint. The hardened concrete must be clean, sound, fairly level, and reasonably rough with some coarse aggregate particles exposed. Any laitance, soft mortar, dirt, wood chips, form oil, or other foreign materials should be removed since they would interfere with proper bonding of the subsequent placement.

The surface of old concrete upon which a concrete overlay is to be placed must be thoroughly roughened and cleaned of all dust, surface films, deposits, loose particles, grease, oil, and other material. In most cases it is necessary to remove the entire surface down to sound concrete. Roughening and cleaning with chipping hammers, waterblasting, scarifiers, sandblasting, shotblasting, hydrojetting, or other mechanical equipment are satisfactory methods for exposing sound concrete (Fig. 9-9). Chemical methods may also be used. Care must be taken to avoid contamination of the clean surface before the bonding grout and overlay concrete are placed.

Partially set or hardened concrete may only require stiff-wire brushing. In some types of construction such as dams, the surface of each concrete lift is cut with a high-velocity air-water jet to expose clean, sound concrete just before final set. This is usually done 4 to 12 hours after placing. The surface must then be protected and continuously cured until concreting is resumed for the next lift.

For two-course floors, the top surface of the base slab can be roughened just before it sets with a steel or stiff-fiber broom. The surface should be level, heavily scored, and free of laitance. It must then be protected until it is thoroughly cleaned just before the grout coat and top course are placed.

Hardened concrete may be left dry or be moistened before new concrete is placed on it; however, its surface should not be wet or have free-standing water. Laboratory studies indicate a slightly better bond is obtained on a dry surface than on a damp surface; however, the increased moisture level in the environment around the concrete reduces water loss from the concrete mixture, especially on hot, dry days.

For a horizontal construction joint in reinforced concrete wall construction, good results have been obtained by constructing the forms to the level of the joint, overfilling the forms an inch or two, and then removing the excess concrete just before setting occurs. The top surface then can be manually roughened with stiff brushes. The procedure is illustrated in Fig. 9-10.

In the case of vertical construction joints cast against a bulkhead, the concrete surface generally is too smooth to permit proper bonding. Particular care should be given to the removal of the smooth surface finish before reerecting the forms for placing freshly mixed concrete against the joint. Stiff-wire brushing may be sufficient if the concrete is less than three days old; otherwise, bushhammering or sandblasting may be needed, followed by washing with clean water to remove all dust and loose particles.*

---

*For additional information, see Reference 9-18.

a. Sandblasting can clean any size or shape surface—horizontal, vertical, or overhead.

c. Scarifiers use steel cutters to remove concrete surfaces.

b. Airless self-contained shotblasting equipment can prepare horizontal surfaces without contaminating the work area with dust or debris.

d. The hydrojetting equipment shown removes surface concrete with fine jets of water at pressures between 6000 and 17,000 psi. Some ultra-high-pressure water equipment can achieve pressures up to 55,000 psi.

Fig. 9-9. Methods commonly used to remove unsound concrete, laitance, scaling, and dirt to expose clean, sound, strong concrete.

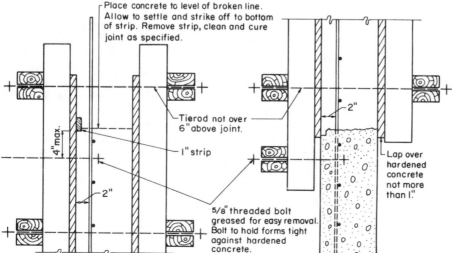

Place concrete to level of broken line. Allow to settle and strike off to bottom of strip. Remove strip, clean and cure joint as specified.

4" max.

2"

Tierod not over 6" above joint.

1" strip

5/8" threaded bolt greased for easy removal. Bolt to hold forms tight against hardened concrete.

2"

Lap over hardened concrete not more than 1".

Fig. 9-10. A straight, horizontal construction joint is provided by this detail.

Bonding New to Previously Hardened Concrete

For horizontal construction joints in wall sections where freshly mixed concrete is to be placed on hardened concrete, field experience indicates that good bond can be obtained by placing a rich concrete (higher cement and sand content than normal) on the bottom of the new lift and thoroughly vibrating it at the joint interface. Often the regular concrete mixture used in the structure is rich enough with proper vibration to form a good bond with clean, prepared hardened concrete.*

For overlays on slabs and for two-course floors, a 1/16- to 1/8-in.-thick layer of grout consisting of one part cement, one part sand, and enough water to make a thick, creamy, paintlike consistency, is brushed onto the clean and dry or damp (no free water present) concrete surface. The grout is placed just a short distance ahead of the overlay or top-course concrete (Fig. 9-11).** The grout should not be allowed to dry prior to the overlay placement, otherwise the dry grout may act as a poor surface for bonding. The old concrete surface should be clean and sound. Cleaning is best done by sandblasting, waterblasting, scarification, shotblasting, and other methods. Overlays are discussed further under "Patching, Cleaning, and Finishing."

Fig. 9-11. Application of a bonding grout a few feet ahead of the overlay concrete. The grout must not dry before the concrete is placed.

MAKING JOINTS IN FLOORS AND WALLS

The following three types of joints are common in concrete construction.

Isolation Joints

Isolation joints (Fig. 9-12) are also sometimes called expansion joints. They permit both horizontal and vertical differential movements at adjoining parts of a structure, for example, around the perimeter of a floor on ground, around columns, and around machine

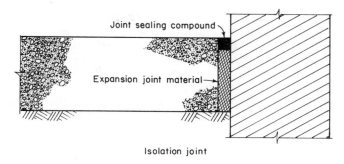

Fig. 9-12. Isolation joints permit horizontal and vertical movements between abutting faces of the slab and fixed parts of a structure.

foundations to separate the slab from the more rigid parts of the structure.

Isolation-joint material (also called expansion-joint material) can be as thin as 1/4 in. or less, but 1/2-in. material is commonly used. Care should be taken to ensure that all the edges for the full depth of the slab are isolated from adjoining construction; otherwise cracking can occur as a result of differential movements.

Columns on separate footings are isolated from the floor slab either with a circular or square-shaped isolation joint. The square shape should be rotated to align its corners with control and construction joints (see Fig. 9-17).

Contraction Joints

Contraction joints (Fig. 9-13) provide for movement in the plane of a slab or wall and induce cracking caused

*For additional information, see Reference 9-7.
**This method may also be applicable to horizontal joints in walls.

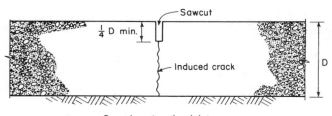

Sawed contraction joint

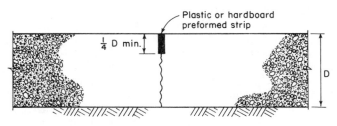

Premolded insert contraction joint

Fig. 9-13. Contraction joints provide for horizontal movement in the plane of a slab or wall and induce controlled cracking caused by drying shrinkage.

by drying and thermal shrinkage at preselected locations. Contraction joints (also sometimes called control joints) should be constructed to permit transfer of loads perpendicular to the plane of the slab or wall. If no contraction joints are used or if they are too widely spaced in slabs on ground or in lightly reinforced walls, random cracks will occur when drying and thermal shrinkage produce tensile stresses in excess of the concrete's tensile strength.

Contraction joints in slabs on ground can be made in several ways. One of the most common methods is to saw a continuous straight slot in the top of the slab (Fig. 9-14). This forms a plane of weakness in which a crack will form. Vertical loads are transmitted across the joint by aggregate interlock between the opposite faces of the crack providing the crack is not too wide

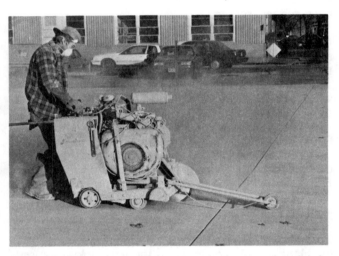

Fig. 9-14. Sawing a continuous cut in the top of a slab is one of the most economical methods for making a contraction joint.

and the spacing between joints is not too great. Steel dowels (Fig. 9-15) may be used to increase load transfer at contraction joints. Sizes and spacing of dowels, which are placed at the center of the slab depth, are shown in Table 9-1.*

Sawing is done as soon as the concrete is strong enough to resist tearing or other damage by the blade. This is normally 4 to 12 hours after the concrete hardens, but proper timing is dictated by the rate of hardening of the concrete. A slight raveling of the sawed edges is acceptable and indicates proper timing of the sawing operation. If sawing is delayed too long, the concrete may crack before it is sawed, or cracking may occur ahead of the saw blade.

Contraction joints also can be formed in the fresh concrete with hand groovers or by placing strips of wood, metal, or preformed joint material at the joint locations. The top of the strips should be flush with the concrete surface. Contraction joints, whether sawed, grooved, or preformed, should extend into the slab to a depth of *one-fourth* the slab thickness.

Contraction joints in walls are also planes of weakness that permit differential movements in the plane of the wall. The thickness of the wall at a contraction

Table 9-1. Dowel and Tiebar Sizes and Spacings

Slab depth, in.	Diameter, in. or bar number	Total length, in.	Spacing, in. center to center
Dowels			
5	⅝	12	12
6	¾	14	12
7	⅞	14	12
8	1	14	12
9	1⅛	16	12
10	1¼	16	12
Tiebars			
5	#4	30	30
6	#4	30	30
7	#4	30	30
8	#4	30	30
9	#5	30	30
10	#5	30	30

Reference 9-22.

Table 9-2. Maximum Spacing of Contraction Joints in Feet

Slab thickness, in.	Slump 4 in. to 6 in.		Slump less than 4 in.*
	Maximum-size aggregate less than ¾ in.	Maximum-size aggregate ¾ in. and larger	
4	8	10	12
5	10	13	15
6	12	15	18
7	14	18	21
8	16	20	24
9	18	23	27
10	20	25	30

Adapted from Table 8, *Concrete Floors on Ground*, EB075D, 2nd ed., Portland Cement Association, 1983.

Note: Given spacings also apply to the distance from contraction joints to parallel isolation joints or to parallel contraction-type construction joints. Spacings greater than 15 ft may show a marked loss in effectiveness of aggregate interlock to provide load transfer across the joint.

*¾ in. and larger maximum size aggregate.

joint should be reduced by at least 20%, preferably 25%. In lightly reinforced walls, half of the horizontal steel bars should be cut at the joint. Care must be taken to cut alternate bars precisely at the joint. At the corners of openings in walls where contraction joints are located, extra diagonal or vertical and horizontal reinforcement should be provided to control cracking. Contraction joints in walls should be spaced not more than about 20 ft apart. In addition, contraction joints should be placed where abrupt changes in thickness or height occur, and near corners—if possible, within 10 to 15 ft. Depending on the structure, these joints may need to be calked to prevent the passage of water through the wall. Instead of calking, a waterstop can be used to prevent water from leaking through the crack that occurs in the joint.

The spacing of contraction joints in floors on ground depends on (1) slab thickness, (2) shrinkage potential of the concrete, (3) subgrade friction, (4) environment,

*References 9-18 and 9-22 discuss further the use of doweled joints.

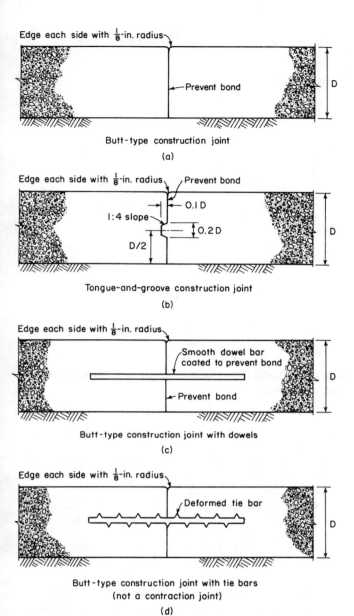

Edge each side with ⅛-in. radius

Prevent bond

D

Butt-type construction joint

(a)

Edge each side with ⅛-in. radius

Prevent bond

0.1 D

1:4 slope

0.2 D

D/2

D

Tongue-and-groove construction joint

(b)

Edge each side with ⅛-in. radius

Smooth dowel bar coated to prevent bond

Prevent bond

D

Butt-type construction joint with dowels

(c)

Edge each side with ⅛-in. radius

Deformed tie bar

D

Butt-type construction joint with tie bars
(not a contraction joint)

(d)

Fig. 9-15. Construction joints are stopping places in the process of construction. Construction-joint types a, b, and c are also used as contraction joints.

and (5) the absence or presence of steel reinforcement. Unless reliable data indicate that more widely spaced joints are feasible, the suggested intervals given in Table 9-2 should be used for well-proportioned concrete with aggregates having normal shrinkage characteristics. Joint spacing should be decreased for concrete suspected of having high shrinkage characteristics. The resulting panels should be approximately square. Panels with excessive length-to-width ratio (more than 1½ to 1) are likely to crack at an intermediate location.

Construction Joints

Construction joints (Fig. 9-15) are stopping places in the process of construction. A true construction joint should bond new concrete to existing concrete and permit no movement. Deformed tiebars (sizes and spacing shown in Table 9-1) are often used in construction joints to restrict movement. Because extra care is needed to make a true construction joint, they are usually designed and built to function as and align with contraction or isolation joints and therefore may purposely be made unbonded. Oils, form-release agents, and paints are used as debonding materials. In thick, heavily loaded floors, unbonded doweled construction joints are commonly used; otherwise unbonded tongue-and-groove joints are satisfactory. For thin slabs, the flat-faced butt-type joint will suffice.

On most structures it is desirable to have wall joints that will not detract from appearance. When properly made, joints can be inconspicuous or hidden by rustication strips. They thus can be an architectural as well as functional feature of the structure. However, if rustication strips are used in structures that may be exposed to deicing salts, such as bridge columns and abutments, care should be taken to ensure that the reinforcing steel has the required depth of concrete cover to prevent corrosion.

Horizontal joints in walls should be made straight, exactly horizontal, and should be placed at suitable locations. A straight horizontal construction joint can be made by nailing a 1-in. wood strip to the inside face of the form near the top (see Fig. 9-10). Concrete should then be placed to a level slightly above the bottom of the strip. After the concrete has settled and before it becomes hard, any laitance that has formed on the top surface should be removed. The strip can then be removed and irregularities in the joint leveled off. The forms are removed and then reerected above the construction joint for the next lift of concrete. To prevent leakage from the concrete, gaskets should be used where forms contact previously placed hardened concrete.

A variation of this procedure makes use of a rustication strip instead of the 1-in. wood strip to form a groove in the concrete for architectural effect (Fig. 9-16). Rustication strips can be V-shaped, rectangular, or slightly beveled. If V-shaped, the joint should be made at the point of the V. If rectangular or beveled,

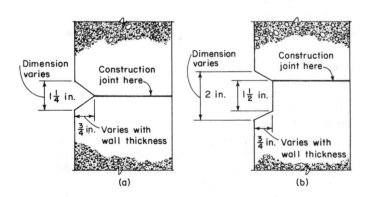

Dimension varies

1¼ in.

Construction joint here

¾ in. Varies with wall thickness

(a)

Dimension varies

2 in.

1½ in.

Construction joint here

¾ in. Varies with wall thickness

(b)

Fig. 9-16. Horizontal construction joints in walls with V-shaped (a) and beveled (b) rustication strips.

the joint should be made at the top edge of the inner face of the strip.

FILLING FLOOR JOINTS

The movement at contraction joints in a floor is generally very small. For some industrial and commercial uses, these joints can be left unfilled. Where there are wet conditions, hygienic and dust-control requirements, or considerable traffic by small, hard-wheel vehicles such as forklifts, joint filling is necessary.

In many places, a resilient material such as an elastomeric sealant is satisfactory, but to provide support to the edges and prevent spalling at saw-cut joints, a good quality, semirigid epoxy filler with a Shore Hardness of A-80 or D-50 (ASTM D 2240) should be used. The material should be installed full depth in the saw cut, without a backer rod, and flush with the floor surface.

Extruded lead strips embedded in contraction joints give good support to the edges and thereby reduce spalling. Lead strips, where permitted, have been highly successful for filling joints in heavy-duty concrete floors where trucking is severe, especially of the small hard-wheel type.

Isolation joints are intended to accommodate movement; thus a flexible, elastomeric sealant should be used.

JOINT LAYOUT FOR FLOORS

A typical joint layout for industrial floors is illustrated in Fig. 9-17. Isolation joints are provided around the perimeter of the floor where it abuts the walls and around all fixed elements that may restrain movement of the slab. This includes columns and machinery bases that pierce the floor slab. With the slab isolated from other building elements, the remaining task is to locate and correctly space contraction joints to eliminate random cracking. Construction-joint locations are coordinated with the floor contractor to accommodate work schedules and crew size. They should coincide with the contraction-joint pattern and act as contraction joints.*

UNJOINTED FLOORS

An unjointed floor or one with a limited number of joints can be constructed when joints are unacceptable. Three methods are suggested:
1. A prestressed floor can be built through use of post-tensioning. With this method, steel strands in ducts are tensioned after the concrete hardens to produce compressive stress in the concrete during stress transfer. This will counteract the development of tensile stresses in the concrete and provide a crack-free floor. Large areas, 10,000 sq ft and more, can be

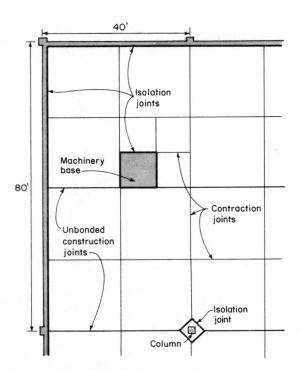

Fig. 9-17. Typical joint layout for an 8-in.-thick concrete floor on ground.

constructed in this manner without intermediate joints.
2. Concrete made with expansive cement can be used to offset the amount of drying shrinkage to be anticipated after curing. Contraction joints are not needed when construction joints are used at intervals of 40 to 120 ft. Large areas, to 20,000 sq ft, have been cast in this manner without joints. Steel reinforcement is needed in order to produce compressive stresses during and after the expansion period, that is, a form of prestressing.
3. Large areas—a single day of slab placement, usually about 8000 to 10,000 sq ft—can be cast without contraction joints when the amount of distributed steel is about one-half of one percent of the cross-sectional area of the slab. Special effort should be made to reduce subgrade friction in floors without contraction joints.**

REMOVING FORMS

It is advantageous to leave forms in place as long as possible to continue the curing period. However, there are times when it is necessary to remove them as soon as possible. For example, where a rubbed finish is specified, forms must be removed early to permit the first rubbing before the concrete becomes too hard.

---

*For more information on floor joints, see References 9-14, 9-18, and 9-22. For joints in walls, see References 9-18, 9-21, 9-23, 9-24, and 9-25.
**Reference 9-22 discusses use of distributed steel in floors.

Furthermore, it is often necessary to remove forms quickly to permit their immediate reuse.

In any case, forms should not be removed until the concrete is strong enough to satisfactorily carry the stresses from both the dead load of the structure and any imposed construction loads. The concrete should be hard enough so that the surfaces will not be injured in any way when reasonable care is used in removing forms. In general, for concrete temperatures above 50°F, the side forms of reasonably thick, supported sections can be removed 12 to 24 hours after concreting. Beam and floor slab forms and supports (shoring) may be removed between 3 and 21 days depending on the size of the member and the strength gain of the concrete.* For most conditions, it is better to rely on the strength of the concrete as determined by test rather than to select arbitrarily the age at which forms may be removed.

The designer should specify the minimum strength requirements of various members for form removal. The age-strength relationship should be determined from representative samples of concrete used in the structure and field-cured under job conditions. Table 9-3 shows the ages typically required to attain certain strengths under average conditions for air-entrained concrete made with a water-cement ratio of about 0.53. It should be remembered, however, that strengths are affected by the materials used, temperature, and other conditions. The time required for form removal, therefore, will vary from job to job (see Chapter 12).

Table 9-3. Typical Age-Strength Relationship of Air-Entrained Concrete

Strength, psi	Age	
	Type I normal cement	Type III high-early-strength cement
500	24 hours	12 hours
750	1½ days	18 hours
1500	3½ days	1½ days
2000	5½ days	2½ days

A pinch bar or other metal tool should not be placed against the concrete to wedge forms loose. If it is necessary to wedge between the concrete and the form, only wooden wedges should be used. Stripping should be started some distance away from and moved toward a projection. This relieves pressure against projecting corners and reduces the chance of edges breaking off.

Recessed forms require special attention. Wooden wedges should be gradually driven behind the form and the form should be tapped lightly to break it away from the concrete. Forms should not be pulled off rapidly after wedging has been started at one end; this is almost certain to break the edges of the concrete.

PATCHING, CLEANING, AND FINISHING

After forms are removed, all bulges, fins, and small projections can be removed by chipping or tooling.

Undesired bolts, nails, ties, or other embedded metal can be removed or cut back to a depth of ½ in. from the concrete surface. When required, the surface can be rubbed or ground to provide a uniform appearance. Any cavities such as tierod holes should be filled unless they are to remain for decorative purposes. Honeycombed areas must be repaired and stains removed to present a concrete surface that is uniform in color. All of these operations can be minimized by exercising care in constructing the formwork and placing the concrete. In general, repairs are easier to make and more successful if they are made as soon as practical, preferably as soon as the forms are removed. However, the procedures discussed below apply to both new and old hardened concrete.

Holes, Defects, and Overlays

Patches usually appear darker than the surrounding concrete; therefore, some white cement should be used in mortar or concrete for patching where appearance is important. Samples should be applied and cured in an inconspicuous location, perhaps a basement wall, several days in advance of patching operations to determine the most suitable proportions of white and gray cements. Steel troweling should be avoided since this darkens the patch.

Bolt holes, tierod holes, and other cavities that are small in area but relatively deep should be filled with a dry-pack mortar. The mortar should be mixed as stiff as practical using 1 part cement, 2½ parts sand passing a No. 16 sieve, and just enough water to form a ball when the mortar is squeezed gently in the hand. The cavity should be cleaned of oil and loose material and kept damp with water for several hours. A bonding grout (1 part cement, 1 part sand, and enough water to form a creamy consistency) should be brushed onto the void surfaces but not allowed to dry before the mortar is placed. The mortar should be tamped into place in layers about ½ in. thick. Vigorous tamping and adequate curing will ensure good bond and minimum shrinkage of the patch.

Large patches and thin-bonded overlays need concrete for the repair material. This concrete should have a low water-cement ratio, often with a cement content equal to or greater than the concrete to be repaired. Cement contents often range from 600 to 850 lb per cubic yard and the water-cement ratio is usually 0.45 or less. The aggregate size should be preferably no more than ⅓ the patch or overlay thickness. A ⅜ in. maximum size coarse aggregate is commonly used. The sand proportion can be higher than usual, often equal to the amount of coarse aggregate, depending on the desired properties and application.

Before the patching concrete is applied, the surrounding concrete should be clean and sound. Abrasive methods of cleaning (sandblasting, hydrojetting, waterblasting, scarification, or shotblasting) are usually required. A grout (1 part cement, 1 part fine sand passing a No. 30 sieve, and sufficient mixing water for

---

*Reference 9-26.

a creamy consistency) should be scrubbed with a brush or broom onto all surfaces to which the new material is to be bonded. The grout should be applied immediately before the new concrete is placed. The grout should not be allowed to dry before the freshly mixed concrete is placed; otherwise bond may be impaired. The concrete may be dry or damp when the grout is applied *but not wet* with free-standing water. The minimum thickness for most patches and overlays is ¾ in. Some structures, like bridge decks, should have a minimum repair thickness of 1½ in. A superplasticizer is often added to overlay or repair concrete to reduce the water-cement ratio and to improve workability and ease of consolidation.*

Honeycombed and other defective concrete should be cut out to expose sound material. If defective concrete is left adjacent to a patch, moisture may get into the voids; in time, weathering action will cause the patch to spall. The edges of the defective area should be cut or chipped straight and at right angles to the surface, or slightly undercut to provide a key at the edge of the patch. No featheredges should be permitted (Fig. 9-18). Based on the size of the patch either a mortar or concrete mixture should be used.

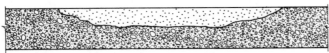

(a) Incorrectly installed patch. The feathered edges will break down under traffic or will weather off.

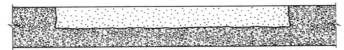

(b) Correctly installed patch. The chipped area should be at least ¾ in. deep with the edges at right angles or undercut to the surface.

Fig. 9-18. Patch installation.

Shallow patches can be filled with stiff mortar as described earlier. This should be placed in layers not more than ½ in. thick, with each layer given a scratch finish to improve bond with the subsequent layer. The final layer can be finished to match the surrounding concrete by floating, rubbing, or tooling, or on formed surfaces by pressing a section of form material against the patch while still plastic.

Deep patches can be filled with concrete held in place by forms. Such patches should be reinforced and doweled to the hardened concrete.** Large, shallow vertical or overhead repairs may best be accomplished by shotcreting (Chapter 15). Several proprietary low-shrinkage cementitious repair products are also available.

Curing Patches

Following patching, good curing is essential (Fig. 9-19). Curing should be started as soon as possible to avoid early drying. Wet burlap, wet sand, plastic sheets, curing

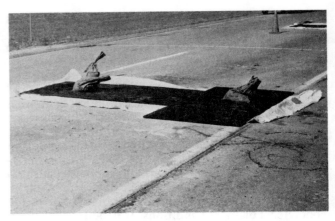

Fig. 9-19. Good curing is essential to successful patching. This patch is covered with polyethylene sheeting plus rigid insulation to retain moisture and heat for rapid hydration and strength gain.

paper, tarpaulins, or a combination of these can be used. In locations where it is difficult to hold these materials in place, an application of two coats of membrane-curing compound is often the most convenient method.

Cleaning Concrete Surfaces

Concrete surfaces are not always uniform in color when forms are removed; they may have a somewhat blotchy appearance and there may be a slight film of form-release agent in certain areas. There may be mortar stains from leaky forms or there may be rust stains. Flatwork can also become discolored during construction. Where appearance is important, all surfaces should be cleaned after construction has progressed to the stage where there will be no discoloration from subsequent construction activities.

There are three techniques for cleaning concrete surfaces: water, chemical, and mechanical (abrasion). Water dissolves dirt and rinses it from the surface. Chemical cleaners, usually mixed with water, react with dirt to separate it from the surface, and then the dirt and chemicals are rinsed off with clean water. Mechanical methods—sandblasting is most common—remove dirt by abrasion.

Before selecting a cleaning method, it should be tried on a test area to be certain that it will be helpful and not harmful. If possible, identify the characteristics of the discoloration because some treatments are more effective than others in removing certain materials.

Water cleaning methods include low-pressure washes, moderate-to-high-pressure waterblasting, and steam. Low-pressure washing is the simplest, requiring only that water run gently down the concrete surface for a day or two. The softened dirt then is flushed off with a

*Reference 9-27.
**See Reference 1-15, page 398.

slightly higher pressure rinse. Stubborn areas can be scrubbed with a nonmetallic-bristle brush and rinsed again. High-pressure waterblasting is used effectively by experienced operators. Steam cleaning must be performed by skilled operators using special equipment. Water methods are the least harmful to concrete, but they are not without potential problems. Serious damage may occur if the concrete surface is subjected to freezing temperatures while it is still wet; and water can bring soluble salts to the surface, forming a chalky, white deposit called efflorescence.

Chemical cleaning is usually done with water-based mixtures formulated for specific materials such as brick, stone, and concrete. An organic compound called a surfactant (surface-active agent), which acts as a detergent to wet the surface more readily, is included in most chemical cleaners. A small amount of acid or alkali is included to separate the dirt from the surface. For example, hydrochloric (muriatic) acid is commonly used to clean masonry walls and remove efflorescence. There can be problems related to the use of chemical cleaners. Their acid or alkaline properties can lead to reaction between cleaner and concrete as well as mortar, painted surfaces, glass, metals, and a host of other building materials. Since chemical cleaners are used in the form of water-diluted solutions, they too can liberate soluble salts from within the concrete to form efflorescence. Some chemicals can also expose the aggregate in concrete. Chemicals commonly used to clean concrete surfaces and remove discoloration include weak solutions (1% to 10% concentration) of hydrochloric, acetic, or phosphoric acid. Diammonium citrate (20% to 30% water solution) is especially useful in removing discoloration stains and efflorescence on formed and flatwork surfaces. Chemical cleaners should be used by skilled operators taking suitable safety precautions.*

Mechanical cleaning (Fig. 9-9) includes sandblasting, shotblasting, scarification, power chipping, and grinding. These methods wear the dirt off the surface rather than separate it from the surface. They, in fact, wear away both the dirt and some of the concrete surface; it is inevitable that there will be some loss of decorative detail, increased surface roughness, and rounding of sharp corners. Abrasive methods may also reveal defects (voids) hidden just beneath the formed surface.

Chemical and mechanical cleaning can each have an abrading effect on the concrete surface that may change the appearance of a surface compared to that of an adjacent uncleaned surface.

Finishing Formed Surfaces

Many off-the-form concrete surfaces require little or no additional treatment when they are carefully constructed with the proper forming materials or form liners. These surfaces are divided into two general classes: smooth and textured or patterned. Smooth surfaces are produced with plastic-coated forms, steel forms, fiberglass-reinforced plastic forms, formica forms, or tempered-hardboard forms. Textured or patterned surfaces are achieved with rough-sawn lumber, special grades and textures of plywood, form liners, or by fracturing the projections of a striated surface.

As-cast finishes may, if specified, require patching of tieholes and defects, but the surfaces otherwise need no further work since texture and finish are imparted by the forms.

For a *smooth off-the-form finish,* it is important to arrange the smooth-facing forming material and tie-rods in a symmetrical pattern. Smooth-finish forms must be supported by studs and wales that are capable of preventing excessive deflections.

A *smooth, rubbed finish* is produced on a newly hardened concrete surface no later than the day following form removal. The forms are removed and necessary patching completed as soon as possible. The surface is then wetted and rubbed with a carborundum brick or other abrasive until a satisfactory uniform color and texture are produced.

A *sand-floated finish* can also be produced on newly hardened concrete surfaces. No later than 5 to 6 hours following form removal, the surface should be thoroughly wetted and rubbed with a wood float in a circular motion, working fine sand into the surface until the resulting finish is even and uniform in texture and color.

A *grout cleandown (sack-rubbed finish)* can be used to impart a uniform color and appearance to a smooth surface. After defects have been repaired, the surface should be saturated thoroughly with water and kept wet during grout operations. A grout consisting of 1 part cement, 1½ to 2 parts of fine sand, and sufficient water for a thick, creamy consistency should be applied uniformly by brush, plasterer's trowel, or rubber float to completely fill air bubbles and holes.

The surface should be vigorously floated with a wood, sponge-rubber, or cork float immediately after applying the grout to fill any small air holes (bugholes) that are left and to remove some excess grout. The remaining excess grout should be scraped off with a sponge-rubber float. If the float pulls grout from holes, a sawing motion of the tool should correct the difficulty. The grout remaining on the surface should be allowed to stand undisturbed until it loses some of its plasticity but not its damp appearance. Then the surface should be rubbed with clean, dry burlap to remove all excess grout. All air holes should remain filled, but no visible film of grout should remain after the rubbing. Any section being cleaned with grout must be completed in one day, since grout remaining on the surface overnight is too difficult to remove.

If possible, work should be done in the shade and preferably during cool, damp weather. During hot or dry weather, the concrete can be kept moist with a fine fog spray.

The completed surface should be moist-cured by keeping the area wet for 36 hours following the cleandown. When completely dry, the surface should have a uniform color and texture.

---

*See References 9-1 and 9-20 for more information.

SPECIAL SURFACE FINISHES

Exposed Aggregate

Methods of exposing aggregate in formed concrete include washing and brushing, using retarders and scrubbing, abrasive blasting, and tooling or grinding.*

In washing and brushing, the surface layer of mortar should be carefully washed away with a light spray of water and brushed until the desired exposure is achieved.

With the retarder method a water-insoluble retarder should generally be used. The retarded surface layer of mortar is scrubbed away by washing with a light spray of water and brushing. When the concrete becomes too hard to produce the required finish with normal scrubbing, a dilute hydrochloric acid can be used.

Abrasive blasting is best applied to a gap-graded-aggregate concrete. The nozzle should be held perpendicular to the surface and the surface removed to a maximum depth of about one-third the diameter of the coarse aggregate.

Waterblasting can also be used to texture the surface of hardened concrete, especially where local ordinances prohibit the use of sandblasting for environmental reasons. High-pressure water jets are used on surfaces that have or have not been treated with retarders.

In tooling or bushhammering, a layer of hardened concrete is removed and the aggregate is fractured at the surface. The surfaces attained can vary from a light scaling to a deep, bold texture obtained by jackhammering with a single-pointed chisel. Combs and multiple points can be used to produce finishes similar to some finishes used on cut stone.

Grinding should be done in several successive steps, each with a finer grit than the preceding one. A polishing compound and buffer can then be used for a honed finish.

Paints and Clear Coatings

Many types of paints and clear coatings can be applied to concrete surfaces. Among the principal paints used are portland cement base, latex-modified portland cement, and latex (acrylic and polyvinyl acetate) paints.**

Portland cement based paints can be used on either interior or exterior exposures. The surface of the concrete should be damp at the time of application and each coat should be dampened as soon as it can be done without disturbing the paint. *Damp curing of conventional portland cement paint is essential.* On open-textured surfaces, such as concrete masonry, the paint should be applied with stiff-bristle brushes such as scrub brushes. Paint should be worked well into the surface. For concrete with a smooth or sandy surface, whitewash or Dutch-type calcimine brushes are best.

The latex materials used in latex-modified portland cement paints retard evaporation, thereby retaining the necessary water for hydration of the portland cement. Moist curing is unnecessary with latex-modified paints.

Most latex paints are resistant to alkali and can be applied to new concrete after 10 days of good drying weather. The preferred method of application is by long-fiber, tapered nylon brushes 4 to 6 in. wide. However, application can also be made by roller or spray. The paints may be applied to damp, but not wet, surfaces, and if the surface is moderately porous or extremely dry conditions prevail, prewetting the surface is advisable.

Clear coatings are frequently used on concrete surfaces to prevent soiling or discoloration of the concrete by air pollution, to facilitate cleaning the surface if it does become dirty, to brighten the color of the aggregates, and to render the surface water-repellent and thus prevent color change due to water absorption. The better coatings often consist of methyl methacrylate forms of acrylic resin, as indicated by a laboratory evaluation of commercial clear coatings.† The methyl methacrylate coatings should have a higher viscosity and solids content when used on smooth concrete, since the original appearance of smooth concrete is more difficult to maintain than the original appearance of exposed-aggregate concrete.

Other materials, such as a new generation of silane and siloxane penetrating sealers, are coming into use as water repellants for architectural concrete.

PLACING CONCRETE UNDERWATER

Concrete should be placed in the air rather than underwater whenever possible. When it must be placed underwater, the work should be done under experienced supervision. The basic principles for normal concrete work in the dry apply, with common sense, to underwater concreting. The following special points, however, should be observed:

The slump of the concrete should not be less than 5 in. and the cement content usually not less than 650 lb per cubic yard. It is important that the concrete flow without segregation; therefore the aim in proportioning should be to obtain a cohesive mixture with high workability. Using rounded aggregates, a higher percentage of fines, and entrained air should help to obtain the desired consistency.

Methods for placing concrete underwater include the following: tremie, concrete pump, bottom-dump buckets, grouting preplaced aggregate, toggle bags, bagwork, and the diving bell.

A tremie is a smooth, straight pipe long enough to reach the lowest point to be concreted from a working platform above the water. A hopper to receive the concrete is attached to the top of the pipe. The lower end of the tremie should be kept buried in the fresh concrete to maintain a seal and to force the concrete to flow into position by pressure. Placing should be continuous with as little disturbance to the previously

---

*References 9-5 and 9-12.
**Reference 9-8.
†Reference 9-3.

placed concrete as possible. The top surface should be kept as level as possible.

Development of the mobile concrete pump with a variable radius boom has made the work of placing concrete underwater easier than before.

Bottom-drop buckets vary in shape and capacity. The bottom-opening doors are operated by a diver or a trip line to the surface. The top of the bucket should be closed with a canvas cover to protect the concrete from damage while it is being lowered.

Grouting preplaced aggregate has advantages when placing concrete in flowing water.

Toggle bags are reusable canvas bags, sausage shaped, that are filled with concrete and lowered to a diver. A slipknot or chain and toggle at the top and bottom permits easy emptying and filling of the bag.

Sand bags half full of plastic concrete can be used for small jobs, filling gaps, or temporary work. The tied end should face away from the outside.

SPECIAL PLACING TECHNIQUES

Concrete may be placed by methods other than the usual cast-in-place method. These methods, such as shotcreting, are described in Chapter 15. No matter what method is used, the basics of mixing, placing, consolidating, and curing apply to all portland cement mortars and concretes.

PRECAUTIONS

When working with freshly mixed concrete, care should be taken to avoid skin irritation or chemical burns that can occur through prolonged contact between such concrete and skin surfaces, eyes, and clothing.

There are three causes of skin irritation or chemical burns through prolonged contact with fresh concrete.
1. Portland cement is alkaline in nature and therefore caustic.
2. Portland cement is hygroscopic—it tends to absorb moisture from the skin.
3. The sand contained in fresh concrete is abrasive to bare skin.

Clothing should not be allowed to become saturated with the moisture from the concrete because saturated clothing can transmit alkaline or hygroscopic effects to the skin. Waterproof gloves, a long-sleeve shirt, and full-length trousers should be worn.

If it is necessary to stand in freshly mixed concrete while it is being placed, rubber boots high enough to prevent concrete from flowing into them should be worn.

When finishing concrete, waterproof pads should be used between concrete surfaces and knees, elbows, hands, and so forth. Clothing areas that become saturated from contact with freshly mixed concrete should be rinsed promptly with clean water to prevent continued contact with skin surfaces. Eyes or skin areas that come in contact with freshly mixed concrete should be washed thoroughly with fresh water. Mild irritation of skin areas can be relieved by applying a lanolin cream to the irritated area after washing. Persistent or severe discomfort should be attended to by a physician.

REFERENCES

9-1. Greening, N. R., and Landgren, R., *Surface Discoloration of Concrete Flatwork*, Research Department Bulletin RX203, Portland Cement Association, 1966.

9-2. Colley, B. E., and Humphrey, H. A., *Aggregate Interlock at Joints in Concrete Pavements*, Development Department Bulletin DX124, Portland Cement Association, 1967.

9-3. Litvin, Albert, *Clear Coatings for Exposed Architectural Concrete*, Development Department Bulletin DX137, Portland Cement Association, 1968.

9-4. Turner, C. D., "Unconfined Free-Fall of Concrete," *Journal of the American Concrete Institute*, American Concrete Institute, Detroit, December 1970, pages 975-976.

9-5. *Bushhammering of Concrete Surfaces*, IS051A, Portland Cement Association, 1972.

9-6. *Subgrades and Subbases for Concrete Pavements*, IS029P, Portland Cement Association, 1975.

9-7. *Bonding Concrete or Plaster to Concrete*, IS139T, Portland Cement Association, 1976.

9-8. *Painting Concrete*, IS134T, Portland Cement Association, 1977.

9-9. Hurd, M. K., *Formwork for Concrete*, SP-4, American Concrete Institute, 1979.

9-10. *Finishing Concrete Slabs, Exposed Aggregate, Patterns, and Colors*, IS206T, Portland Cement Association, 1979.

9-11. *Bridge Deck Renewal with Thin-Bonded Concrete Resurfacing*, IS207E, Portland Cement Association, 1980.

9-12. *Color and Texture in Architectural Concrete*, SP021A, Portland Cement Association, 1980.

9-13. *Roller Compacted Concrete*, ACI 207.5R-80, ACI Committee 207 Report, American Concrete Institute, 1980.

9-14. *Guide for Concrete Floor and Slab Construction*, ACI 302.1R-80, ACI Committee 302 Report, American Concrete Institute, 1980.

9-15. *Concrete Basements for Residential and Light Building Construction*, IS208B, Portland Cement Association, 1980.

9-16. *Joint Design for Concrete Highway and Street Pavements*, IS059P, Portland Cement Association, 1980.

9-17. *Resurfacing Concrete Floors*, IS144T, Portland Cement Association, 1981.

9-18. *Building Movements and Joints,* EB086B, Portland Cement Association, 1982.

9-19. *Standard Practice for Consolidation of Concrete,* ACI 309-72, revised 1982, ACI Committee 309 Report, American Concrete Institute.

9-20. *Removing Stains and Cleaning Concrete Surfaces,* IS214T, Portland Cement Association, 1982.

9-21. *Joints in Walls Below Ground,* CR059T, Portland Cement Association, 1982.

9-22. *Concrete Floors on Ground,* EB075D, Portland Cement Association, 1983.

9-23. "Why Concrete Walls Crack," *Concrete Technology Today,* PL842B, Portland Cement Association, June 1984.

9-24. "Joints to Control Cracking in Walls," *Concrete Technology Today,* PL843B, Portland Cement Association, September 1984.

9-25. "Sealants for Joints in Walls," *Concrete Technology Today,* PL844B, December 1984.

9-26. *Recommended Practice for Concrete Formwork,* ACI 347-78, reaffirmed 1984, ACI Committee 347 Report, American Concrete Institute.

9-27. Kosmatka, Steven H., "Repair with Thin-Bonded Overlay," *Concrete Technology Today,* PL851B, Portland Cement Association, March 1985.

9-28. *Guide for Measuring, Mixing, Transporting and Placing Concrete,* ACI 304-85, ACI Committee 304 Report, American Concrete Institute, 1985.

9-29. Kosmatka, Steven H., "Floor-Covering Materials and Moisture in Concrete," *Concrete Technology Today,* PL853B, Portland Cement Association, September 1985.

9-30. *Specifications for Structural Concrete for Buildings,* ACI 301-84, revised 1985, ACI Committee 301 Report, American Concrete Institute.

9-31. *Concrete Slab Surface Defects: Causes, Prevention, Repair,* IS177T, Portland Cement Association, 1987.

9-32. *Cement Mason's Guide,* PA122H, Portland Cement Association, 1971, revised 1987.

CHAPTER 10
Curing Concrete

Curing is the maintenance of a satisfactory moisture content and temperature in concrete during some definite period immediately following placing and finishing so that the desired properties may develop. The need for adequate curing of concrete cannot be over-emphasized. Curing has a strong influence on the properties of hardened concrete such as durability, strength, watertightness, abrasion resistance, volume stability, and resistance to freezing and thawing and deicer salts. Exposed slab surfaces are especially sensitive to curing as the top surface strength development can be reduced significantly when curing is defective.

When portland cement is mixed with water, a chemical reaction called hydration takes place. The extent to which this reaction is completed influences the strength, durability, and density of the concrete. Most freshly mixed concrete contains considerably more water than is required for complete hydration of the cement; however, any appreciable loss of water by evaporation or otherwise will delay or prevent complete hydration. If temperatures are favorable, hydration is relatively rapid the first few days after concrete is placed; it is important for the water to be retained during this period, that is, for evaporation to be prevented or at least reduced. The objectives of curing, therefore, are to

1. Prevent (or replenish) the loss of moisture from concrete
2. Maintain a favorable concrete temperature for a definite period of time

With proper curing, concrete will become stronger, more impermeable, more resistant to stress, abrasion, and freezing and thawing. The improvement is rapid at early ages but continues more slowly thereafter for an indefinite period. Fig. 10-1 shows the strength gain of concrete with age for different curing periods.

The most effective method for curing concrete depends upon the circumstances. For most jobs, normal curing is adequate, but in some cases, such as in hot and cold weather, special care is needed.

When moist curing is interrupted, the development of strength continues for a short period and then stops after the concrete's internal relative humidity drops to about 80%. However, if moist curing is resumed, strength development will be reactivated. Although it can be done in a laboratory, it is difficult to resaturate

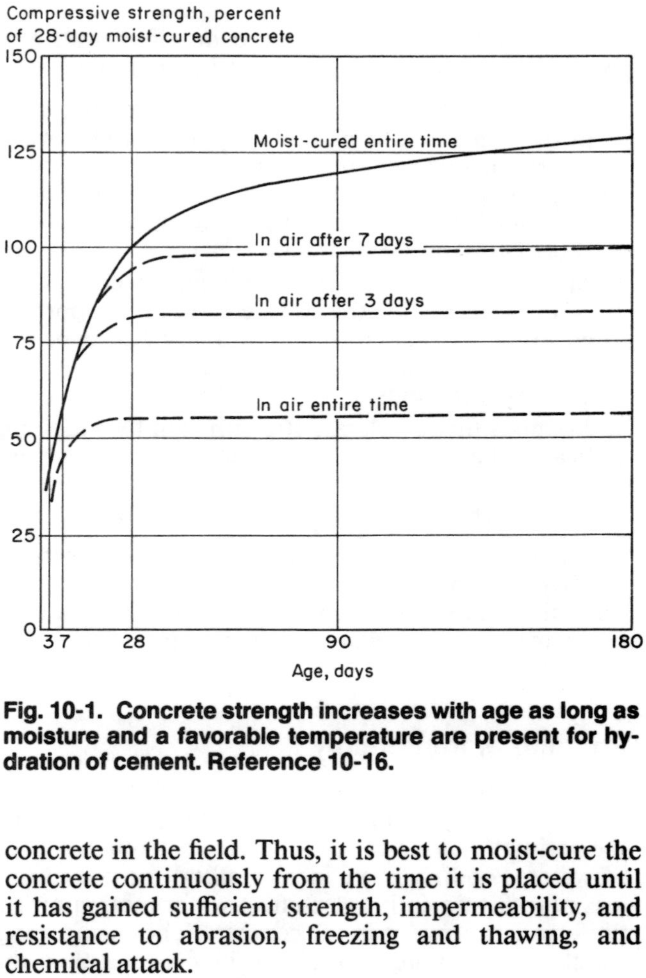

Fig. 10-1. Concrete strength increases with age as long as moisture and a favorable temperature are present for hydration of cement. Reference 10-16.

concrete in the field. Thus, it is best to moist-cure the concrete continuously from the time it is placed until it has gained sufficient strength, impermeability, and resistance to abrasion, freezing and thawing, and chemical attack.

Loss of water will also cause the concrete to shrink, thus creating tensile stresses within the concrete. If these stresses develop before the concrete has attained adequate tensile strength, surface cracking can result. All exposed surfaces, including exposed edges and joints, must be protected against moisture evaporation.

Hydration proceeds at a much slower rate when the concrete temperature is low. Temperatures below 50°F are unfavorable for the development of early strength; below 40°F the development of early strength is greatly retarded; and at or below freezing temperatures, down to 14°F, little or no strength develops.

In recent years, a maturity concept has been introduced to evaluate the development of strength when there is variation in the curing temperature of the concrete. *Maturity* is defined as the product of the age of the concrete and its average curing temperature above a certain base temperature. Refer to Chapter 12 for more information on the maturity concept. It follows that concrete should be protected so that its temperature remains favorable for hydration and moisture is not lost during the early hardening period.

CURING METHODS AND MATERIALS

Concrete can be kept moist (and in some cases at a favorable temperature) by three curing methods:

1. Methods that maintain the presence of mixing water in the concrete during the early hardening period. These include ponding or immersion, spraying or fogging, and saturated wet coverings. These methods afford some cooling through evaporation, which is beneficial in hot weather.

2. Methods that prevent loss of mixing water from the concrete by sealing the surface. This can be done by covering the concrete with impervious paper or plastic sheets, or by applying membrane-forming curing compounds.

3. Methods that accelerate strength gain by supplying heat and additional moisture to the concrete. This is usually accomplished with live steam, heating coils, or electrically heated forms or pads.

The method or combination of methods chosen depends on factors such as availability of curing materials, size and shape of concrete, production facilities (in place or in a plant), esthetic appearance, and economics.

Ponding or Immersion

On flat surfaces such as pavements and floors, concrete can be cured by ponding. Earth or sand dikes around the perimeter of the concrete surface can retain a pond of water. Ponding is an ideal method for preventing loss of moisture and is also effective for maintaining a uniform temperature in the concrete. The curing water should not be more than about 20°F cooler than the concrete to prevent thermal stresses that could result in cracking. Since ponding requires considerable labor and supervision, the method is generally used only for small jobs.

The most thorough method of curing with water consists of total immersion of the finished concrete element. This method is commonly used in the laboratory for curing concrete test specimens. Where appearance of the concrete is important, the water used for curing by ponding or immersion must be free of substances that will stain or discolor the concrete. The material used for dikes may also discolor the concrete.

Spraying or Fogging

Continuous spraying or fogging with water is an excellent method of curing when the ambient temperature is well above freezing and the humidity is very low. A fine mist should be applied continuously through a system of nozzles or sprayers. Ordinary lawn sprinklers are effective if good coverage is provided and water runoff is of no concern. Soil-soaker hoses are useful on surfaces that are vertical or nearly so.

The cost of spraying or fogging may be a disadvantage. The method requires an ample water supply and careful supervision. If spraying or fogging is done at intervals, the concrete must be prevented from drying between applications of water because alternate cycles of wetting and drying can cause surface crazing or cracking. Care must also be taken that water erosion of the newly finished concrete does not occur.

Wet Coverings

Fabric coverings saturated with water, such as burlap, cotton mats, rugs, or other moisture-retaining fabrics, are commonly used for curing (Fig. 10-2). Treated burlaps that reflect light and are resistant to rot and fire are available. The requirements for burlap are described in the Specification for Burlap Cloths Made from Jute or Kenaf (American Association of State Highway and Transportation Officials M182), and those for white burlap-polyethylene sheeting are described in ASTM C171.

Burlap must be free of sizing or any substance that is harmful to concrete or causes discoloration. New burlap should be thoroughly rinsed in water to remove soluble substances and make the burlap more absorbent.

Wet, moisture-retaining fabric coverings should be placed as soon as the concrete has hardened sufficiently to prevent surface damage. Care should be taken to cover the entire surface, including the edges of slabs. The coverings should be kept continuously moist so

Fig. 10-2. Wet burlap kept saturated with water is an effective method for moist-curing concrete.

that a film of water remains on the concrete surface throughout the curing period. Use of polyethylene film over burlap will eliminate the need for continuous watering of the covering. Alternate cycles of wetting and drying during the early curing period may cause crazing of the surface.

Wet coverings of earth, sand, or sawdust are effective for curing and are often useful on small jobs.* A layer about 2 in. thick should be evenly distributed over the previously moistened surface of the concrete and kept continuously wet.

Wet hay or straw can be used to cure flat surfaces. If used, it should be placed in a layer at least 6 in. thick and held down with wire screen, burlap, or tarpaulins to prevent its being blown off by wind.

A major disadvantage of moist earth, sand, sawdust, hay, or straw coverings is the possibility of discoloring the concrete.

Impervious Paper

Impervious paper for curing concrete consists of two sheets of kraft paper cemented together by a bituminous adhesive with fiber reinforcement. Such paper, conforming to ASTM C171, is an efficient means of curing horizontal surfaces and structural concrete of relatively simple shapes. An important advantage of this method is that periodic additions of water are not required. Curing with impervious paper assures suitable hydration of cement by preventing loss of moisture from the concrete (Fig. 10-3).

As soon as the concrete has hardened sufficiently to prevent surface damage, it should be thoroughly wetted and the widest paper available applied. Edges of adjacent sheets should be overlapped about 6 in. and tightly sealed with sand, wood planks, pressure-sensitive tape, mastic, or glue. The sheets must be weighted

to maintain close contact with the concrete surface during the entire curing period.

Impervious paper can be reused if it effectively retains moisture. Tears and holes can easily be repaired with curing-paper patches. When the condition of the paper is questionable, additional use can be obtained by using it in double thickness.

Impervious paper provides some protection to the concrete against damage from subsequent construction activity as well as protection from the direct sun. It should be light in color and nonstaining to the concrete. Paper with a white upper surface is preferable for curing exterior concrete during hot weather.

Plastic Sheets

Plastic sheet materials such as polyethylene film can be used to cure concrete (Fig. 10-4). Polyethylene film is a lightweight, effective moisture barrier and is easily applied to complex as well as simple shapes. Its application is the same as described for impervious paper.

Curing with polyethylene film (or impervious paper) can cause patchy discoloration, especially if the concrete contains calcium chloride and has been finished by hard-steel troweling. This discoloration is more pronounced when the film becomes wrinkled, but it is difficult and time consuming on a large project to place sheet materials without wrinkles. Discoloration may be prevented by flooding the surface under the covering, but other means of curing should be used when uniform color is important.

Polyethylene film should conform to ASTM C171, which specifies a 4-mil thickness for curing concrete and lists only clear and white opaque film. However,

---

*Sawdust from most woods is acceptable, but oak and other woods that contain tannic acid should not be used since deterioration of the concrete may occur.

Fig. 10-3. Impervious curing paper is an efficient means of curing horizontal surfaces.

Fig. 10-4. Polyethylene film is an effective moisture barrier for curing concrete and easily applied to complex as well as simple shapes. To minimize discoloration, the film should be kept as flat as possible on the concrete surface.

black film is available and is satisfactory under some conditions. White film should be used for exteriors during hot weather to reflect the sun's rays. Black film can be used during cool weather or for interior locations. Clear film has little effect on heat absorption.

ASTM C171 also includes a sheet material consisting of burlap impregnated on one side with white opaque polyethylene film. Combinations of polyethylene film bonded to absorbent fabric such as burlap help retain moisture on the concrete surface.

Polyethylene film may also be placed over wet burlap or other wet covering materials to retain the water in the wet covering material. This procedure eliminates the labor-intensive need for continuous watering of wet covering materials.

Membrane-Forming Curing Compounds

Liquid membrane-forming compounds consisting of waxes, resins, chlorinated rubber, and solvents of high volatility can be used to retard or reduce evaporation of moisture from concrete. They are suitable not only for curing freshly placed concrete but also for extending curing of concrete after removal of forms or after initial moist curing.

Membrane-forming curing compounds are of two general types: clear, or translucent; and white pigmented. Clear or translucent compounds may contain a fugitive dye that makes it easier to check visually for complete coverage of the concrete surface when the compound is applied. The dye fades away soon after application. On hot, sunny days, use of white-pigmented compounds reduces solar-heat gain, thus reducing the concrete temperature. Pigmented compounds should be kept agitated in the container to prevent pigment from settling out.

Curing compounds should be applied by hand-operated or power-driven spray equipment immediately after final finishing of the concrete (Fig. 10-5). The concrete surface normally should be damp when the coating is applied. On dry, windy days or during periods when adverse weather conditions could result in plastic shrinkage cracking, application of curing compound immediately after final finishing and before all free water on the surface has evaporated will help prevent the formation of cracks. Power-driven spray equipment is recommended for uniform application of curing compounds on large paving projects. Spray nozzles and windshields on such equipment should be arranged to prevent wind-blown loss of curing compound.

Normally only one smooth, even coat is applied at a rate of 150 to 200 sq ft per gallon, but two coats may be necessary to ensure complete coverage and effective protection. The second coat, when used, should be applied at right angles to the first. Complete coverage of the surface must be attained because even small pinholes in the membrane will permit evaporation of some moisture from the concrete.

Curing compounds can prevent bond between hardened and freshly placed concrete or other floor surfacing materials. Consequently, they should not be used

Fig. 10-5. Liquid membrane-forming curing compounds should be applied with uniform and adequate coverage over the entire surface for effective, extended curing of concrete.

when subsequent bonding is necessary. For example, a curing compound should not be applied to the base slab of a two-course floor. Similarly, some curing compounds affect the adhesion of paint or resilient flooring materials to concrete floors. Curing compound manufacturers should be consulted to determine if their product is in this category.

Curing compounds should be uniform and easy to maintain in a thoroughly mixed solution. They should not sag, run off peaks, or collect in grooves. They should form a tough film to withstand early construction traffic without damage, be nonyellowing, and have good moisture-retention properties.

Curing compounds should conform to ASTM C 309. A method for determining the efficiency of curing compounds, waterproof paper, and plastic sheets is described in ASTM C156.

Forms Left in Place

Forms provide satisfactory protection against loss of moisture if the top exposed concrete surfaces are kept wet. A soil-soaker hose is excellent for this. The forms should be left on the concrete as long as practical.

Wood forms left in place should be kept moist by sprinkling, especially during hot, dry weather. If this can not be done, they should be removed as soon as practical and another curing method started without delay.

Steam Curing

Steam curing is advantageous where early strength gain in concrete is important or where additional heat is required to accomplish hydration, as in cold weather.

Two methods of steam curing are used: live steam at atmospheric pressure (for enclosed cast-in-place structures and large precast concrete units) and high-pressure steam in autoclaves (for small manufactured units).

A steam-curing cycle consists of (1) an initial delay prior to steaming, (2) a period for increasing the temperature, (3) a period for holding the maximum temperature constant, and (4) a period for decreasing the temperature. An optimum atmospheric steam-curing cycle is shown in Fig. 10-6.

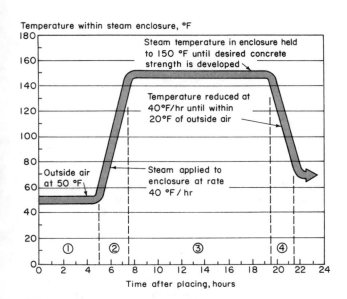

① Initial delay prior to steaming 2 to 5 hours
② Temperature increase period   2 1/2 hours
③ Constant temperature period   6 to 12 hours*
④ Temperature decrease period   2 hours
 * Type III or high-early-strength cement, longer for other types

Fig. 10-6. A typical atmospheric steam-curing cycle (idealized).

Steam curing at atmospheric pressure is generally done in an enclosure to minimize moisture and heat losses. Tarpaulins are frequently used to form the enclosure. Application of steam to the enclosure should be delayed at least 2 hours after final placement of concrete to allow for some hardening of the concrete. However, a 3- to 5-hour delay period prior to steaming will achieve maximum early strength, as shown in Fig. 10-7. Steam temperature in the enclosure should be kept at about 150°F until the desired concrete strength has developed. Strength will not increase significantly if the maximum steam temperature is raised from 150°F to 175°F. Steam-curing temperatures above 180°F should be avoided; they are uneconomical and may result in undue reduction in ultimate strength. Besides early strength gain, other advantages of curing concrete at temperatures of around 150°F are reduced drying shrinkage and creep as compared to concrete cured at 70°F for 28 days.*

Excessive rates of heating and cooling should be avoided to prevent damaging volume changes. Temperatures in the enclosure surrounding the concrete should not be increased or decreased more than 40°F

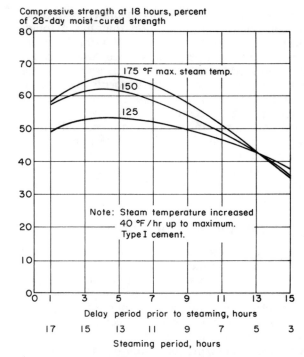

Fig. 10-7. Relationship between strength at 18 hours and delay period prior to steaming. In each case, the delay period plus the steaming period totaled 18 hours. Reference 10-5.

to 60°F per hour depending on the size and shape of the concrete element. The temperature of the concrete prior to casting can be increased up to 140°F by injecting live steam into the mixture along with the addition of mixing water, obviating the need for a delay and temperature increase period. Handling time, however, may be reduced.

The maximum curing temperature in the enclosure should be held until the concrete has reached the desired strength. The time required will depend on the concrete mixture and steam temperature in the enclosure.**

High-pressure steam curing in autoclaves takes advantage of temperatures ranging from 325°F to 375°F and corresponding pressures of 80 to 170 lb per square inch gage. Hydration is greatly accelerated and the elevated temperatures and pressures can produce additional beneficial chemical reactions between aggregates and cementitious materials not occurring under atmospheric steam curing. Autoclaving can produce within a few hours concrete strengths equal to those obtained in concrete moist-cured for 28 days at 70°F.

Insulating Blankets or Covers

Layers of dry, porous material such as straw or hay can be used to provide insulation against freezing of concrete when temperatures fall below 32°F.

---

*Reference 10-4.
**See Reference 10-14 for information on specific steam-curing applications.

Formwork can be economically insulated with commercial blanket or batt insulation that has a tough moistureproof covering. Suitable insulating blankets are manufactured of fiberglass, sponge rubber, cellulose fibers, mineral wool, vinyl foam, and open-cell polyurethane foam. When insulated formwork is used, care should be taken to ensure that concrete temperatures do not become excessive.

Framed enclosures of canvas, reinforced polyethylene film, or other materials can be placed around the structure and heated by space heaters or steam.

Curing concrete in cold weather should follow the recommendations in Chapter 12 and ACI 306, *Cold-Weather Concreting*.

Electrical, Oil, and Infrared Curing

Electrical, hot oil, and infrared curing methods have been available for accelerated and normal curing of concrete for many years. Electrical curing methods include a variety of techniques: use of the concrete itself as the electrical conductor, use of reinforcing steel as the heating element, use of a special wire as the heating element, electric blankets, and the use of electrically heated steel forms (presently the most popular method). Electrical heating is especially useful in cold-weather concreting. Hot oil may be circulated through steel forms to heat the concrete. Infrared rays have had limited use in accelerated curing of concrete. Concrete that is cured by infrared methods is usually under a covering or is enclosed in steel forms. Electrical, oil, and infrared curing methods are used primarily in the precast concrete industry.

CURING PERIOD AND TEMPERATURE

The period of time that concrete should be protected against loss of moisture depends upon the type of cement, mixture proportions, required strength, size and shape of the concrete member, ambient weather, and future exposure conditions. The period may be 3 weeks or longer for lean concrete mixtures used in massive structures such as dams; conversely, it may be only a few days for rich mixes, especially if Type III cement is used. Steam-curing periods are normally much shorter, ranging from 3 hours to 3 days; but generally about 24-hour cycles are used.

Since all the desirable properties of concrete are improved by curing, the curing period should be as long as practical. For concrete slabs on ground (floors, pavements, canal linings, parking lots, driveways, sidewalks) and for structural concrete (cast-in-place walls, columns, slabs, beams, small footings, piers, retaining walls, bridge decks), the length of the curing period for ambient temperatures above 40°F should be a minimum of 7 days or the time necessary to attain 70% of the specified compressive or flexural strength. A higher curing temperature provides earlier strength gain in concrete than a lower temperature but it may decrease 28-day strength as shown in Fig. 10-8. If strength tests are made to establish the time when curing can cease

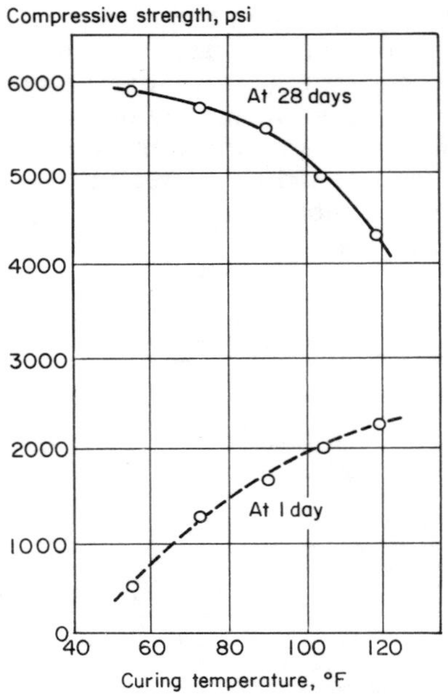

Fig. 10-8. One-day strength increases with increasing curing temperature but 28-day strength decreases with increasing curing temperature. Reference 10-13.

or forms can be removed, representative concrete cylinders or beams should be made in the field, kept adjacent to the structure or pavement they represent, and cured by the same methods. Cores, cast-in-place removable cylinders, and nondestructive testing methods may also be used to determine the strength of the concrete member. Natural curing above 50°F (rain, mist, high humidity, moist backfill, and so on) may be ample if it is equivalent to keeping Type I cement concrete moist for 7 days, Type II cement concrete moist for 14 days, and Type III moist for 3 days.

Since the rate of hydration is influenced by cement composition and fineness, the curing period should be prolonged for concretes made with cements possessing slow-strength-gain characteristics. For mass concrete (in large piers, locks, abutments, dams, heavy footings, and massive columns and transfer girders) containing no pozzolan as part of cementitious material, curing of unreinforced sections should continue for at least 2 weeks. If the mass concrete contains a pozzolan, minimum curing time for unreinforced sections should be extended to 3 weeks. Heavily reinforced mass concrete sections should be cured for a minimum of 7 days.

During cold weather, additional heat is often required to maintain favorable curing temperatures of 50°F to 70°F. This can be supplied by vented gas- or oil-fired heaters, heating coils, or live steam. In all cases, care must be taken to avoid loss of moisture from the concrete. Exposure of fresh concrete to heater or engine exhaust gases must be avoided as this can result in surface deterioration and dusting (rapid carbonation).

High-early-strength concrete can be used in cold weather to speed setting time and strength development. This can reduce the curing period from 7 to 3 days, but a minimum temperature of 50°F must be maintained in the concrete for 3 days.

For adequate scale resistance of concrete to chemical deicers, the minimum curing period generally corresponds to the time required to develop the design strength of the concrete. A period of air drying, which enhances resistance to scaling, should then elapse before application of deicing salts. This drying period should be at least 1 month.

REFERENCES

10-1. Powers, T. C., *A Discussion of Cement Hydration in Relation to the Curing of Concrete,* Research Department Bulletin RX025, Portland Cement Association, 1948.

10-2. Lerch, William, *Plastic Shrinkage,* Research Department Bulletin RX081, Portland Cement Association, 1957.

10-3. Klieger, Paul, *Curing Requirements for Scale Resistance of Concrete,* Research Department Bulletin RX082, Portland Cement Association, 1957.

10-4. Klieger, Paul, *Some Aspects of Durability and Volume Change of Concrete for Prestressing,* Research Department Bulletin RX118, Portland Cement Association, 1960.

10-5. Hanson, J. A., *Optimum Steam Curing Procedure in Precasting Plants,* with discussion, Development Department Bulletins DX062 and DX062A, Portland Cement Association, 1963.

10-6. *Curing of Concrete, 1925-1960,* Bibliography 32, Highway Research Board, Washington, D.C., 1963.

10-7. *Curing of Concrete Pavements,* Current Road Problems No. 1-2R, Highway Research Board, May 1963.

10-8. Hanson, J. A., *Optimum Steam Curing Procedures for Structural Lightweight Concrete,* Development Department Bulletin DX092, Portland Cement Association, 1965.

10-9. ACI Committee 516 Report, "High Pressure Steam Curing: Modern Practice and Properties of Autoclaved Products," *Proceedings of the American Concrete Institute,* American Concrete Institute, Detroit, August 1965, pages 869-908.

10-10. Proudley, C. E., "Curing Materials," *Significance of Tests and Properties of Concrete and Concrete-Making Materials,* STP 169-A, American Society for Testing and Materials, Philadelphia, 1966, pages 522-529.

10-11. Greening, N. R., and Landgren, R., *Surface Discoloration of Concrete Flatwork,* Research Department Bulletin RX203, Portland Cement Association, 1966.

10-12. McCoy, W. J., "Mixing and Curing Water for Concrete," *Significance of Tests and Properties of Concrete and Concrete-Making Materials,* STP 169-A, American Society for Testing and Materials, 1966, pages 515-521.

10-13. Verbeck, George J., and Helmuth, R. A., "Structures and Physical Properties of Cement Pastes," *Proceedings, Fifth International Symposium on the Chemistry of Cement,* vol. III, The Cement Association of Japan, Tokyo, 1968, page 9.

10-14. *Accelerated Curing of Concrete at Atmospheric Pressure—State of the Art,* ACI 517.2R-80, ACI Committee 517 Report, American Concrete Institute, 1980.

10-15. *Hot-Weather Concreting,* ACI 305R-77, revised 1982, ACI Committee 305 Report, American Concrete Institute.

10-16. *Cold-Weather Concreting,* ACI 306R-78, revised 1983, ACI Committee 306 Report, American Concrete Institute.

10-17. *Standard Practice for Curing Concrete,* ACI 308-81, revised 1986, ACI Committee 308 Report, American Concrete Institute.

CHAPTER 11
Hot-Weather Concreting

Weather conditions at a jobsite—hot or cold, windy or calm, dry or humid—may be vastly different from the optimum conditions assumed at the time a concrete mix is specified, designed, or selected. Hot weather can create difficulties in fresh concrete, such as

—increased water demand
—accelerated slump loss
—increased rate of setting
—increased tendency for plastic cracking
—difficulties in controlling entrained air
—critical need for prompt early curing

Adding water to the concrete at the jobsite can adversely affect properties and serviceability of the hardened concrete, resulting in

—decreased strength
—decreased durability and watertightness
—nonuniform surface appearance
—increased tendency for drying shrinkage

Only by taking precautions to alleviate these difficulties in anticipation of hot-weather conditions can concrete work proceed smoothly.

WHEN TO TAKE PRECAUTIONS

The most favorable temperature for freshly mixed concrete is lower during hot weather than can usually be obtained without artificial cooling. A concrete temperature of 50°F to 60°F is desirable but not always practical. Many specifications require only that concrete when placed should have a temperature of less than 85°F or 90°F.

For most work it is impractical to limit the maximum temperature of concrete as placed because circumstances vary widely. A limit that would serve successfully at one jobsite could be highly restrictive at another. For example, flatwork done under a roof with exterior walls in place could be completed at a concrete temperature that would cause difficulty were the same concrete placed outdoors on the same day.

The effects of a high concrete temperature should be anticipated and the concrete placed at a temperature limit that will allow best results in hot-weather conditions, probably somewhere between 75°F and 100°F. The limit should be established for conditions at the jobsite based on trial-batch tests at the limiting temperature rather than at ideal temperatures.

EFFECTS OF HIGH CONCRETE TEMPERATURES

As concrete temperature increases there is a loss in slump that is often unadvisedly compensated for by adding more water at the jobsite. At higher temperatures a greater amount of water is required to hold slump constant than is needed at lower temperatures. Adding water *without* adding cement results in a higher water-cement ratio, thereby lowering the strength at all ages and adversely affecting other desirable properties of the hardened concrete. This is in addition to the adverse effect on strength at later ages due to the higher temperature even without the addition of water.

As shown in Fig. 11-1, if the temperature of freshly mixed concrete is increased from 50°F to 100°F, about 33 lb of additional water is needed per cubic yard of concrete to maintain the same 3-in. slump. This addi-

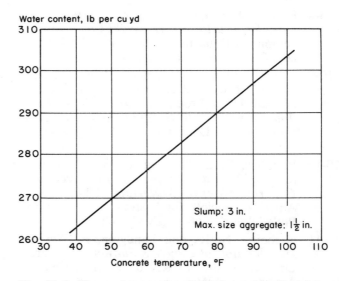

Fig. 11-1. The water requirement of a concrete mixture increases with an increase in concrete temperature. Reference 11-6.

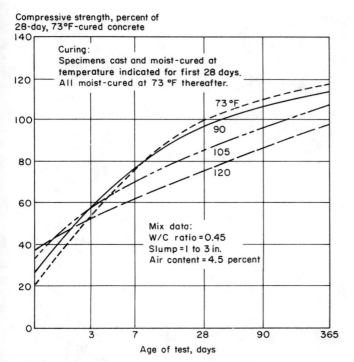

Fig. 11-2. Effect of high concrete temperatures on compressive strength at various ages. Reference 11-3.

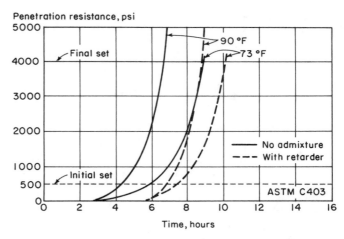

Fig. 11-3. Effect of concrete temperature and retarder on setting time. Reference 11-5.

tional water could reduce strength by 12% to 15% and produce a compressive strength cylinder test break of less than the specified compressive strength (f'_c).

Fig. 11-2 shows the effect of high initial concrete temperatures on compressive strength. The concrete temperatures at time of mixing, casting, and curing were 73°F, 90°F, 105°F, and 120°F. After 28 days, the specimens were all moist-cured at 73°F until the 90-day and one-year test ages. The tests, using identical concretes of the same water-cement ratio, show that while higher concrete temperatures give higher early strength than concrete at 73°F, at later ages concrete strengths are lower. If the water content had been increased to maintain the same slump (without increasing cement content), the reduction in strength would have been even greater than shown.

Besides reducing strength and increasing the mixing-water requirement, high temperatures of freshly mixed concrete have other harmful effects. Setting time is significantly reduced—high temperatures increase the rate of concrete hardening and shorten the length of time within which the concrete can be transported, placed, and finished. Setting time can be reduced by 2 or more hours with a 20°F increase in concrete temperature (Fig. 11-3). Concrete should remain plastic sufficiently long so that each layer can be placed without development of cold joints or discontinuities in the concrete. Retarding admixtures, ASTM C 494 Type B, can be beneficial in offsetting the accelerating effects of high temperature.

In hot weather, the tendency for cracks to form is increased both before and after hardening. Rapid evaporation of water from freshly placed concrete can cause plastic-shrinkage cracks before the surface has hard-

ened (discussed in more detail later in this chapter). Cracks may also develop in the hardened concrete because of increased drying shrinkage due to a higher water content or thermal volume changes at the surface due to cooling.

Air entrainment is also affected in hot weather. At elevated temperatures, an increase in the amount of air-entraining admixture is required to produce a given air content.

Because of the detrimental effects of high concrete temperatures, operations in hot weather should be directed toward keeping the concrete as cool as is practicable.

COOLING CONCRETE MATERIALS

The usual method of cooling concrete is to lower the temperature of the concrete materials before mixing. One or more of the ingredients should be cooled. In hot weather the aggregates and water should be kept as cool as practicable, as these materials have a greater influence on temperature after mixing than other components.

The contribution of each material in a concrete mixture to the initial temperature of concrete is related to the temperature, specific heat, and quantity of each material. Fig. 11-4 shows graphically the effect of temperature of materials on the temperature of fresh concrete. It is evident that although concrete temperature is primarily dependent upon the aggregate temperature, it can be effectively lowered by cooling the mixing water.

The approximate temperature of concrete can be calculated from the temperatures of its ingredients by using the following equation:*

$$T = \frac{0.22(T_a W_a + T_c W_c) + T_w W_w + T_{wa} W_{wa}}{0.22(W_a + W_c) + W_w + W_{wa}}$$

---

*Reference 11-4.

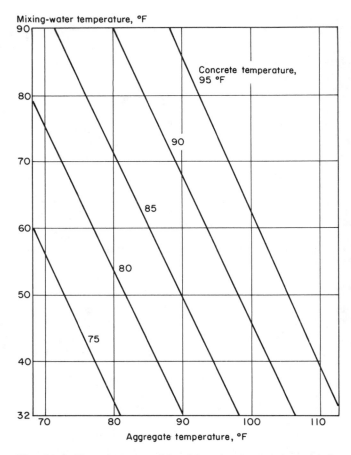

Mixing-water temperature, °F

Concrete temperature, 95 °F

Aggregate temperature, °F

Fig. 11-4. Temperature of freshly mixed concrete as affected by temperature of its ingredients. Although the chart is based on the following mixture, it is reasonably accurate for other typical mixtures:

Aggregate	3000 lb
Moisture in aggregate	60 lb
Added mixing water	240 lb
Cement at 150°F	564 lb

where

T = temperature of the freshly mixed concrete

T_a, T_c, T_w, and T_{wa} = temperature of aggregates, cement, added mixing water, and free water on aggregates, respectively

W_a, W_c, W_w, and W_{wa} = weight of aggregates, cement, added mixing water, and free water on aggregates, respectively

Example calculations for initial concrete temperature are shown in Table 11-1A.

Of the materials in concrete, water is the easiest to cool. Even though it is used in smaller quantities than the other ingredients, cold water will effect a moderate reduction in the concrete temperature. Mixing water from a cool source should be used. It should be stored in tanks that are not exposed to the direct rays of the sun. Tanks and pipelines carrying the mixing water should be buried, insulated, shaded, or painted white to keep water at the lowest practical temperature.

Fig. 11-5. Liquid nitrogen water-cooling installation at a ready mixed concrete plant.

Water can be cooled by refrigeration, liquid nitrogen, or ice. Fig. 11-5 shows a liquid nitrogen water-cooling installation at a ready mixed concrete plant. Liquid nitrogen can also be added directly into a central mixer drum or the drum of a truck mixer to lower concrete temperature. Ice can be used as part of the mixing water provided it is completely melted by the time mixing is completed. When using crushed ice, care must be taken to store it at a temperature that will prevent the formation of lumps.

When ice is added as part of the mixing water, the effect of the heat of fusion of the ice must be considered, so the equation for temperature of fresh concrete is modified as follows:

$$T = \frac{0.22(T_a W_a + T_c W_c) + T_w W_w + T_{wa} W_{wa} - 112 W_i}{0.22(W_a + W_c) + W_w + W_{wa} + W_i}$$

where W_i is the weight in pounds of ice.*

The heat of fusion of ice in British thermal units is 144 Btu per pound. Calculations in Table 11-1B show the effect of 75 lb of ice in reducing the temperature of concrete. Crushed or flaked ice is more effective than chilled water in reducing concrete temperature. The

*Reference 11-4.

132

Table 11-1.

A. Effect of Temperature of Materials on Initial Concrete Temperatures

Material	Weight, W, lb/cu yd	Specific heat	Btu to vary temperature 1°F	Initial temperature of material, T, °F	Total Btu's in material
	(1)	(2)	(3) Col. 1 × Col. 2	(4)	(5) Col. 3 × Col. 4
Cement	564 (W_c)	0.22	124	150 (T_c)	18,600
Water	282 (W_w)	1.00	282	80 (T_w)	22,560
Total aggregate	3100 (W_a)	0.22	682	80 (T_a)	54,560
			1088		95,720

Initial concrete temperature $= \dfrac{95,720}{1088} = 88.0°F$

To achieve 1°F reduction in initial concrete temperature:

Cement temperature must be lowered $= \dfrac{1088}{124} = 9°F$

Or water temperature dropped $= \dfrac{1088}{282} = 3.8°F$

Or aggregate temperature cooled $= \dfrac{1088}{682} = 1.6°F$

B. Effect of Ice (75 lb) on Temperature of Concrete

Material	Weight, W, lb/cu yd	Specific heat	Btu to vary temperature 1°F	Initial temperature of material, T, °F	Total Btu's in material
	(1)	(2)	(3) Col. 1 × Col. 2	(4)	(5) Col. 3 × Col. 4
Cement	564 (W_c)	0.22	124	150 (T_c)	18,600
Water	207 (W_w)	1.00	207	80 (T_w)	16,560
Total aggregate	3100 (W_a)	0.22	682	80 (T_a)	54,560
Ice*	75 (W_i)	1.00	75	32 (T_i)	2,400
			1088		
minus	75 (W_i) × heat of fusion, 144 Btu/lb =				−10,800
					81,320

Concrete temperature $= \dfrac{81,320}{1088} = 74.7°F$

*$32\,W_i - 144\,W_i = -112\,W_i$.

amount of water and ice must not exceed the total mixing-water requirements. Fig. 11-6 shows crushed ice being charged into a truck mixer prior to the addition of other materials.

Aggregates have a pronounced effect on the fresh concrete temperature because they represent 70% to 85% of the total weight of concrete. To lower the temperature of concrete 10°F requires only a 15°F reduction in the temperature of the aggregates.

There are several simple methods of keeping aggregates cool. Stockpiles should be shaded from the sun and kept moist by sprinkling. Since evaporation is a cooling process, sprinkling provides effective cooling, especially when the relative humidity is low.

Sprinkling of coarse aggregates should be adjusted to avoid producing excessive variations in the free moisture content and thereby causing a loss of slump uniformity. Refrigeration is another method of cooling materials. Aggregates can be immersed in cold-water tanks, or cooled air can be circulated through storage bins. Vacuum cooling can reduce aggregate temperatures to as low as 34°F.

Fig. 11-6. Substituting ice for part of the mixing water will substantially lower concrete temperature. A crusher delivers finely crushed ice to a truck mixer reliably and quickly.

Cement temperature has only a minor effect on the temperature of the freshly mixed concrete because of cement's low specific heat and the relatively small amount of cement in the mixture. A cement temperature change of 10°F generally will change the concrete temperature by only 1°F. Because cement loses heat slowly during storage, it may be warm when delivered. (This heat is produced in grinding the cement clinker during manufacture.) Since the temperature of cement does affect the temperature of the fresh concrete to some extent, some specifications place a limit on its temperature at the time of use. The ACI 305 recommended maximum temperature at which cement should be used in concrete is 170°F.* However, it is preferable to specify a maximum temperature for freshly mixed concrete rather than place a temperature limit on individual ingredients.**

PREPARATION BEFORE CONCRETING

Before concrete is placed, certain precautions should be taken during hot weather to maintain or reduce concrete temperature. Mixers, chutes, belts, hoppers, pump lines, and other equipment for handling concrete should be shaded, painted white, or covered with wet burlap to reduce solar heat.

Forms, reinforcing steel, and subgrade should be fogged or sprinkled with cool water just before the concrete is placed. Fogging the area during placing and finishing operations cools the contact surfaces and surrounding air and increases its relative humidity. This reduces the temperature rise of the concrete and minimizes the rate of evaporation of water from the concrete after placement. For slabs on ground, it is a good practice to moisten the subgrade the evening before concreting. There should be no standing water or puddles on forms or subgrade at the time concrete is placed.

During extremely hot periods, improved results can be obtained by restricting concrete placement to early morning, evening, or nighttime hours, especially in arid climates. This practice has resulted in less thermal shrinkage and cracking of thick slabs and pavements.

TRANSPORTING, PLACING, FINISHING

Transporting and placing concrete should be done as quickly as is practical during hot weather. Delays contribute to loss of slump and an increase in concrete temperature. Sufficient labor and equipment must be available at the jobsite to handle and place concrete immediately upon delivery.

Prolonged mixing, even at agitating speed, should be avoided. If delays occur, the heat generated by mixing can be minimized by stopping the mixer and then agitating intermittently. The Standard Specification for Ready Mixed Concrete (ASTM C 94) requires that discharge of concrete be completed within 1½ hours or before the drum has revolved 300 times, whichever

occurs first. During hot weather the time limit can be reasonably reduced to 1 hour or even 45 minutes.

Since concrete hardens more rapidly in hot weather, extra care must be taken with placement techniques to avoid cold joints. For placement of walls, shallower layers can be specified to assure enough time for consolidation with the previous lift. Temporary sunshades and windbreaks help to minimize cold joints.

Floating should be done promptly after the water sheen disappears from the surface or when the concrete can support the weight of a finisher. Finishing on dry and windy days requires extra care. Rapid drying of the concrete at the surface may cause plastic shrinkage cracking.

PLASTIC SHRINKAGE CRACKING

Plastic shrinkage cracks are cracks that sometimes occur in the surface of freshly mixed concrete soon after it has been placed and while it is being finished (Fig. 11-7). These cracks appear mostly on horizontal surfaces and can be substantially eliminated if preventive measures are taken.

Plastic shrinkage cracking is usually associated with hot-weather concreting; however, it can occur any time ambient conditions produce rapid evaporation of moisture from the concrete surface. Such cracks occur when water evaporates from the surface faster than it can appear at the surface during the bleeding process. This creates rapid drying shrinkage and tensile stresses in the surface that often result in short, irregular cracks. The following conditions, singly or collectively, increase evaporation of surface moisture and increase the possibility of plastic shrinkage cracking:

1. High air and concrete temperature
2. Low humidity
3. High winds

Fig. 11-7. Typical plastic shrinkage cracks.

---

*Reference 11-7, page 8.
**See Reference 11-1 for more information.

The crack length is generally from a few inches to a few feet in length and they are usually spaced in an irregular pattern from a few inches to 2 ft apart. Fig. 11-8 is useful for determining when precautionary measures should be taken. There is no way to predict with certainty when plastic shrinkage cracking will occur. However, when the rate of evaporation exceeds 0.2 lb per square foot per hour, precautionary measures are almost mandatory. Cracking is possible if the rate of evaporation exceeds 0.1 lb per square foot per hour.

The simple precautions listed below can minimize the possibility of plastic shrinkage cracking. They should be considered while planning for hot-weather-concrete construction or while dealing with the problem after construction has started. They are listed in the order in which they should be done during construction.

1. Moisten the subgrade and forms.
2. Moisten concrete aggregates that are dry and absorptive.
3. Erect temporary windbreaks to reduce wind velocity over the concrete surface.
4. Erect temporary sunshades to reduce concrete surface temperatures.
5. Keep the freshly mixed concrete temperature low by cooling the aggregates and mixing water.

6. Protect the concrete with temporary coverings, such as polyethylene sheeting, during any appreciable delay between placing and finishing. Evaporation retarders (usually polymers) can be spray-applied immediately after screeding to retard water evaporation before final finishing operations and curing commence. These materials are floated and troweled into the surface during finishing and should have no adverse effect on the concrete or inhibit the adhesion of membrane-curing compounds.
7. Reduce time between placing and start of curing by eliminating delays during construction.
8. Protect the concrete immediately after final finishing to minimize evaporation. This is most important to avoid cracking. Use of a fog spray to raise the relative humidity of the ambient air is an effective means of preventing evaporation from the concrete. Fogging should be continued until a suitable curing material such as a curing compound, wet burlap, or curing paper can be applied.

If plastic shrinkage cracks should appear during finishing, the cracks can be closed by striking each side of the crack with a float and refinishing. However, the cracking may recur unless the causes are corrected.

CURING AND PROTECTION

Curing and protection are more critical in hot and cold weather than in temperate periods. Retaining forms in place cannot be considered a satisfactory substitute for curing in hot weather; they should be loosened as soon as practical without damage to the concrete. Water should then be applied at the top exposed concrete surfaces, for example, with a soil-soaker hose, and allowed to run down inside the forms. On hardened concrete and on flat concrete surfaces in particular, curing water should not be excessively cooler than the concrete. This will minimize cracking caused by thermal stresses due to temperature differentials between the concrete and curing water.

The need for moist curing is greatest during the first few hours after finishing. To prevent the drying of exposed concrete surfaces, moist curing should commence as soon as the surfaces are finished and continue for at least 24 hours. In hot weather, continuous moist curing for the entire curing period is preferred. However, if moist curing can not be continued beyond 24 hours, the concrete surfaces should be protected from drying with curing paper, heat-reflecting plastic sheets, or membrane-forming curing compounds while the surfaces are still damp. Moist-cured surfaces should dry out slowly after the curing period to reduce the possibility of surface crazing* and cracking.

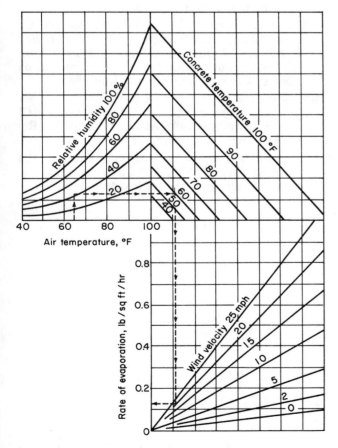

Fig. 11-8. Effect of concrete and air temperatures, relative humidity, and wind velocity on rate of evaporation of surface moisture from concrete. Reference 11-2.

---

*Crazing, a network pattern of fine cracks that do not penetrate much below the surface, is caused by minor surface shrinkage. Crazing cracks are very fine and barely visible except when the concrete is drying after the surface has been wet. The cracks encompass small concrete areas less than 2 in. in dimension, forming a chicken-wire pattern.

White-pigmented curing compounds can be used on horizontal surfaces. Application of a curing compound during hot weather should be preceded by 24 hours of moist curing. If this is not practical, the compound should be applied immediately after final finishing. The concrete surfaces should be moist.

ADMIXTURES

For unusual cases in hot weather and where careful inspection is maintained, a retarding admixture may be beneficial in delaying the setting time (Fig. 11-3), despite the somewhat increased rate of slump loss generally resulting from their use.

Retarding admixtures should conform to the requirements of ASTM C 494 Type B. Admixtures should be tested with job materials under job conditions before construction begins in order to determine their compatibility with the basic concrete ingredients and their ability under the particular conditions to produce the desired properties.

HEAT OF HYDRATION

Heat generated during cement hydration raises the temperature of the concrete to a greater or lesser extent depending on the size of the concrete placement, its surrounding environment, and the amount of cement in the concrete. ACI 211.1 states that as a general rule a 10°F to 15°F temperature rise per 100 lb of cement can be expected from the heat of hydration.* There may be instances in hot-weather-concrete work and massive concrete placements when measures must be taken to cope with the generation of heat and attendant thermal volume changes to control cracking (see Chapters 13 and 15).

---

*Reference 11-9.

REFERENCES

11-1. Lerch, William, *Hot Cement and Hot Weather Concrete Tests*, TA015T, Portland Cement Association, 1955.

11-2. Menzel, Carl A., "Causes and Prevention of Crack Development in Plastic Concrete," *Proceedings of the Portland Cement Association*, 1954, pages 130–136.

11-3. Klieger, Paul, *Effect of Mixing and Curing Temperature on Concrete Strength*, Research Department Bulletin RX103, Portland Cement Association, 1958.

11-4. *Cooling Ready Mixed Concrete*, NRMCA Publication No. 106, National Ready Mixed Concrete Association, Silver Spring, Maryland, 1962.

11-5. Sprouse, J. H., and Peppler, R. B., "Setting Time," *Significance of Tests and Properties of Concrete and Concrete-Making Materials*, STP169B, American Society for Testing and Materials, Philadelphia, 1978, pages 105-121.

11-6. *Concrete Manual*, 8th ed., U.S. Bureau of Reclamation, Denver, revised 1981.

11-7. *Hot Weather Concreting*, ACI 305R-77, revised 1982, ACI Committee 305 Report, American Concrete Institute, Detroit.

11-8. Gaynor, Richard D.; Meininger, Richard C.; and Khan, Tarek S., *Effect of Temperature and Delivery Time on Concrete Proportions*, NRMCA Publication No. 171, National Ready Mixed Concrete Association, June 1985.

11-9. *Standard Practice for Selecting Proportions for Normal, Heavyweight, and Mass Concrete*, ACI 211.1-81, revised 1985, ACI Committee 211 Report, American Concrete Institute.

11-10. *Standard Practice for Curing Concrete*, ACI 308-81, revised 1986, ACI Committee 308 Report, American Concrete Institute.

CHAPTER 12
Cold-Weather Concreting

Concrete can be placed safely throughout the winter months in cold climates if certain precautions are taken. Cold weather is defined by ACI 306 as "a period when for more than 3 successive days the mean daily temperature drops below 40°F." Normal concreting practices can be resumed once the ambient temperature is above 50°F for more than half a day.

During cold weather, the concrete mixture and its temperature should be adapted to the construction procedure and ambient weather conditions. Preparations should be made to protect the concrete; enclosures, windbreaks, portable heaters, insulated forms, and blankets should be ready to maintain the concrete temperature. Forms, reinforcing steel, and embedded fixtures must be clear of snow and ice at the time concrete is placed. Thermometers and proper storage facilities for test cylinders should be available to verify that precautions are adequate.

Fig. 12-1. When suitable preparations have been made, cold weather is no obstacle to concrete construction.

EFFECT OF FREEZING FRESH CONCRETE

Concrete gains very little strength at low temperatures. Freshly mixed concrete must be protected against the disruptive effects of freezing until the degree of saturation of the concrete has been sufficiently reduced by the process of hydration. The time at which this reduction is accomplished corresponds roughly to the time required for the concrete to attain a compressive strength of 500 psi.* At normal temperatures, this occurs within the first 24 hours after placement. Significant ultimate strength reductions, up to about 50%, can occur if concrete is frozen within a few hours after placement or before it attains a compressive strength of 500 psi.**

Concrete that has been frozen just once at an early age can be restored to nearly normal strength by providing favorable subsequent curing conditions. Such concrete, however, will not be as resistant to weathering nor as watertight as concrete that had not been frozen. The critical period after which concrete is not seriously damaged by one or two freezing cycles is dependent upon the concrete ingredients and conditions of mixing, placing, curing, and subsequent drying. For example, air-entrained concrete is less susceptible to damage by early freezing than non-air-entrained

concrete. See Chapter 5 on "Freeze-Thaw Resistance" and "Resistance to Deicers and Salts" for more information.

STRENGTH GAIN OF CONCRETE AT LOW TEMPERATURES

Temperature affects the rate at which hydration of cement occurs—low temperatures retard hydration and consequently retard the hardening and strength gain of concrete.

If concrete is frozen and kept frozen above about 14°F, it will gain strength slowly. Below that temperature, cement hydration and concrete strength gain cease. Fig. 12-3 illustrates the effect of cool temperatures on setting time. Figs. 12-4 and 12-5 show the age-compressive strength relationship for concrete that has been

*Reference 12-7.
**Reference 12-1.

137

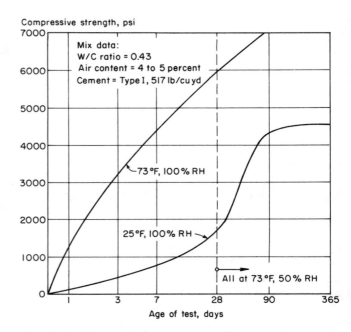

Fig. 12-4. Effect of temperature conditions on the strength development of concrete. Specimens for the lower curve were made at 40°F and placed immediately in a curing room at 25°F. Both curves represent 100% relative-humidity curing for first 28 days followed by 50% relative-humidity curing. Reference 12-5.

Fig. 12-2. Insulated column forms permit concreting even when the air temperature is well below the freezing mark. For more details see Figs. 12-20 and 12-21.

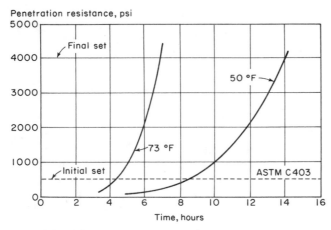

Fig. 12-3. Effect of cold temperature on rate of hardening. Reference 12-11.

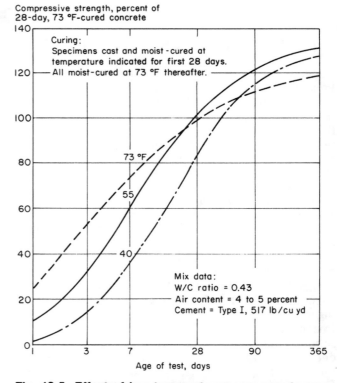

Fig. 12-5. Effect of low temperatures on concrete compressive strength at various ages. Reference 12-5.

cast and cured at various temperatures. Note in Fig. 12-5 that concrete cast and cured at 40°F and 55°F had relatively low strengths for the first week; but after 28 days—when all specimens were moist-cured at 73°F—strengths for the 40°F and 55°F concretes grew faster than the 73°F concrete and at one year they were slightly higher.

Higher early strengths can be achieved through use of Type III, high-early-strength, cement, as illustrated in Fig. 12-6. Principal advantages occur during the first 7 days. At a 40°F curing temperature, the advantages of Type III cement are more pronounced and persist longer than at higher temperatures.

HEAT OF HYDRATION

Concrete generates heat during hardening as a result of the chemical process by which cement reacts with water to form a hard, stable paste. The heat generated is called heat of hydration; it varies in amount and rate for different portland cements. Heat generation and buildup are affected by dimensions of the concrete, ambient air temperature, initial concrete temperature, water-cement ratio, cement composition and fineness, amount of cement, and admixtures.

Heat of hydration is particularly useful in winter concreting as it often generates enough heat to provide a satisfactory curing temperature without other temporary heat sources, particularly in more massive elements.

Concrete must be delivered at the proper temperature and account must be taken of the temperature of forms, reinforcing steel, the ground, or other concrete on which the concrete is cast. Concrete should not be cast on frozen concrete or on frozen ground.

Fig. 12-7 shows a concrete pedestal on a spread footing being covered with a tarpaulin just after the concrete was placed. Thermometer readings of the concrete's temperature will tell whether the covering is adequate. The heat liberated during hydration will offset to a considerable degree the loss of heat during placing, finishing, and early curing operations.

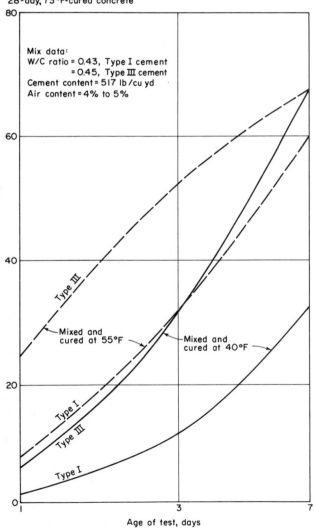

Compressive strength, percent of 28-day, 73°F-cured concrete

Mix data:
W/C ratio = 0.43, Type I cement
        = 0.45, Type III cement
Cement content = 517 lb/cu yd
Air content = 4% to 5%

Type III
Mixed and cured at 55°F
Mixed and cured at 40°F
Type I
Type III
Type I

Age of test, days

Fig. 12-6. Early-age compressive-strength relationships for Type I and Type III portland cements at low curing temperatures. Reference 12-5.

Fig. 12-7. Concrete footing pedestal being covered with a tarpaulin to retain the heat of hydration.

SPECIAL CONCRETE MIXTURES

High strength at an early age is desirable in winter construction to reduce the length of time temporary protection is required. The additional cost of high-early-strength concrete is often offset by earlier reuse of forms and removal of shores, savings in the shorter duration of temporary heating, earlier finishing of flatwork, and earlier use of the structure. High-early-strength concrete can be obtained by using one or a combination of the following:

1. Type III or IIIA high-early-strength cement
2. Additional portland cement (100 to 200 lb per cu yd)
3. Chemical accelerators

Small amounts of an accelerator such as calcium chloride (at a maximum dosage of 2% by weight of portland cement) can be used to accelerate the setting and early-age strength development of concrete in cold weather. Accelerators containing chlorides should not be used where there is an in-service potential for corrosion, such as in prestressed concrete or where aluminum or galvanized inserts will be used. Chlorides are not recommended for concretes exposed to soil or water containing sulfates or for concrete susceptible to alkali-aggregate reactions. Other restrictions on accelerators containing chlorides are discussed in detail in Chapter 6.

Accelerators must not be used as a substitute for proper curing and frost protection. Also, the use of so-called antifreeze compounds to lower the freezing point of concrete must not be permitted. The quantity of these materials needed to appreciably lower the freezing point of concrete is so great that strength and other properties are seriously affected by their use.

Low water-cement ratio and low-slump concrete is particularly desirable for cold-weather flatwork. Evaporation is minimized and finishing can be accomplished quicker (Fig. 12-8).

AIR-ENTRAINED CONCRETE

Entrained air is particularly desirable in any concrete placed during freezing weather. Concrete that is not air entrained will suffer a strength loss as a result of freezing and thawing (Fig. 12-9). Air entrainment provides the capacity to absorb stresses due to ice formation within the concrete. See Chapter 5, "Air-Entrained Concrete."

*Cold-Weather Concreting** implies that air entrainment should always be used for construction during the freezing months. The exception is concrete work done under roof where there is no chance that rain, snow, or water from other sources can saturate the concrete and where there is no chance of freezing.

The likelihood of water saturating a concrete floor during construction is very real. Fig. 12-10 shows conditions in the upper story of an apartment building during winter construction. Snow fell on the top deck. When heaters were used below to warm the deck, the

Fig. 12-8. Finishing the concrete surface can proceed because a windbreak has been provided, there is adequate heat under the slab, and the concrete has low slump.

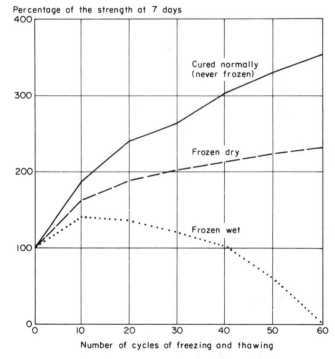

Fig. 12-9. Effect of freezing on strength of concrete that does not contain entrained air (cured 7 days before first freeze). Reference 12-3.

snow melted. Water ran through floor openings down to a level that was not being heated. The water-saturated concrete froze, which caused a strength loss, particularly at the floor surface. This could also result in greater deflection of the floor and a surface that is less abrasion-resistant than it might have been.

---

*Reference 12-13.

Fig. 12-10. Example of a concrete floor that was saturated and then frozen, showing the need for air entrainment.

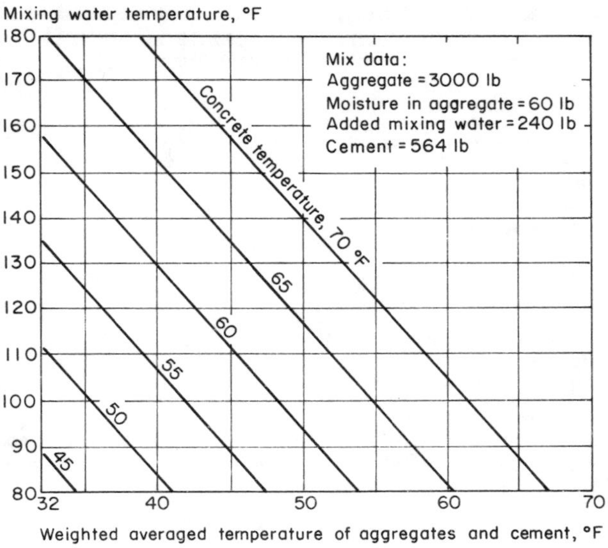

Mixing water temperature, °F

Mix data:
Aggregate = 3000 lb
Moisture in aggregate = 60 lb
Added mixing water = 240 lb
Cement = 564 lb

Weighted averaged temperature of aggregates and cement, °F

Fig. 12-11. Temperature of mixing water needed to produce heated concrete of required temperature. Temperatures are based on the mixture shown but are reasonably accurate for other typical mixtures.

TEMPERATURE OF CONCRETE

Temperature of Concrete As Mixed

The temperature of fresh concrete *as mixed* should not be less than shown in Lines 1, 2, or 3 of Table 12-1. Note that lower concrete temperatures are recommended for more massive concrete sections because heat generated during hydration is dissipated less rapidly in heavier sections. Also note that at lower air temperatures more heat is lost from concrete during transporting and placing; hence the recommended concrete temperatures are higher for colder weather.

There is little advantage in using fresh concrete at a temperature much above 70°F. Higher concrete temperatures do not afford proportionately longer protection from freezing because the rate of heat loss is greater. Also, high concrete temperatures are undesirable since they increase thermal shrinkage after hardening, require more mixing water for the same slump, and contribute to the possiblity of plastic-shrinkage cracking (caused by rapid moisture loss through evaporation). Therefore, the temperature of the concrete *as mixed* should not be more than 10°F above the minimums recommended in Table 12-1.

Aggregate temperature. The temperature of aggregates varies with weather and type of storage. Aggregates usually contain frozen lumps and ice when the temperature is below freezing. Frozen aggregates must be thawed to avoid pockets of aggregate in the concrete after batching, mixing, and placing. If thawing takes place in the mixer, excessively high water contents due to the ice melting must be avoided.

At temperatures above freezing it is seldom necessary to heat aggregates. At temperatures below freezing, often only the fine aggregate needs to be heated to produce concrete of the required temperature, provided the coarse aggregate is free of frozen lumps. If

aggregate temperatures are above freezing, the desired concrete temperature can usually be obtained by heating only the mixing water.

Circulating steam through pipes over which aggregates are stockpiled is a recommended method for heating aggregates. Stockpiles can be covered with tarpaulins to retain and distribute heat and to prevent formation of ice. Live steam, preferably at pressures of 75 to 125 psi, can be injected directly into the aggregate pile to heat it, but the resultant variable moisture content in aggregates might result in erratic mixing-water control.

On small jobs aggregates can be heated by stockpiling over metal culvert pipes in which fires are maintained. Care should be taken to prevent scorching the aggregates.

Mixing-water temperature. Of the ingredients used to make concrete, mixing water is the easiest and most practical to heat. The weight of aggregates and cement in concrete is much greater than the weight of water; however, water can store about five times as much heat as can cement and aggregate of the same weight. For cement and aggregate the average specific heat (that is, heat units required to raise the temperature of 1 lb of material 1°F) is 0.22 Btu per pound per degree Fahrenheit compared to 1.0 for water.

Equation for concrete temperature. Fig. 12-11 shows graphically the effect of temperature of materials on temperature of freshly mixed concrete. The chart is based on the equation

$$T = \frac{0.22(T_a W_a + T_c W_c) + T_w W_w + T_{wa} W_{wa}}{0.22(W_a + W_c) + W_w + W_{wa}}$$

Table 12-1. Recommended Concrete Temperature for Cold-Weather Construction—Air-Entrained Concrete*

Line	Condition		Thickness of sections, in.:			
			Less than 12	12 to 36	36 to 72	Over 72
1	Minimum temperature of fresh concrete *as mixed* for weather indicated, °F	Above 30°F	60	55	50	45
2		0°F to 30°F	65	60	55	50
3		Below 0°F	70	65	60	55
4	Minimum temperature of fresh concrete *as placed and maintained*, °F**		55	50	45	40
5	Maximum allowable *gradual* drop in temperature in first 24 hours after end of protection, °F		50	40	30	20

*Adapted from Table 1.4.1 of Reference 12-13.
**Placement temperatures listed are for normal-weight concrete. Lower temperatures can be used for lightweight concrete if justified by tests. For recommended duration of temperatures in Line 4, see Table 12-2.

where

$$T = \text{temperature of the freshly mixed concrete}$$

$T_a, T_c, T_w, \text{and } T_{wa} = $ temperature of aggregates, cement, added mixing water, and free moisture on aggregates, respectively

$W_a, W_c, W_w, \text{and } W_{wa} = $ weight of aggregates, cement, added mixing water, and free moisture on aggregates, respectively

If the weighted average temperature of aggregates and cement is above 32°F, the proper mixing-water temperature for the required concrete temperature can be selected from Fig. 12-11. The range of concrete temperatures in the chart corresponds with the recommended values given in Lines 1, 2, and 3 of Table 12-1.

To avoid the possibility of a quick or flash set of the concrete when either water or aggregates are heated to above 100°F, they should be combined in the mixer first before the cement is added. If this mixer-loading sequence is followed, water temperatures up to the boiling point can be used, provided the aggregates are cold enough to reduce the final temperature of the aggregates and water mixture to appreciably less than 100°F.

Fluctuations in mixing-water temperature from batch to batch should be avoided. The temperature of the mixing water can be adjusted by blending hot and cold water.

Temperature of Concrete As Placed and Maintained

There will be some temperature loss after mixing while the truck mixer is traveling to the construction site and waiting to discharge its load. The concrete should be placed in the forms before its temperature drops below that given on Line 4 of Table 12-1, and that concrete temperature should be maintained for the duration of the protection period given in Table 12-2.

CONTROL TESTS

Thermometers are needed to check the concrete temperatures as delivered, as placed, and as maintained. An inexpensive pocket thermometer is shown in Fig. 12-12.

After the concrete has hardened, temperatures can be checked with special surface thermometers or with an ordinary thermometer that is kept covered with insulating blankets. A simple way to check temperature below the concrete surface is shown in Fig. 12-13. Instead of filling the hole shown in Fig. 12-13 with a fluid, it can be fitted with insulation except at the bulb.

Concrete test cylinders must be maintained at a temperature between 60°F and 80°F at the jobsite for 24 hours until they are taken to a laboratory for curing. During this 24-hour period cylinders should be kept in a curing box with the temperature accurately controlled by a thermostat (Fig. 12-14). When stored in an insulated curing box outdoors, cylinders are less likely to be jostled by vibrations than if they were left on the floor of a trailer. If kept in a trailer or shanty where the heat may be turned off at night or over a weekend or holiday, the cylinders would not be at the prescribed curing temperatures during this critical period.

Fig. 12-12. A bimetallic pocket thermometer with a metal sensor suitable for checking concrete temperatures.

Table 12-2.

A. Recommended Duration of Concrete Temperature in Cold Weather— Air-Entrained Concrete*

Service category††	For durability		For safe stripping strength	
	Conventional concrete,** days	High-early-strength concrete,† days	Conventional concrete,** days	High-early-strength concrete,† days
No load, not exposed, favorable moist-curing	2	1	2	1
No load, exposed, but later has favorable moist-curing	3	2	3	2
Partial load, exposed	3	2	6	4
Fully stressed, exposed	3	2	See Table B below	

B. Recommended Duration of Concrete Temperature for Fully Stressed, Exposed, Air-Entrained Concrete

Required percentage of design strength, f'_c	Days at 50°F			Days at 70°F		
	Type of portland cement			Type of portland cement		
	I	II	III	I	II	III
50	6	9	3	4	6	3
65	11	14	5	8	10	4
85	21	28	16	16	18	12
95	29	35	26	23	24	20

*Adapted from Table 1.4.2 of Reference 12-13. Cold weather is defined as that in which average daily temperature is less than 40°F for 3 successive days except that if temperatures above 50°F occur during at least 12 hours in any day, the concrete should no longer be regarded as winter concrete and normal curing practice should apply. For recommended concrete temperatures, see Table 12-1. For concrete that is *not* air entrained, ACI 306 states that protection for durability should be at least twice the number of days listed in Table A.

Part B was adapted from Table 7.7 of Reference 12-13. The values shown are approximations and will vary according to the thickness of concrete, mix proportions, etc. They are intended to represent the ages at which supporting forms can be removed. For recommended concrete temperatures, see Table 12-1.

**Made with ASTM Type I or II portland cement.

†Made with ASTM Type III portland cement, or an accelerator, or an extra 100 lb of cement per cu yd.

††"Exposed" means subject to freezing and thawing.

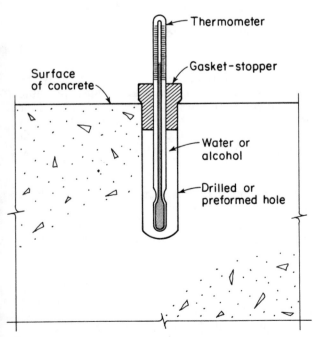

Fig. 12-13. Scheme for measuring concrete temperatures below the surface with a glass thermometer.

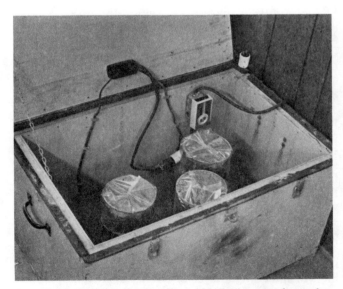

Fig. 12-14. Insulated curing box with thermostat for curing test cylinders. Heat is supplied by electric rubber heating mats on the bottom. A wide variety of designs are possible for curing boxes.

In addition to laboratory-cured cylinders, it is useful to field-cure some test cylinders in order to monitor actual curing conditions on the job in cold weather. It is sometimes difficult to find the right locations for field curing. A preferred location is in a boxout in a floor slab or wall with thermal insulation for cover. When placed on a formwork ledge just below a heated, suspended floor, possible high temperatures there will not duplicate the average temperature in the slab, nor the lowest temperature on top of the slab. Still, field-cured cylinders are more indicative of actual concrete strength than are laboratory-cured cylinders.

Molds should be stripped from the cylinders after the first 24 ±4 hours and the cylinders wrapped tightly in plastic bags. When cylinders are picked up for delivery to the laboratory, they must be maintained at 60°F to 80°F temperatures until they are placed in the laboratory curing room.

Cast-in-place cylinders (ASTM C 873) and nondestructive testing methods discussed in Chapter 14, as well as maturity techniques discussed later in this chapter, are helpful in monitoring in-place concrete strength.

Fig. 12-15. Even in the dead of winter, an outdoor swimming pool can be constructed if a heated enclosure is used.

CONCRETING ON GROUND

Concreting on ground during cold weather involves some extra effort and expense, but many contractors find that it more than pays for itself. In winter, the site may be frozen rather than a morass of mud. The concrete will furnish some if not all of the heat needed for proper curing. Insulated blankets or simple enclosures are easily provided. Embankments are frozen and require less bracing. With a good start during the winter months, construction gets above the ground before warmer weather arrives.

Placing concrete on the ground involves different procedures from those used at an upper level: (1) the ground must be thawed before placing concrete; (2) cement hydration will furnish some of the curing heat; (3) construction of enclosures is much simpler and use of insulating blankets may be sufficient; and (4) in the case of a floor slab, a *vented* heater is required if the area is enclosed.

Footings can be started if the ground is thawed with temporary heat and the footings backfilled as soon as possible. Concrete should never be placed on a frozen subgrade because when the subgrade thaws, uneven settlement may occur, causing cracking. Also, on a frozen subgrade, heat will migrate rapidly away from the bottom of the concrete and its rate of hardening will be retarded. Ideally, the temperature of the subgrade should be as close as practicable to the temperature of the concrete to be placed on it.

When the subgrade is frozen to a depth of only a few inches, this surface region can be thawed by (1) steaming; (2) spreading a layer of hot sand, gravel, or other granular material where the grade elevations allow it; (3) burning straw or hay if local air pollution ordinances permit it; or (4) covering the subgrade with insulation for a few days. Placing concrete for floor slabs and exposed footings should be delayed until the ground thaws and warms sufficiently to ensure that it will not freeze again during the curing period.

For slabs cast on ground at 35°F, ACI 306* gives two sets of tables and graphs showing required resistance to heat transfer (R) of insulating blankets. The data extends to a protection period of only one week and states that insulating blankets must be supplemented with heat for slabs 12 in. or less in thickness.

CONCRETING ABOVEGROUND

Working aboveground in cold weather usually involves several different approaches compared to working at ground level:

1. The concrete mixture need not necessarily be changed to generate more heat because portable heaters can be used to heat the undersides of floor and roof slabs. Nevertheless, there are advantages to having a mix that will produce a high strength at an early age; for example, artificial heat can be cut off sooner (see Table 12-2), and forms can be recycled faster.
2. Enclosures must be constructed to retain the heat under floor and roof slabs.
3. Portable heaters used to warm the underside of concrete can be direct-fired heating units (without venting).

Before concreting, the heaters under a formed deck should be turned on to preheat the forms and melt any snow or ice remaining on top. When slab finishing is completed, insulating blankets or other insulation must be placed on top of the slab to ensure that proper

*Reference 12-13.

144

curing temperatures are maintained. The insulation R values necessary to maintain an adequate curing temperature above 50°F for 3 or 7 days may be estimated from Figs. 12-16 and 12-17. To maintain a temperature for longer periods, more insulation is required. Figs. 12-16 and 12-17 also apply to concrete walls. Insulation can be selected based on R values provided by insulation manufacturers or by using the information in Table 12-3. ACI 306 has additional graphs and tables for slabs on or above ground.

The period for the maintained temperature may be based on Table 12-2. However, the actual amount of insulation and length of the protection period should be determined from the monitored in-place concrete temperature and the desired strength. A correlation between curing temperature, curing time, and compressive strength can be determined from laboratory testing of the particular concrete mix used in the field (see maturity concept discussed in "Duration of Heating" in this chapter).

Since square and rectangular columns have twice as many heat-flow paths as a long, high wall, twice as much insulation is needed to maintain the same heat-loss characteristics. If the ambient temperature rises much above the temperature assumed in selecting insulation values, the temperature of the concrete may become excessive. This increases the probability of thermal shock and cracking when forms are removed. Temperature readings of insulated concrete should therefore be taken at regular intervals and should not be allowed to rise much above 80°F. In case of a sudden increase in concrete temperature, up to say 95°F, it may be necessary to remove some of the insulation or loosen the formwork. The maximum temperature differential between the concrete interior and the concrete surface should be about 35°F to minimize cracking. The weather forecast should be checked and appropriate action taken for expected temperature changes.

Columns and walls should not be cast on frozen foundations, because chilling of concrete in the bottom of the column or wall will retard strength development.

ENCLOSURES

Heated enclosures can be used for protecting concrete in cold weather. Enclosures can be of wood, canvas tarpaulins, or polyethylene film (Fig. 12-18). Prefabricated, rigid-plastic enclosures are also available. Plastic enclosures that admit daylight are the most popular (Fig. 12-18b).

When enclosures are being constructed below a deck, the framework can be extended above the deck to serve as a windbreak. A height of 6 ft will give good protection against biting winds.

Enclosures can be made to be moved with flying forms; more often, though, they must be removed so that the wind will not interfere with maneuvering the forms into position. Similarly, enclosures can be built in large panels like gang forms with the windbreak included (Fig. 12-1).

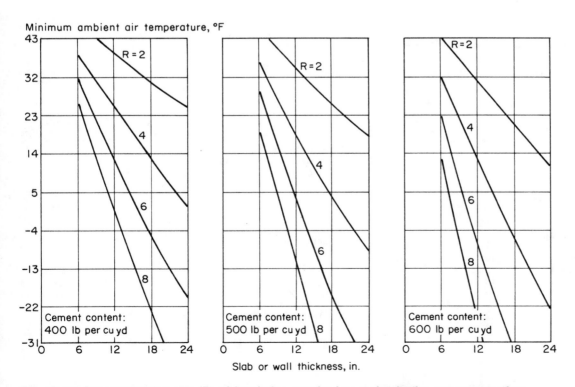

Minimum ambient air temperature, °F

Slab or wall thickness, in.

Fig. 12-16. Thermal resistance (R) of insulation required to maintain the concrete surface temperature of walls and slabs aboveground at 50°F or above for 3 days. Concrete temperature as placed: 50°F. Maximum wind velocity: 15 mph. Reference 12-13.

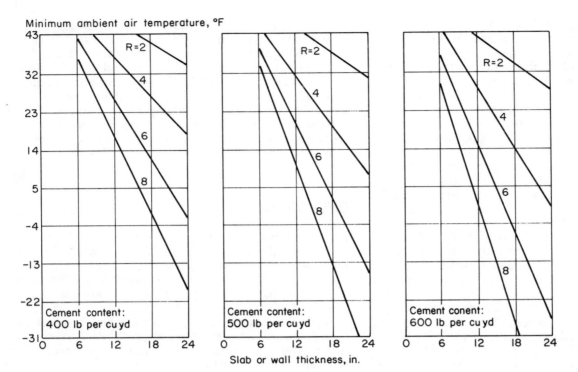

Fig. 12-17. Thermal resistance (*R*) of insulation required to maintain the concrete surface temperature of walls and slabs aboveground at 50°F or above for 7 days. Concrete temperatures as placed: 50°F. Maximum wind velocity: 15 mph. Note that in order to maintain a certain minimum temperature for a longer period of time, more insulation or a higher *R* value is required. Reference 12-13.

Table 12-3. Insulation Values of Various Materials

A.

Material	Density, pcf	Thermal resistance, *R*, for 1-in. thickness of material,* (°F · hr · ft²)/Btu
Board and Slabs		
Expanded polyurethane	1.5	6.25
Expanded polystyrene extruded smooth-skin surface	1.8 to 3.5	5.00
Expanded polystyrene extruded cut-cell surface	1.8	4.00
Glass fiber, organic bonded	4 to 9	4.00
Expanded polystyrene, molded beads	1	3.85
Mineral fiber with resin binder	15	3.45
Mineral fiberboard, wet felted	16 to 17	2.94
Vegetable fiberboard sheathing	18	2.64
Cellular glass	8.5	2.86
Laminated paperboard	30	2.00
Particle board (low density)	37	1.85
Plywood	34	1.24
Loose fill		
Wood fiber, soft woods	2.0 to 3.5	3.33
Perlite (expanded)	5.0 to 8.0	2.70
Vermiculite (exfoliated)	4.0 to 6.0	2.27
Vermiculite (exfoliated)	7.0 to 8.2	2.13
Sawdust or shavings	8.0 to 15.0	2.22

B.

Material	Material thickness, in.		Thermal resistance, *R*, for 1-in. thickness of material,* (°F · hr · ft²)/Btu
Mineral fiber blanket, fibrous form (rock, slag, or glass) 0.3 to 2 pcf	2	to 2.75	7
	3	to 3.5	11
	5.5	to 6.5	19
	6	to 7	22
	8.5	to 9	30
	12		38
Mineral fiber loose fill (rock, slag, or glass) 0.6 to 2 pcf	3.75	to 5	11
	6.5	to 8.75	19
	7.5	to 10	22
	10.25	to 13.75	30

*Values are from *ASHRAE Handbook of Fundamentals*, 1977 and 1981, American Society of Heating, Refrigerating, and Air-Conditioning Engineers, Inc., New York.

R values are the reciprocal of U values (conductivity).

a.

b.

Fig. 12-18. Heated enclosures maintain an adequate temperature for proper curing. a. Tarpaulin enclosure provides protection for a bridge during severe and prolonged winter weather. b. Polyethylene plastic sheets admitting daylight are used to fully enclose a building frame. The temperature inside is maintained at 50°F with space heaters.

Fig. 12-19. Insulating blankets trap heat and moisture in the concrete, providing beneficial curing.

Fig. 12-20. Blanket or batt insulation may be applied to the outside of job-built or prefabricated forms.

INSULATING MATERIALS

Heat and moisture can be retained in the concrete by covering it with commercial insulating blankets or batt insulation (Fig. 12-19). The effectiveness of insulation can be determined by placing a thermometer under it and in contact with the concrete. If the temperature falls below the minimum required, additional insulating material, or material with a higher R value, should be applied. Corners and edges of concrete are most vulnerable to freezing and the temperature should be checked there particularly.

The thermal resistance values for common insulating materials are given in Table 12-3. For maximum efficiency, insulating materials should be kept dry and in close contact with concrete or formwork.

Concrete pavements can be protected from cold weather by spreading 12 in. or more of dry straw or hay on the surface for insulation. Tarpaulins, polyethylene film, or waterproof paper should be used as a protective cover over the straw or hay to make the insulation more effective and prevent it from blowing away. The

straw or hay should be kept dry or its insulation value will drop considerably.

Forms built for repeated use often can be economically insulated with commercial blanket or batt insulation (Fig. 12-20). The insulation should have a tough moistureproof covering to withstand handling abuse and exposure to the weather. Rigid insulation can also be used (Fig. 12-21).

Insulating blankets for construction are made of fiberglass, sponge rubber, open-cell polyurethane foam, vinyl foam, mineral wool, or cellulose fibers. The outer covers are made of canvas, woven polyethylene, or other tough fabrics that will take rough handling.

Fig. 12-21. With air temperatures down to −10°F, concrete was cast in this insulated column form made of ¾-in. high-density plywood inside, 1-in. rigid polystyrene in the middle, and ½-in. rough plywood outside. R value: 5.6 (°F · hr · ft²)/Btu.

The *R* value for a typical insulating blanket is about 7(°F · hr · ft²)/Btu, but since *R* values are not marked on the blankets, their effectiveness should be checked with a thermometer. If necessary, they can be used in two or three layers to attain the desired insulation.

HEATERS

Two types of heaters are used in cold-weather concrete construction: direct fired and indirect fired (Fig. 12-22). Indirect-fired heaters are vented to remove the products of combustion. Where heat is to be supplied to the top of fresh concrete—for example, a floor slab—vented heaters are required. Carbon dioxide (CO_2) in the exhaust will be vented to the outside and prevented from reacting with the fresh concrete (Fig. 12-23). Direct-fired units can be used to heat enclosures beneath a floor or a roof deck (Fig. 12-24).

Any heater burning a fossil fuel produces CO_2 that will combine with calcium hydroxide on the surface of fresh concrete to form a weak layer of calcium carbonate that interferes with cement hydration.* The result is a soft, chalky surface that will dust under traffic. Depth and degree of carbonation depend on concentration of CO_2, curing temperature, humidity, porosity of the concrete, length of exposure, and method of curing. Direct-fired heaters, therefore, should not be permitted to heat the air over concreting operations—at least until 24 hours have elapsed.

Carbon monoxide (CO), another product of combustion, is not usually a problem unless the heater is using recirculated air. Four hours of exposure to 200 parts per million of CO will produce headaches and nausea. Three hours of exposure to 600 ppm can be fatal. The American National Standard Safety Requirements for Temporary and Portable Space Heating Devices and Equipment Used in the Construction Industry (ANSI

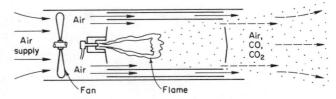

a) Direct-fired heater

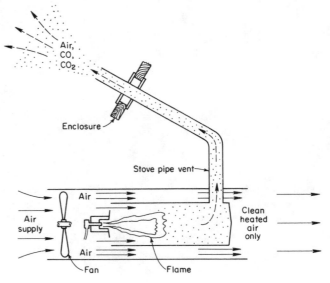

b) Indirect-fired heater

Fig. 12-22. Two basic types of heaters.

Fig. 12-23. An indirect-fired heater. Notice vent pipe that carries combustion gases outside the enclosure.

*Reference 12-2.

Fig. 12-24. A direct-fired heater mounted outside the enclosure, thus using a fresh air supply.

A10.10) limits concentrations of CO to 50 ppm at worker breathing levels. The standard also establishes safety rules for ventilation and the stability, operation, fueling, and maintenance of heaters.

A salamander is an inexpensive combustion heater without a fan that discharges its combustion products into the surrounding air; heating is accomplished by radiation from its metal casing. Salamanders are fueled by coke, oil, wood, or liquid propane. They are but one form of a direct-fired heater. A primary disadvantage of salamanders is the high temperature of their metal casing, a definite fire hazard. Salamanders should be placed so as not to overheat formwork or enclosure materials. When placed on floor slabs they should be elevated to avoid scorching the concrete.

Some heaters burn more than one type of fuel. The approximate heat values of fuels are as follows:

No. 1 fuel oil	135,000 Btu/gal
Kerosene	134,000 Btu/gal
Gasoline	128,000 Btu/gal
Liquid-propane gas	91,500 Btu/gal
Natural gas	1,000 Btu/ft^3

The Btu rating of a portable heater is usually the heat content of the fuel consumed per hour. A rule of thumb is that about 36,000 Btu are required for each 10,000 cu ft of air to develop a 20°F heat rise.

Electricity can also be used to cure concrete in winter. The use of large electric blankets equipped with thermostats is one method. The blankets can also be used to thaw subgrades or concrete foundations.

Use of electric resistance wires that are cast into the concrete is another method. The power supplied is under 50 volts, and from 1.5 to 5 kilowatt-hours of electricity per cubic yard of concrete is required, depending on the circumstances. The method has been used in the Montreal, Quebec, area for many years. Where electrical resistance wires are used, insulation should be included during the initial setting period. If insulation is removed before the recommended time, the concrete should be covered with an impervious sheet and the power continued for the required time.

Steam is another source of heat for winter concreting. Live steam can be piped into an enclosure or supplied through radiant heating units. For accelerated steam curing at elevated temperatures, see Chapter 10, "Curing Concrete."

In choosing a heat source, it must be remembered that the concrete itself supplies heat through hydration of cement, and this is often enough for curing needs if the heat can be retained within the concrete with insulation.

DURATION OF HEATING

After concrete is in place, it should be protected and kept at a favorable curing temperature until it gains sufficient strength to withstand exposure to low temperatures, anticipated environment, and construction and service loads. Recommended periods of protection are given in Table 12-2. The columns headed "for durability" list the length of time in days required to provide adequate durability against exposure to freezing and thawing. The columns headed "for safe stripping strength" list the length of time in days after which bottom forms or reshores can be removed. This is based on the assumption that construction loads will be less than those provided for in the design and that normal curing conditions will enable the concrete to reach its full design strength before being put into service.

Maturity Concept

The maturity concept is based on the principle that strength gain is a function of curing time and temperature. The maturity concept, as described in ACI 306 and ASTM C1074 and C918, can be used to evaluate strength development when the prescribed curing temperatures have not been maintained for the required time or when curing temperatures have fluctuated. The concept is expressed by the equation

$$M = \Sigma(F - 14)\Delta t$$

where

M = maturity factor
Σ = summation
F = temperature, degrees Fahrenheit
Δt = duration of curing at temperature F, usually in hours

The equation is based on the premise that concrete gains strength (or that cement continues to hydrate) at temperatures as low as 14°F.

Before construction begins, a calibration curve is drawn plotting the relationship between compressive strength and the maturity factor for a series of test cylinders (of the particular concrete mixture proportions) cured in a laboratory and tested for strength at successive ages.

The maturity concept is not precise and has some limitations. The concept is useful only in checking the curing of concrete and estimating strength in relation to time and temperature. It presumes that all other factors affecting concrete strength have been properly controlled. Thus the maturity concept is another method for monitoring temperatures, but it is no substitute for quality control and proper concreting practices.*

Moist Curing

Strength gain stops when moisture required for curing is no longer available. Concrete retained in forms or covered with insulation seldom loses enough moisture at 40°F to 55°F to impair curing. However, a positive means of providing moist curing is needed to offset drying when heated enclosures are used during cold weather.

Live steam exhausted into an enclosure around the concrete is an excellent method of curing because it provides both heat and moisture. Steam is especially practical in extremely cold weather because the moisture provided offsets the rapid drying that occurs when very cold air is heated.

Liquid membrane-forming compounds can be used for early curing of concrete surfaces within heated enclosures. These compounds are also helpful in reducing carbonation of the surface by unvented heaters.

Terminating the Heating Period

Rapid cooling of concrete at the end of the heating period should be avoided. Sudden cooling of the concrete surface while the interior is still warm may cause thermal cracking, especially in massive sections such as bridge piers, abutments, dams, and large structural members; thus cooling should be gradual. The maximum uniform drop in temperature throughout the first 24 hours after the end of protection should not be more than the amounts given in Line 5 of Table 12-1. Gradual cooling can be accomplished by lowering the heat or by simply shutting off the heat and allowing the enclosure to cool to outside air temperature.

FORM REMOVAL AND RESHORING

It is good practice in cold weather to leave forms in place as long as possible. Even within heated enclosures, forms serve to distribute heat more evenly and help prevent drying and local overheating.

Table 12-2A can be used to determine the length of time in days that vertical support for forms should be left in place. The time span is also the length of time that the required concrete temperature should be maintained. Table 12-2B lists length of time in days for heating and shoring or reshoring of members that will be fully stressed. The engineer in charge must determine what percentage of the design strength is required.

Side forms can be removed sooner than shoring and temporary falsework.**

REFERENCES

12-1. McNeese, D. C., "Early Freezing of Non-Air-Entrained Concrete," *Journal of the American Concrete Institute Proceedings,* vol. 49, American Concrete Institute, Detroit, December 1952, pages 293-300.

12-2. Kauer, J. A., and Freeman, R. L., "Effect of Carbon Dioxide on Fresh Concrete," *Journal of the American Concrete Institute Proceedings,* vol. 52, December 1955, pages 447-454. Discussion: December 1955, Part II, pages 1299-1304, American Concrete Institute.

12-3. Powers, T. C., *Resistance of Concrete to Frost at Early Ages,* Research Department Bulletin RX071, Portland Cement Association, 1956.

12-4. Klieger, Paul, *Curing Requirements for Scale Resistance of Concrete,* Research Department Bulletin RX082, Portland Cement Association, 1957.

12-5. Klieger, Paul, *Effect of Mixing and Curing Temperature on Concrete Strength,* Research Department Bulletin RX103, Portland Cement Association, 1958.

12-6. Copeland, L. E.; Kantro, D. L.; and Verbeck, George, *Chemistry of Hydration of Portland Cement,* Research Department Bulletin RX153, Portland Cement Association, 1960, page 452.

12-7. Powers, T. C., *Prevention of Frost Damage to Green Concrete,* Research Department Bulletin RX148, Portland Cement Association, 1962.

12-8. Brewer, Harold W., *General Relation of Heat Flow Factors to the Unit Weight of Concrete,* Development Department Bulletin DX114, Portland Cement Association, 1967.

12-9. *Cold Weather Ready Mixed Concrete,* Publication No. 130, National Ready Mixed Concrete Association, Silver Spring, Maryland, 1968.

12-10. Malhotra, V. M., "Maturity Concept and the Estimation of Concrete Strength: A Review," Parts I and II, *Indian Concrete Journal,* vol. 48, Associated Cement Companies, Ltd., Bombay, April and May 1974.

12-11. Sprouse, J. H., and Peppler, R. B., "Setting Time," *Significance of Tests and Properties of Concrete and Concrete-Making Materials,* STP169B, American Society for Testing and Materials, Philadelphia, 1978, pages 105-121.

12-12. *Concrete Manual,* 8th ed., revised 1981, U.S. Bureau of Reclamation, Denver.

12-13. *Cold-Weather Concreting,* ACI 306R-78, revised 1983, ACI Committee 306 Report, American Concrete Institute.

12-14. *Recommended Practice for Concrete Formwork,* ACI 347-78, reapproved 1984, ACI Committee 347, American Concrete Institute.

---

*See References 12-10 and 12-13 for more information.
**See Reference 12-14.

CHAPTER 13
Volume Changes of Concrete

Concrete changes slightly in volume for various reasons, and understanding the nature of these changes is useful in planning or analyzing concrete work. If concrete were free of any restraints to deform, normal volume changes would be of little consequence; but since concrete in service is usually restrained by foundations, subgrades, reinforcement, or connecting members, significant stresses can develop. This is particularly true of tensile stresses.

Cracks sometimes develop, since concrete is relatively weak in tension but quite strong in compression. High stresses and cracking can be prevented or minimized by controlling the variables that affect volume changes. When reinforced concrete is allowed to crack or cracking cannot be avoided, the tolerable crack widths of Table 13-1 should be considered in the structural design.

Volume change is defined merely as an increase or decrease in volume. Most commonly, the subject of concrete volume changes deals with expansion and contraction due to temperature and moisture cycles. But volume changes are also caused by chemical effects such as carbonation shrinkage, sulfate attack, and the disruptive expansion of alkali-aggregate reactions. Also, creep is a volume change or deformation caused by sustained stress or load. Equally important is the elastic or inelastic change in dimensions or shape that occurs instantaneously under applied load.

For convenience, the magnitude of volume changes is generally stated in linear rather than volumetric units. Changes in length are often expressed in millionths and are applicable to any length unit (e.g., in./in. or ft/ft). For example, one millionth is 0.000001 in./in. and 600 millionths is 0.000600 in./in. Change of length can also be expressed as a percentage; thus 0.06% is the same as 0.000600, which incidentally is approximately the same as ¾ in. per 100 ft. The volume changes that ordinarily occur in concrete are small, ranging in length change from perhaps 10 millionths up to about 1000 millionths.

TEMPERATURE CHANGES

Concrete expands slightly as temperature rises and contracts as temperature falls, although it can expand

Table 13-1. Tolerable Crack Widths for Reinforced Concrete

Exposure condition	Tolerable crack width, in.
Dry air or protective membrane	0.016
Humidity, moist air, soil	0.012
Deicing chemicals	0.007
Seawater and seawater spray; wetting and drying	0.006
Water retaining structures*	0.004

*Excluding nonpressure pipes.
Reference 13-39.

Table 13-2. Effect of Aggregate Type on Thermal Coefficient of Expansion of Concrete

Aggregate type (from one source)	Coefficient of expansion, millionths per °F
Quartz	6.6
Sandstone	6.5
Gravel	6.0
Granite	5.3
Basalt	4.8
Limestone	3.8

Source: Reference 13-1. Note: Coefficients of concretes made with aggregates from different sources may vary widely from these values, especially those for gravels, granites, and limestones. The coefficient for structural lightweight concrete varies from 3.9 to 6.1 millionths per degree Fahrenheit, depending on the aggregate type and the amount of natural sand.

slightly as free water in the concrete freezes. Temperature changes may be caused by environmental conditions or by cement hydration. An average value for the coefficient of thermal expansion of concrete is about 5.5 millionths per degree Fahrenheit, although values ranging from 3.2 to 7.0 have been observed. This amounts to a length change of ⅔ in. for 100 ft of concrete subjected to a rise or fall of 100°F. The coefficient of thermal expansion for structural lightweight concrete varies from 3.6 to 6.1 millionths per degree Fahrenheit.

Thermal expansion and contraction of concrete varies with factors such as aggregate type, cement content, water-cement ratio, temperature range, concrete age, and relative humidity. Of these, aggregate type has the greatest influence.

Table 13-2 shows some experimental values of the thermal coefficient of expansion of concretes made with aggregates of various types. These data were obtained from tests on small concrete specimens in which all factors were the same except aggregate type. The fine aggregate was of the same material as the coarse aggregate.

The thermal coefficient of expansion for steel is about 6.5 millionths per degree Fahrenheit, which is comparable to that for concrete. The coefficient for reinforced concrete can be assumed as 6 millionths per degree Fahrenheit, the average for concrete and steel.

Temperature changes that result in shortening will crack concrete members that are highly restrained by another part of the structure or by ground friction. Consider a long restrained concrete section cast without joints that, after moist curing, is allowed to drop in temperature. As the temperature drops, the concrete tends to shorten, but cannot as it is restrained longitudinally. The resulting tensile stresses cause the concrete to crack. Since tensile strength and modulus of elasticity of concrete both may be assumed proportional to the square root of concrete strength, calculations show that a large enough temperature drop (depending upon type of aggregate) will crack concrete regardless of its age or strength, provided that the coefficient of expansion does not vary with the temperature and the concrete is fully restrained.

An example of these calculations follows:*

Assume the coefficient of thermal expansion for concrete, α, is $5.5 \times 10^{-6}/°F$. If restrained tightly at its ends against contraction, the tensile stress in the concrete is $f = \alpha E$ and

$$E = 57,000 \sqrt{f'_c} = 3,120,000 \text{ when } f'_c = 3000 \text{ psi}$$

Thus tensile stress

$$f = \frac{5.5 \times 3.12 \times 1,000,000}{1,000,000} = 17.2 \text{ psi/°F}$$

Tensile strength of concrete $f_t = 7.5 \sqrt{f'_c}$ and

$$f_t = 410 \text{ psi when } f'_c = 3000 \text{ psi}$$

The temperature drop necessary to crack concrete is Δ_t and

$$\Delta_t = \frac{410}{17.2} = 24°F \text{ when concrete has 3000 psi}$$
compressive strength

The general equation for Δ_t at any age, assuming a constant coefficient of thermal contraction, is

$$\Delta_t = \frac{f_t}{\alpha E} = \frac{7.5 \sqrt{f'_c}}{\alpha \times 57,000 \sqrt{f'_c}} = \frac{1}{7600\alpha}$$

When $\alpha = 5.5 \times 10^{-6}$, $\Delta_t = 24°F$ for any age or strength concrete

Precast wall panels and slabs and pavements on ground are susceptible to bending and curling caused by temperature gradients that develop when concrete is cool on one side and warm on the other. The calculated amount of curling in a wall panel is illustrated in Fig. 13-1.

For the effect of temperature changes in mass concrete due to heat of hydration, see Chapter 15.

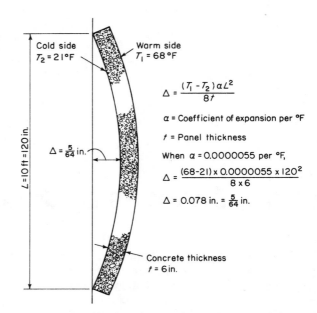

Fig. 13-1. Curling of a plain concrete wall panel due to temperature that varies uniformly from inside to outside. Reference 13-40.

Low Temperatures

Concrete continues to contract as the temperature is reduced below freezing. The amount of volume change at subfreezing temperatures is greatly influenced by the moisture content, behavior of the water (physical state—ice or liquid), and type of aggregate in the concrete. In one study, the coefficient of thermal expansion for a temperature range of 75°F to −250°F varied from 3.3×10^{-6} per °F for a lightweight aggregate concrete to 4.5×10^{-6} per °F for a sand and gravel mix. Subfreezing temperatures can significantly increase the compressive and tensile strength and modulus of elasticity of moist concrete. Dry concrete properties are not as affected by low temperatures. In the same study, moist concrete with an original compressive strength of 5000 psi at 75°F achieved over 17,000 psi at −150°F, a 240% increase. The same concrete tested ovendry or at a 50% internal relative humidity had strength increases of only about 20%. The modulus of elasticity for sand and gravel concrete with 50% relative humidity was only 8% higher at −250°F than at 75°F, whereas the moist concrete had a 50% increase in modulus of elasticity. Going from 75°F to −250°F, the thermal conductivity of normal-weight concrete is also increased, especially for moist concrete. The thermal conductivity of lightweight aggregate concrete is little affected.**

High Temperatures

Temperatures greater than 200°F that are sustained for several months or even several hours can have significant effects on concrete. The total amount of volume

*Reference 13-40, page 19.
**References 13-13 and 13-23.

change of concrete is the sum of volume changes of the cement paste and aggregate. At high temperatures, the paste shrinks due to dehydration while the aggregate expands. For normal-aggregate concrete, the expansion of the aggregate exceeds the paste shrinkage resulting in overall expansion of the concrete. Some aggregates such as expanded shale, andesite, or pumice with low coefficients of expansion can produce a very volume-stable concrete in high temperature environments (see Chapter 4 and Fig. 13-2). On the other hand, some aggregates undergo extensive and abrupt volume changes at a particular temperature, causing distress in the concrete. For example, in one study a dolomitic limestone aggregate contained an iron sulfide impurity that caused severe expansion, cracking, and disintegration in concrete exposed to a temperature of 302°F for four months; at temperatures above and below 302°F there was no detrimental expansion.* The coefficient of thermal expansion tends to increase with temperature rise.

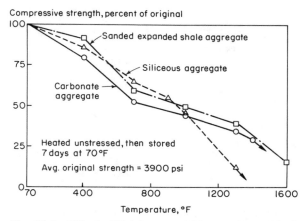

Fig. 13-3. Effect of high temperatures on the residual compressive strength of concretes containing various types of aggregate. Reference 13-33.

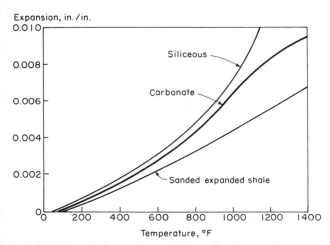

Fig. 13-2. Thermal expansion of concretes containing various types of aggregate. Reference 13-36.

Besides volume change, sustained high temperatures can also have other, usually irreversible, effects such as a reduction in strength, modulus of elasticity, and thermal conductivity. Creep increases with temperature. Above 212°F, the paste begins to dehydrate (lose chemically combined water of hydration) resulting in significant strength losses. Strength decreases with increases in temperature until the concrete loses essentially all its strength. The effect of high-temperature exposure on compressive strength of particular unrestrained concretes is illustrated in Fig. 13-3. Several factors including concrete moisture content, aggregate type and stability, cement content, exposure time, rate of temperature rise, age of concrete, restraint, and existing stress all influence the behavior of concrete at high temperatures.

If stable aggregates are used and strength reduction and the effects on other properties are accounted for in

the mix design, high quality concrete can be exposed to temperatures of 200°F to 400°F for long periods. Although some concrete elements have been exposed to temperatures up to 500°F or even 600°F for long periods, special steps should be taken or special materials (such as heat-resistant high-alumina cement) should be considered for exposure temperatures greater than 400°F. Before any structural concrete is exposed to high temperatures (greater than 200°F), laboratory testing should be performed to determine the particular concrete's thermal properties to avoid unexpected distress.**

MOISTURE CHANGES (DRYING SHRINKAGE)

Concrete expands slightly with a gain in moisture and contracts with a loss in moisture. The effects of these moisture movements are illustrated schematically in Fig. 13-4. Specimen A represents concrete stored continuously in water from time of casting; Specimen B represents the same concrete exposed first to drying in air and then to alternate cycles of wetting and drying. For comparative purposes it can be noted that the swelling that occurs during continuous wet storage over a period of several years is usually less than 150 millionths or about one-fourth of the shrinkage of air-dried concrete for the same period.

Tests indicate that the drying shrinkage of small, plain concrete specimens (without reinforcement) ranges from about 400 to 800 millionths when exposed to air at 50% humidity. Concrete with a unit shrinkage of 550 millionths shortens about the same amount as the thermal contraction caused by a decrease in temperature of 100°F. Preplaced aggregate concrete has a

*Reference 13-42.
**References 13-8, 13-9, 13-20, 13-22, 13-25, 13-29, 13-30, 13-31, 13-32, 13-33, 13-36, 13-37, 13-38, 13-42, and 13-45.

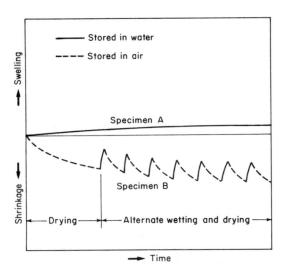

Fig. 13-4. Schematic illustration of moisture movements in concrete. If concrete is kept continuously wet, a slight expansion occurs. However, drying usually takes place, causing shrinkage. Further wetting and drying causes alternate swelling and shrinkage. Reference 13-12.

drying shrinkage of 200 to 400 millionths, considerably less than normal concrete due to point-to-point contact of aggregate particles. The drying shrinkage of structural lightweight concrete ranges from slightly less than to 30 percent more than that of normal-weight concrete, depending on the type of aggregate used.

The shrinkage of reinforced concrete is less than that for plain concrete, the difference depending on the amount of reinforcement. Steel reinforcement restricts but does not prevent drying shrinkage. In reinforced concrete structures with normal amounts of reinforcement, drying shrinkage is commonly assumed to be 200 to 300 millionths.

For many outdoor applications, concrete reaches its maximum moisture content in winter; so in winter the volume changes due to increase in moisture content and the decrease in average temperature generally offset each other.

The amount of moisture in concrete is affected by the relative humidity of the surrounding air. The free moisture content of concrete elements after drying in air at relative humidities of 50% to 90% for several months is about 1% to 2% by weight of the concrete depending on the concrete's constituents, original water content, drying conditions, and the size and shape of the concrete element.

After concrete has dried to constant moisture content at one humidity condition, a decrease in humidity causes it to lose moisture and an increase causes it to gain moisture. The concrete shrinks or swells with each such change in moisture content due primarily to the response of the cement paste to moisture loss. Most aggregates show little response to changes in moisture content, although there are a few aggregates that swell or shrink in response to such changes.

As drying takes place, concrete near the surface dries and shrinks faster than the inner concrete, causing

tensile stress and possible cracks. Major random cracks may develop if joints are not properly provided and the concrete element is restrained from shortening (Fig. 13-5). Fig. 13-6 illustrates the relationship between drying rate at different depths, drying shrinkage, and weight loss for normal-weight concrete.*

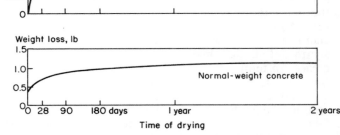

Fig. 13-5. Drying-shrinkage cracks like these often result from improper joint spacing.

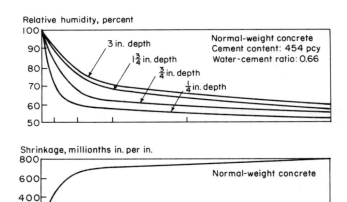

Fig.13-6. Relative humidity distribution, drying shrinkage, and weight loss of 6x12-in. cylinders moist-cured 7 days followed by drying at 73°F and 50% RH. Reference 13-28.

*Reference 13-28 also gives relationships for lightweight concrete.

154

Shrinkage may continue for a number of years, depending on the size and shape of the concrete mass. The rate and ultimate amount of shrinkage are usually smaller for large masses of concrete than for small masses, although shrinkage continues longer for large masses. Higher volume-to-surface ratios (larger elements) yield lower shrinkage as shown in Fig. 13-7.

The rate and amount of drying shrinkage for small concrete specimens are shown in Fig. 13-8. Specimens were initially moist-cured for 14 days at 70°F, then stored for 38 months in air at the same temperature and 50% relative humidity. Shrinkage recorded at the age of 38 months ranged from 600 to 790 millionths. An average of 34% of this shrinkage occurred within the first month. At the end of 11 months an average of 90% of the 38-month shrinkage had taken place. The general uniformity of shrinkages of all the types of cement at the different ages can be noted.

If the tensile stress that results from restrained drying shrinkage exceeds the tensile strength of the concrete, cracks will develop. Where there is no restraint, movement occurs freely and no stresses or cracks develop. Recommendations for proper joint spacing to control cracks are discussed in Chapter 9. See "Plastic Shrinkage Cracking" in Chapter 11 for information on early volume changes.

Effect of Concrete Ingredients on Drying Shrinkage

The most important controllable factor affecting shrinkage is the amount of water per unit volume of concrete. The results of tests illustrating the water content-shrinkage relationship are shown in Fig. 13-9. Shrinkage can be minimized by keeping the water

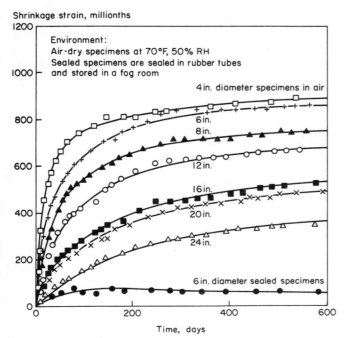

Fig. 13-7. Drying shrinkage of various sizes of cylindrical specimens of Elgin gravel concrete. Reference 13-24.

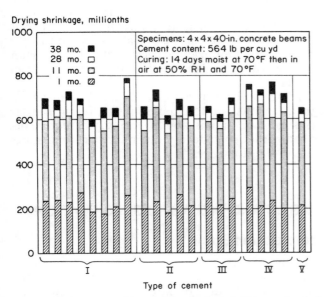

Fig. 13-8. Results of long-term drying shrinkage tests by the U.S. Bureau of Reclamation. Shrinkage ranged from 600 to 790 millionths after 38 months of drying. The shrinkage of concretes made with air-entraining cements was similar to that for non-air-entrained concretes in this study. References 13-4 and 13-6.

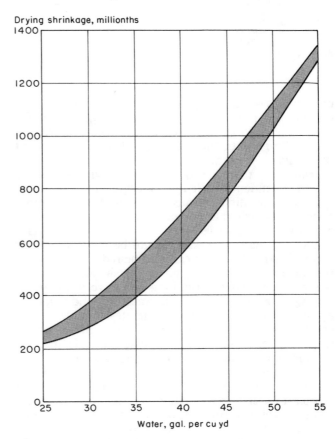

Fig. 13-9. Relationship of total water content and drying shrinkage. A large number of mixtures with various proportions is represented within the area of the band curves. Drying shrinkage increases with increasing water contents. 1 gal = 8.34 lb of water.

content of concrete as low as possible. This is achieved by keeping the total coarse aggregate content of the concrete as high as possible. Use of low slumps and placing methods that minimize water requirements are thus major factors in controlling concrete shrinkage. Any practice that increases the water requirement of the cement paste, such as the use of high slumps, excessively high freshly mixed concrete temperatures, high fine-aggregate contents, or small-size coarse aggregate, will increase shrinkage. Work performed at the Massachusetts Institute of Technology showed that for each 1% increase in mixing water, concrete shrinkage increased about 2%.*

The type of cement, cement fineness and composition, and cement content have relatively little effect on the drying shrinkage of normal-strength concrete. Table 13-3 illustrates that within the range of practical con-

shrinkage. Drying shrinkage can be evaluated in accordance with ASTM C 157.

CURLING (WARPING)

In addition to horizontal movement caused by changes in moisture and temperature, curling of slabs on ground can be caused by differences in moisture content and temperature between the top and bottom of slabs.

The edges at the joints tend to curl upward when the surface of a slab is drier or cooler than the bottom. A slab will assume a reverse curl when the surface is wetter or warmer than the bottom, but enclosed slabs, such as floors on ground, curl only upward. When the edges of an industrial floor slab are curled upward, lift-

Table 13-3. Effect of Cement Content on Drying Shrinkage of Concrete*

| Cement content, bags/cu yd** | Concrete composition by absolute volume† | | | | | Water + Air | Water-cement ratio by wt. | Slump, in. | Shrinkage†† |
	Cement	Water	Air	Total, paste	Aggregate				
4.99	0.089	0.202	0.017	0.308	0.692	0.219	0.72	3.3	0.0330
5.99	0.107	0.207	0.016	0.330	0.670	0.223	0.62	3.6	0.0330
6.98	0.124	0.210	0.014	0.348	0.652	0.224	0.54	3.8	0.0289
8.02	0.143	0.207	0.015	0.365	0.635	0.223	0.46	3.8	0.0300

*American River sand and gravel graded to 1-in. maximum.
**One bag equals 94 lb.
†Average of 3 batches.
††Average of nine 3x3x10-in. prisms cured wet for 7 days, then dried for 14 days.
Reference 13-17.

crete mixes—5 to 8 bags of cement per cubic yard—cement content has little effect on shrinkage of concrete.

Aggregates in concrete, especially coarse aggregate, physically restrain the shrinkage of hydrated cement paste. Paste content affects the drying shrinkage of mortar more than that of concrete. Drying shrinkage is also dependent on the type of aggregate. Hard, rigid aggregates are difficult to compress and provide more restraint to shrinkage of cement paste. As an extreme example, if steel balls were substituted for ordinary coarse aggregate, shrinkage would be reduced 30% or more. Drying shrinkage can also be reduced by avoiding aggregates that have high drying shrinkage properties and aggregates containing excessive amounts of clay. Quartz, granite, feldspar, limestone, and dolomite aggregates generally produce concretes with low drying shrinkages.** Steam curing will also reduce drying shrinkage. (See Chapter 10, "Curing Concrete.")

Some admixtures require an increase in the unit water content of concrete and for this reason they can be expected to increase drying shrinkage. The use of accelerators such as calcium chloride results in increased drying shrinkage of concrete. Despite reductions in water content, many chemical admixtures of the water-reducing type increase drying shrinkage substantially, particularly those that contain an accelerator to counteract the retarding effect of the admixture. Air entrainment and some finely divided mineral admixtures such as fly ash, have little or no effect on drying

truck traffic passing over joints causes a repetitive vertical deflection that creates a great potential for fatigue cracking in the slab. The amount of vertical curl upward (curling) is small for a short, thick slab.

Even though designers are aware of moisture effects and the existence of curling of slabs, they tend to use only temperature-range changes as the design factor when determining slab length and the size of joint openings.†

Curling can be reduced or eliminated by using design and construction techniques that minimize shrinkage differentials and by using techniques described earlier to reduce temperature and moisture-related volume changes.††

ELASTIC AND INELASTIC DEFORMATION

Compression Strain

The series of curves in Fig. 13-10 illustrate the amount of compressive stress and strain that results instantaneously due to loading of unreinforced concrete. With

---

*Reference 13-2, page 425.
**Reference 13-39.
†Reference 13-40.
††See References 13-46 and 13-47 for more information.

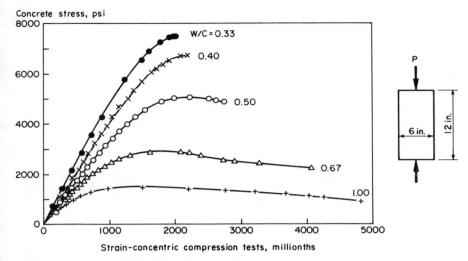

Fig. 13-10. Stress-strain curves for compression tests on 6x12-in. concrete cylinders at an age of 28 days. Reference 13-7.

water-cement ratios of 0.50 or less and strains up to 1500 millionths, the upper three curves show that strain is closely proportional to stress; in other words the concrete is almost elastic. The upper portions of the curves and beyond show that the concrete is inelastic. The curves for high-strength concrete have sharp peaks, whereas those for lower-strength concretes have long and relatively flat peaks. Fig. 13-10 also shows the sudden failure characteristics of higher strength concrete cylinders.

When load is removed from concrete in the inelastic zone, the recovery line usually is not parallel to the original line for the first load application. Therefore, the amount of permanent set may differ from the amount of inelastic deformation (Fig. 13-11).

The term "elastic" is not favored for general discussion of concrete behavior because frequently the strain may be in the inelastic range. For this reason, the term "instantaneous strain" is often used.

Modulus of Elasticity

The ratio of stress to strain in the elastic range of a stress-strain curve for concrete defines the modulus of elasticity (E) of that concrete (Fig. 13-11). Normal-weight concrete has a modulus of elasticity of 2,000,000 psi to 6,000,000 psi, depending on factors such as compressive strength and aggregate type. For normal-weight concrete with compressive strengths (f_c') between 3000 psi and 5000 psi, the modulus of elasticity can be estimated as 57,000 times the square root of f_c'. The modulus of elasticity for structural lightweight concrete is between 1,000,000 psi and 2,500,000 psi. E for any particular concrete can be determined in accordance with ASTM C 469.

Deflection

Deflection of concrete beams and slabs is one of the more common and obvious building movements. The

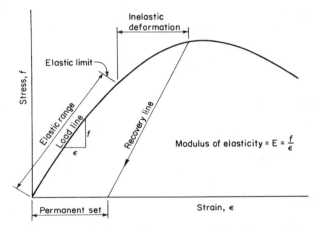

Fig. 13-11. Generalized stress-strain curve for concrete. Reference 13-40.

deflections are the result of flexural strains that develop under dead load and live loads and that may result in cracking in the tensile zone of concrete members. Reinforced concrete structural design anticipates these tension cracks. ACI 318R-83 Section 10.6. Commentary on the Building Code Requirements for Reinforced Concrete limits flexural crack width to 0.016 in. for interior concrete and 0.013 in. for exterior concrete. Table 13-1 lists tolerable crack widths for reinforced concrete with respect to various exposure conditions.*

Concrete members are often cambered, that is, built with an upward bow, to compensate for the expected deflection.

Poisson's Ratio

When a block of concrete is loaded in uniaxial compression, as in Fig. 13-12, it will shorten and at the

---

*Flexural deflection and crack width calculations are given in Reference 13-40.

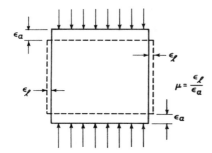

Fig. 13-12. Ratio of lateral to axial strain is Poisson's ratio, μ. Reference 13-40.

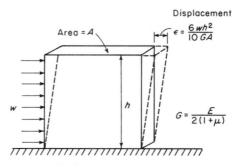

Fig. 13-13. Strain that results from shear forces on a body. G = shear modulus. μ = Poisson's ratio. Strain resulting from flexure is not shown. Reference 13-40.

same time develop a lateral strain or bulging. The ratio of lateral to axial strain is called Poisson's ratio, μ. A common value used is 0.20 to 0.21, but the value may vary from 0.15 to 0.25 depending upon the aggregate, moisture content, concrete age, and compressive strength. Poisson's ratio (ASTM C 469) is generally of no concern to the structural designer, but is used in advanced structural analysis of flat-plate floors, shell roofs, arch dams, and mat foundations.

Shear Strain

Concrete, like other materials, deforms under shear forces. The shear strain produced is important in determining the load paths or distribution of forces in indeterminate structures—for example where shear-walls and columns both participate in resisting horizontal forces in a concrete building frame. The amount of movement, while not large, is significant in short, stubby members; in larger members it is overshadowed by flexural strains.

The shear modulus (modulus of rigidity), G, in Fig. 13-13, varies with the strength and temperature of the concrete. When the compressive strength is between 4000 and 5000 psi, G at 75°F is approximately 42% of the elastic modulus, E.*

Torsional Strain

Plain rectangular concrete members can also fail in torsion, that is, a twisting action caused by bending about an axis parallel to the wider face and inclined at an angle of about 45 degrees to the longitudinal axis of a member. Microcracks develop at low torque; however, concrete behaves reasonably elastically up to the maximum limit of the elastic torque.**

CREEP

When concrete is loaded, the deformation caused by the load can be divided into two parts: a deformation that occurs immediately (such as elastic strain) and a time-dependent deformation that begins immediately but continues at a decreasing rate for as long as the concrete is loaded. This latter deformation is called creep.

The amount of creep is dependent upon (1) the magnitude of stress, (2) the age and strength of the concrete when stress is applied, and (3) the length of time the concrete is stressed. It is also affected by other factors related to the quality of the concrete and conditions of exposure, such as the type, amount, and maximum size of aggregate; type of cement; amount of cement paste; size and shape of concrete mass; volume to surface ratio; amount of steel reinforcement; prior curing conditions; and the ambient temperature and humidity.

Within normal stress ranges, creep is proportional to stress. In relatively young concrete, the change in volume or length due to creep is largely unrecoverable; in older or drier concrete it is largely recoverable.

The creep curves shown in Fig. 13-14 are based on tests conducted under laboratory conditions in accordance with ASTM C 512. Cylinders were loaded to almost 40% of their compressive strength. Companion cylinders not subject to load were used to measure drying shrinkage that was then deducted from the total deformation of the loaded specimens to determine creep. Cylinders were allowed to dry while under load except for those marked "sealed." The two 28-day curves for each concrete strength in Fig. 13-14 illustrate that creep of concrete loaded under drying conditions is greater than creep of concrete sealed against drying. Concrete specimens loaded at a late age will creep less than those loaded at an early age. It can be seen that as concrete strength decreases, creep increases. Fig. 13-15 illustrates recovery from the elastic and creep strains after load removal.

A combination of strains occurring in a reinforced column is illustrated in Fig. 13-16. The curves represent deformations and volume changes in a 14th-story column of a 76-story reinforced concrete building while under construction. The 16x48-in. column contained 2.08% vertical reinforcement and was designed for 9000-psi concrete.

The method of curing prior to loading has a marked effect on the amount of creep in concrete. The effects on creep of three different methods of curing are shown in Fig. 13-17. Note that very little creep occurs in

---

*References 13-22 and 13-40.
**Reference 13-27.

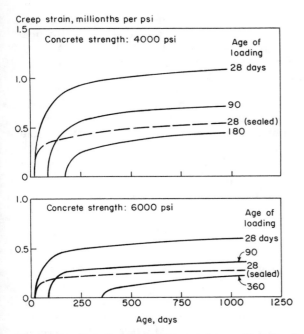

Fig. 13-14. Relationship of time and age of loading to creep of two different strength concretes. Specimens were allowed to dry during loading, except for those labeled as sealed. Reference 13-34, updated.

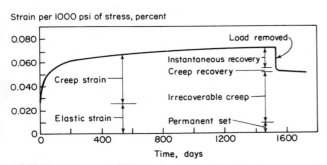

Fig. 13-15. Combined curve of elastic and creep strains showing amount of recovery. Specimens (cylinders) were loaded at 8 days immediately after removal from fog room and then stored at 70°F and 50% RH. The applied stress was 25% of the compressive strength at 8 days. References 13-24 and 13-40.

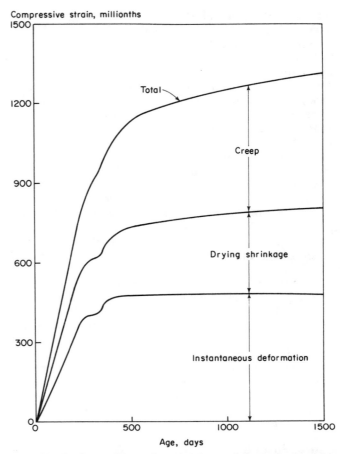

Fig. 13-16. Summation of strains in a reinforced concrete column during construction of a tall building. Reference 13-34.

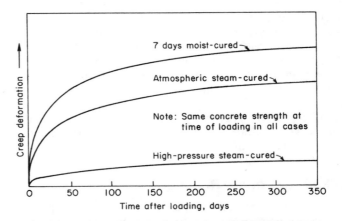

Fig.13-17. Effect of curing method on magnitude of creep for typical normal-weight concrete. Reference 13-19.

concrete that is cured by high-pressure steam (autoclaving). Note also that atmospheric steam-cured concrete has considerably less creep than 7-day moist-cured concrete. The two methods of steam curing shown in Fig. 13-17 reduce drying shrinkage of concrete about half as much as they reduce creep.

CHEMICAL CHANGES AND EFFECTS

Some volume changes of concrete result from chemical reactions that may take place shortly after placing and finishing or from later reactions within the hardened concrete in the presence of water or moisture.

Carbonation

Hardened concrete containing some moisture reacts with carbon dioxide present in air, a reaction that results in a slight shrinkage of the concrete. The effect, known as carbonation, is not destructive but actually increases the chemical stability and strength of the concrete. However, carbonation also reduces the pH of concrete and if steel is present in the carbonated area, steel corrosion can occur in absence of the protective oxide film normally provided by concrete's high pH. The expansive rust formation results in expansion, cracking, and spalling of the concrete.* The depth of the carbonation reaction is very shallow (usually less than 0.1 in. depending on age and exposure conditions) in dense, quality concrete, but can penetrate deeply in porous, poor-quality concrete. During manufacture some concrete masonry units are deliberately exposed to carbon dioxide after reaching 80% of their rated strength to induce carbonation shrinkage and make the units more dimensionally stable. Future drying shrinkage is reduced 30% or more.**

One of the causes of surface crazing of concrete is the shrinkage that accompanies natural air carbonation of young concrete.

Carbonation of another kind also can occur in freshly placed, unhardened concrete. This carbonation, the outcome of which is a soft, chalky surface called dusting, usually takes place during cold-weather concreting when there is an unusual amount of carbon dioxide in the air due to use of unvented heaters or gasoline-powered equipment in an enclosure. It is not accompanied by significant movement or cracking, and after the concrete is 24 hours old, the danger no longer exists.

Sulfate Attack

Sulfate attack of concrete can occur where soil and groundwater have a high sulfate content and when measures to reduce sulfate attack, such as the use of sulfate-resistant cement, have not been taken. The attack is greater in concrete that is exposed to moisture, such as foundations and slabs on ground. It usually results in an expansion of the concrete because of the formation of solids as a result of the chemical action. The amount of expansion in severe circumstances has often been significantly higher than 0.1%, and the disruptive effect within the concrete can result in extensive cracking and disintegration. The amount of expansion cannot be accurately predicted. Procedures to follow when sulfate attack is anticipated are discussed in Chapters 2, 5, and 7.

Alkali-Aggregate Reactions

In most parts of North America, concrete aggregates are considered more or less chemically inert in concrete. However, in some areas the aggregates react with the alkalies in cement, causing expansion and cracking over a period of years. The reaction is greater in those parts of a structure that are exposed to moisture.

A knowledge of the characteristics of local aggregates is essential. There are two types of alkali-reactive aggregates: siliceous and carbonate. Alkali-silica reaction expansion may exceed 1.5% in mortar or 0.5% in concrete and can cause the concrete to fracture and break apart.

The amount of expansion due to to alkali-carbonate reactions, which are less widespread than alkali-silica reactions, has been measured as high as 0.18% in 9 months for 3x4-in. concrete prisms exposed to 100% relative humidity at 73°F.

Structural design techniques cannot counter the effects of alkali-aggregate expansion nor can the expansion be controlled by jointing. In areas where deleteriously reactive aggregates are known to exist, special measures must be taken to prevent the occurrence of alkali-aggregate reaction.†

---

*See the carbonation-corrosion discussions in Chapter 6 for more information.
**References 13-10, 13-11, 13-14, and 13-15.
†See Chapters 4 and 6 and Reference 13-5.

REFERENCES

13-1. Davis, R. E., "A Summary of the Results of Investigations Having to Do with Volumetric Changes in Cements, Mortars, and Concretes Due to Causes Other Than Stress," *Proceedings of the American Concrete Institute,* American Concrete Institute, Detroit, vol. 26, 1930, pages 407-443.

13-2. Carlson, Roy W., "Drying Shrinkage of Concrete as Affected by Many Factors," *Proceedings of the Forty-First Annual Meeting of the American Society for Testing and Materials,* vol. 38, part II, Technical Papers, American Society for Testing and Materials, Philadelphia, 1938, pages 419-440.

13-3. Pickett, Gerald, *The Effect of Change in Moisture Content on the Creep of Concrete Under a Sustained Load,* Research Department Bulletin RX020, Portland Cement Association, 1947.

13-4. "Long-Time Study of Cement Performance in Concrete—Tests of 28 Cements Used in the Parapet Wall of Green Mountain Dam," *Materials Laboratories Report No. C-345,* U.S. Department of the Interior, Bureau of Reclamation, Denver, 1947.

13-5. Lerch, William, *Studies of Some Methods of Avoiding the Expansion and Pattern Cracking Associated with the Alkali-Aggregate Reaction,* Research Department Bulletin RX031, Portland Cement Association, 1950.

13-6. Jackson, F. H., *Long-Time Study of Cement Performance in Concrete—Chapter 9. Correlation of the Results of Laboratory Tests with Field Performance Under Natural Freezing and Thawing Conditions,* Research Department Bulletin RX060, Portland Cement Association, 1955.

13-7. Hognestad, E.; Hanson, N. W.; and McHenry, D., *Concrete Stress Distribution in Ultimate Strength Design,* Development Department Bulletin DX006, Portland Cement Association, 1955.

13-8. Malhotra, M. L., "The Effect of Temperature on the Compressive Strength of Concrete," *Magazine of Concrete Research,* Cement and Concrete Association, Wexham Springs, Slough, England, vol. 8., no. 23, August 1956, pages 85-94.

13-9. Philleo, Robert, *Some Physical Properties of Concrete at High Temperatures,* Research Department Bulletin RX097, Portland Cement Association, 1958.

13-10. Verbeck, G. J., *Carbonation of Hydrated Portland Cement,* Research Department Bulletin RX087, Portland Cement Association, 1958.

13-11. Steinour, Harold H., "Some Effects of Carbon Dioxide on Mortars and Concrete—Discussion," *Proceedings of the American Concrete Institute,* vol. 55, American Concrete Institute, 1959, pages 905-907.

13-12. Roper, Harold, *Volume Changes of Concrete Affected by Aggregate Type,* Research Department Bulletin RX123, Portland Cement Association, 1960.

13-13. Monfore, G. E., and Lentz, A. E., *Physical Properties of Concrete at Very Low Temperatures,* Research Department Bulletin RX145, Portland Cement Association, 1962.

13-14. Powers, T. C., *A Hypothesis on Carbonation Shrinkage,* Research Department Bulletin RX146, Portland Cement Association, 1962.

13-15. Toennies, H. T., and Shideler, J. J., *Plant Drying and Carbonation of Concrete Block—NCMA-PCA Cooperative Program,* Development Department Bulletin DX064, Portland Cement Association, 1963.

13-16. Monfore, G. E., *A Small Probe-Type Gage for Measuring Relative Humidity,* Research Department Bulletin RX160, Portland Cement Association, 1963.

13-17. Tremper, Bailey, and Spellman, D. L., "Shrinkage of Concrete—Comparison of Laboratory and Field Performance," *Highway Research Record Number 3, Properties of Concrete,* Transportation Research Board, National Research Council, Washington, D.C., 1963.

13-18. *Symposium on Creep of Concrete,* SP-9, American Concrete Institute, 1964.

13-19. Hanson, J. A., *Prestress Loss As Affected by Type of Curing,* Development Department Bulletin DX075, Portland Cement Association, 1964.

13-20. Abrams, M. S., and Orals, D. L., *Concrete Drying Methods and Their Effects on Fire Resistance,* Research Department Bulletin RX181, Portland Cement Association, 1965.

13-21. Brewer, H. W., *Moisture Migration—Concrete Slab-on-Ground Construction,* Development Department Bulletin DX089, Portland Cement Association, 1965.

13-22. Cruz, Carlos R., *Elastic Properties of Concrete at High Temperatures,* Research Department Bulletin, RX191, Portland Cement Association, 1966.

13-23. Lentz, A. E., and Monfore, G. E., *Thermal Conductivities of Portland Cement Paste, Aggregate, and Concrete Down to Very Low Temperatures,* Research Department Bulletin RX207, Portland Cement Association, 1966.

13-24. Hansen, Torben C., and Mattock, Alan H., *Influence of Size and Shape of Member on the Shrinkage and Creep of Concrete,* Development Department Bulletin DX103, Portland Cement Association, 1966.

13-25. Petersen, R. H., "Resistance to High Temperature," *Significance of Tests and Properties of Concrete and Concrete-Making Materials,* STP 169-A, American Society for Testing and Materials, 1966.

13-26. Brewer, Harold W., *"General Relation of Heat Flow Factors to the Unit Weight of Concrete,* Development Department Bulletin DX114, Portland Cement Association, 1967.

13-27. Hsu, Thomas T. C., *Torsion of Structural Concrete—Plain Concrete Rectangular Sections,* Development Department Bulletin DX134, Portland Cement Association, 1968.

13-28. Hanson, J. A., *Effects of Curing and Drying Environments on Splitting Tensile Strength of Concrete,* Development Department Bulletin DX141, Portland Cement Association, 1968.

13-29. *Effect of Long Exposure of Concrete to High Temperature,* ST32, Portland Cement Association, 1969.

13-30. "Effect of High Temperature on Hardened Concrete," *Concrete Construction,* Concrete Construction Publications, Inc., Addison, Illinois, November 1971, pages 477-479.

13-31. *Temperature and Concrete,* SP-25, American Concrete Institute, 1971.

13-32. *Concrete for Nuclear Reactors,* vol. 1, SP-34, American Concrete Institute, 1972.

13-33. Abrams, M. S., *Compressive Strength of Concrete at Temperatures to 1600°F,* Research and Development Bulletin RD016T, Portland Cement Association, 1973.

13-34. Russell, H. G., and Corley, W. G., *Time-Dependent Behavior of Columns in Water Tower Place,* Research and Development Bulletin RD052B, Portland Cement Association, 1977.

13-35. Kaar, P. H.; Hanson, N. W.; and Capell, H. T., *Stress-Strain Characteristics of High-Strength Concrete,* Research and Development Bulletin RD051D, Portland Cement Association, 1977.

13-36. Abrams, Melvin S., *Performance of Concrete Structures Exposed to Fire,* Research and Development Bulletin RD060D, Portland Cement Association, 1977.

13-37. Abrams, M. S., *Behavior of Inorganic Materials in Fire,* Research and Development Bulletin RD067M, Portland Cement Association, 1979.

13-38. Cruz, C. R., and Gillen, M., *Thermal Expansion of Portland Cement Paste, Mortar, and Concrete at High Temperatures,* Research and Development Bulletin RD074T, Portland Cement Association, 1980.

13-39. *Control of Cracking in Concrete Structures,* ACI 224R-80 ACI Committee 224 Report, American Concrete Institute, 1980.

13-40. *Building Movements and Joints,* EB086B, Portland Cement Association, 1982.

13-41. *Prediction of Creep, Shrinkage, and Temperature Effects in Concrete Structures,* ACI 209R-82, ACI Committee 209 Report, American Concrete Institute, 1982.

13-42. Carette, G. G.; Painter, K. E.; and Malhotra, V. M., "Sustained High Temperature Effect on Concretes Made with Normal Portland Cement, Normal Portland Cement and Slag, or Normal Portland Cement and Fly Ash," *Concrete International,* American Concrete Institute, July 1982.

13-43. *Causes, Evaluation, and Repair of Cracks in Concrete Structures,* ACI 224.1R-84, ACI Committee 224 Report, American Concrete Institute, 1984.

13-44. Kosmatka, Steven H., "Floor-Covering Materials and Moisture in Concrete," *Concrete Technology Today,* PL853B, Portland Cement Association, 1985.

13-45. *Temperature Effects on Concrete,* STP 858, American Society for Testing and Materials, 1985.

13-46. *Concrete Slab Surface Defects: Causes, Prevention, Repair,* IS177T, Portland Cement Association, 1987.

13-47. Ytterberg, Robert F., "Shrinkage and Curling of Slabs on Grade, Part I—Drying Shrinkage, Part II—Warping and Curling, and Part III—Additional Suggestions," *Concrete International,* American Concrete Institute, April 1987, pages 22-31; May 1987, pages 54-61; and June 1987, pages 72-81.

CHAPTER 14

Control Tests for Quality Concrete

Satisfactory concrete construction and performance requires concrete possessing specific properties. To assure that these properties are obtained, quality control and acceptance testing are indispensable parts of the construction process. Past experience and sound judgment must be relied on in evaluating tests and assessing their significance in the ultimate performance of the concrete.

CLASSES OF TESTS

In general, specifications for concrete and its component materials give detailed requirements for limits of acceptability. The requirements may affect (1) characteristics of the mixture, such as maximum size of aggregate or minimum cement content; (2) characteristics of the cement, water, aggregates, and admixtures; and (3) characteristics of the freshly mixed and hardened concrete, such as temperature, slump, air content, or compressive strength.

Cements are tested for their compliance with established standards to avoid any abnormal performance such as early stiffening, delayed setting, or low strengths in concrete.

Tests of aggregates have two major purposes: First, to determine the suitability of the material itself for use in concrete, including tests for abrasion, soundness, specific gravity, and petrographic and chemical analysis; second, to assure uniformity, such as tests for moisture control and gradation of aggregates. Some tests are used for both purposes.

Tests of concrete to evaluate the performance of available materials, to establish mixture proportions, and to control concrete quality in the field include slump, air content, unit weight, and strength. Slump, air content, and strength tests are usually required in project specifications for concrete quality control, whereas unit weight is used more in mixture proportioning.

Following is a discussion of frequency of testing and descriptions of the major control tests to ensure uniformity of materials, desired properties of freshly mixed concrete, and required strength of hardened concrete. Special tests are also described.

FREQUENCY OF TESTING

Frequency of testing is a significant factor in the effectiveness of quality control of concrete.

The frequency of testing aggregates and concrete for typical batch-plant procedures depends largely upon the uniformity of materials, including the moisture content of aggregates. Initially it is advisable to make tests several times a day, but as work progresses the frequency often can be reduced.

Usually, moisture tests are made once or twice a day. The first batch of fine aggregate in the morning is often overly wet since moisture will migrate overnight to the bottom of the storage bin. As fine aggregate is drawn from the bottom, the moisture content should become stabilized at a lower level and the first moisture test can be made. After a few tests, changes in moisture content can be judged fairly accurately by sight and by feel. Subsequent tests are usually necessary only when a change is readily apparent.

Slump tests should be made for the first batch of concrete each day, whenever consistency of concrete appears to vary, and whenever strength-test cylinders are made at the jobsite.

Air-content tests should be made often enough at the point of delivery to ensure proper air content, particularly if temperature and aggregate grading change. An air-content test is desirable for each sample of concrete from which cylinders are made; a record of the temperature of each sample of concrete should also be kept.

The number of strength tests made will depend on the job specifications and the occurrence of variations. Building Code Requirements for Reinforced Concrete (ACI 318) specifies that strength tests of each class of concrete placed each day should be taken not less than once a day, nor less than once for each 150 cu yd of concrete, nor less than once for each 5000 sq ft of surface area for slabs or walls. The average strength of two cylinders is required for each test. Additional specimens may be required when high-strength concrete is involved or where structural requirements are critical. The specimens should be laboratory cured. Specifications may require that additional specimens be made and field-cured, as nearly as practical in the same manner as the concrete in the structure. A 7-day test cylinder, along with the two 28-day test cylinders, is often made and tested to provide an early indication of strength development. As a rule of thumb, the 7-day strength is about 60% to 75% of the 28-day strength, depending upon the type and amount of cement, water-cement ratio, curing temperature, and other variables.

TESTING AGGREGATES

Sampling Aggregates

Methods for obtaining representative samples of aggregates are given in ASTM D 75. Accurate sampling is important. Reducing large field samples to small quantities for individual tests must be done with care so that the final samples will be truly representative. For coarse aggregate, this is usually done by the quartering method: The sample, thoroughly mixed, is spread on a piece of canvas in an even layer 3 or 4 in. thick. It is divided into four equal parts. Two opposite parts are then discarded. This process is repeated until the desired size of sample remains. A similar procedure is sometimes used for moist, fine aggregate. Sample splitters are desirable for dry fine aggregate (Fig. 14-1).

Organic Impurities

Organic impurities in fine aggregate should be determined in accordance with ASTM C 40. A sample of fine aggregate is placed in a sodium hydroxide solution and shaken. The following day the color of the solution is compared with a standard color solution. If the color is darker than the standard, the fine aggregate should

Fig. 14-1. Sample splitter commonly used to reduce sand samples.

not be used for important work without further investigation. Some fine aggregates contain small quantities of coal or lignite that give the liquid a dark color. The quantity may be insufficient to reduce the strength of the concrete appreciably and the fine aggregate may be acceptable otherwise. In such cases, mortar strength tests (ASTM C 87) using the fine aggregate in question will indicate the effect of the impurities present. It should be noted that appreciable quantities of coal or lignite in aggregates can cause popouts and staining of the concrete and can reduce durability when concrete is exposed to weathering. Local experience is often the best indication of the durability of concrete made with such aggregates.

Objectionable Fine Material

Large amounts of clay and silt in aggregates can adversely affect durability, increase water requirements, and increase shrinkage. Specifications usually limit the amount of material passing the No. 200 sieve to 2% or 3% in fine aggregate and to 1% or less in coarse aggregate. Testing for material finer than the No. 200 sieve should be done in accordance with ASTM C 117. Testing for clay lumps should be in accordance with ASTM C 142.

Grading

Gradation of aggregates significantly affects concrete mixture proportioning and workability. Hence, gradation tests are an important element in the assurance of concrete quality. The grading of an aggregate is determined by a sieve analysis test in which the particles are divided into their various sizes by standard sieves. The analysis should be made in accordance with ASTM C 136.

Results of sieve analyses are used in three ways: (1) to determine whether or not the materials meet specifications; (2) to select the most suitable material if several aggregates are available; and (3) to detect variations in grading that are sufficient to warrant blending selected sizes or an adjustment of concrete mix proportions.

The grading requirements for concrete aggregate are shown in Chapter 4 and ASTM C 33. Materials containing too much or too little of any one size should be avoided. Some specifications require that mixture proportions be adjusted if the average fineness modulus of fine aggregate changes by more than 0.20. Other specifications require an adjustment in mixture proportions if the amount retained on any two consecutive sieves changes by more than 10% by weight of the total fine-aggregate sample. A small quantity of clean particles that pass a No. 100 sieve but are retained on a No. 200 sieve is desirable for workability. For this reason most specifications permit up to 10% of this material in fine aggregate.

Moisture Content of Aggregates

Several methods can be used for determining the

amount of moisture in aggregate samples. The total moisture content for fine or coarse aggregate can be tested in accordance with ASTM C 566. In this method a measured sample of damp aggregate is dried either in an oven or over a hotplate or open fire. From the weights before and after drying, the total and surface (free) moisture contents can be calculated. The total moisture content can be calculated as follows:

$$P = 100(W - D)/D$$

where

P = moisture content of sample, percent
W = weight of original sample
D = weight of dried sample

The surface moisture content is equal to the total moisture content minus the absorption. Absorption can be assumed as 1% for average aggregates or, for greater accuracy, it should be determined in accordance with the methods given in ASTM C 127 for coarse aggregate and ASTM C 128 for fine aggregate. Only the surface moisture, not the absorbed moisture, becomes part of the mixing water in concrete.

Another method to determine moisture content, which is not as accurate, is to evaporate the moisture by burning alcohol. In this method, a measured sample of damp fine aggregate is placed in a shallow pan; alcohol (about 5 oz for each pound) is poured over the sample; the mixture is stirred with a rod and spread in a thin layer over the bottom of the pan. The alcohol is then ignited and allowed to burn until the sample is dry. After burning, the sand is cooled for a few minutes and weighed. The percentage of moisture is then computed.

A test for surface (free) moisture in fine aggregate can also be made in accordance with ASTM C 70. The same procedure can be used for coarse aggregate with appropriate changes in the size of sample and dimensions of the container. This test depends on displacement of water by a known weight of moist aggregate; therefore, the relative density (specific gravity) of the aggregate must be known accurately.

Electric moisture meters are used in many concrete batching plants to check the moisture content of fine aggregates. They operate on the principle that the electrical resistance of damp fine aggregate decreases as moisture content increases, within the range of dampness normally encountered. The meters measure the electrical resistance of the fine aggregate between electrodes protruding into the batch hopper or bin. Such meters require periodic calibration and must be maintained properly. They measure moisture content accurately and rapidly, but only at the level of the electrodes.

Table 14-1 illustrates a method of adjusting batch weights for moisture in aggregates.

TESTING FRESHLY MIXED CONCRETE

Sampling Freshly Mixed Concrete

The importance of obtaining truly representative samples of freshly mixed concrete for control tests must be emphasized. Unless the sample is representative, test results will be misleading. Samples should be obtained and handled in accordance with ASTM C 172. Except for routine slump and air-content tests, this method requires that the sample be at least 1 cu ft, used within 15 minutes of the time it was taken, and protected from sunlight, wind, and other sources of rapid evaporation during this period. The sample should not be taken from the very first or last portion of the batch discharge.

Consistency

The slump test, ASTM C 143, is the most generally accepted method used to measure the consistency of concrete (Fig. 14-2). The test equipment consists of a slump cone (a metal conical mold 12 in. high, with an 8-in.-diameter base and 4-in.-diameter top) and a steel rod (⅝ in. in diameter, 24 in. long) with a hemispherically shaped tip. The dampened slump cone, placed upright on a flat, solid surface, should be filled in three layers of approximately equal volume. Therefore, the cone should be filled to a depth of about 2½ in. (after rodding) for the first layer, to about 6 in. for the second layer, and overfilled for the third layer. Each layer is rodded 25 times. Following rodding, the last layer is struck off and the cone is slowly removed vertically as the concrete subsides or settles to a new height. The empty slump cone is then placed next to the settled concrete. The slump is the vertical distance the con-

Table 14-1. Example of Adjustment in Batch Weights for Moisture in Aggregates

Concrete ingredients	Batch weight (aggregates in SSD condition),* lb	Surface (free) moisture content of aggregates, percentage above SSD	Correction for surface moisture in aggregates, lb	Adjusted batch weight, lb
Cement	600	—	—	600
Fine aggregate	1200	6	+72	1272
Coarse aggregate	1800	1	+18	1818
Water	300	—	−90	210
Total	3900			3900

*An aggregate in a saturated, surface-dry (SSD) condition is one with its permeable voids filled with water and with no free water on its surface.

A

B

Fig. 14-2. Slump test for consistency of concrete. Figure A illustrates a low slump, Figure B a high slump.

test is especially applicable to stiff and extremely dry mixes, and the flow table is especially applicable to flowing concrete.

Temperature Measurement

Because of the important influence concrete temperature has on the properties of freshly mixed and hardened concrete, many specifications place limits on the temperature of fresh concrete. Glass or armored thermometers are available (Figs. 14-3 and 12-12). The thermometer should be accurate to plus or minus 1°F and should remain in a representative sample of concrete for a minimum of 2 minutes or until the reading stabilizes. A minimum of 3 in. of concrete should surround the sensing portion of the thermometer. Electronic temperature meters with precise digital readouts are also available. The temperature measurement (ASTM C 1064) should be completed within 5 minutes after obtaining the sample.

Fig. 14-3. A thermometer is used to take the temperature of fresh concrete.

crete settles, measured to the nearest ¼ in., from the top of the slump cone (mold) to the displaced original center of the subsided concrete.

A high slump value is indicative of a wet or fluid concrete. The slump test should be started within 5 minutes after the sample has been obtained and the test should be completed in 2½ minutes, as concrete loses slump with time.

Another test method involves the use of the K-Slump Tester. This is a probe-type instrument that is thrust into the concrete in any location where there is a minimum of 6 in. of concrete around the tester. The amount of mortar flowing into openings in the tester is a measure of consistency. The test has not been standardized by ASTM.

Additional consistency tests include the British compacting factor test, Powers remolding test, German flow table test (DIN 1048), Vebe test, ball penetration test (ASTM C 360), and the inverted slump cone (ASTM C 995 for fiber-reinforced concrete). The Vebe

Unit Weight and Yield

The unit weight and yield of freshly mixed concrete are determined in accordance with ASTM C 138. The results can be sufficiently accurate to determine the quantity of concrete produced per batch (see Chapter 7). The test also can give indications of air content provided the specific gravities of the ingredients are known. A balance or scale sensitive to 0.1 lb is required. The size of the container used to determine unit weight and yield varies with the size of aggregate; the 0.5 cu ft container is commonly used with aggregates up to 2 in. Care is needed to consolidate the concrete adequately and strike off the surface so that the container is filled properly. The container should be calibrated periodically. The unit weight is expressed in pounds per cubic foot and the yield (volume of the batch) in cubic feet.

The unit weight of unhardened as well as hardened concrete can also be determined by nuclear methods, ASTM C 1040.

Air Content

A number of methods for measuring air content of freshly mixed concrete can be used. ASTM standards include the pressure method (C231), the volumetric method (C173), and the gravimetric method (C138). Variations of the first two methods can also be used.

The pressure method (Fig. 14-4) is based on Boyle's law, which relates pressure to volume. Many commercial air meters of this type are calibrated to read air content directly when a predetermined pressure is applied. The applied pressure compresses the air within the concrete sample, including the air in the pores of aggregates. For this reason, tests by this method are not suitable for determining the air content of concretes made with some lightweight aggregates or other very porous materials. Correction factors for normal-weight aggregates are relatively constant and, though small, should be applied to obtain the correct amount of entrained air. The instrument should be calibrated for various elevations above sea level if it is to be used in localities having considerable differences in elevation. Some meters utilize change in pressure of a known volume of air and are not affected by changes in elevation. Pressure meters are widely used because the mix proportions and specific gravities of the materials need not be known. Also a test can be conducted in less time than is required for other methods.

The volumeteric method (Fig. 14-5) requires removal of air from a known volume of concrete by agitating the concrete in an excess of water. This method can be used for concrete containing any type of aggregate including lightweight or porous materials. The test is not affected by atmospheric pressure, and specific gravity of the materials need not be known. Care must be taken to agitate the sample sufficiently to remove all air.

The gravimetric method utilizes the same test equipment as that for unit weight of concrete. The measured unit weight of concrete is subtracted from the theoretical unit weight as determined from the absolute volumes of the ingredients, assuming no air is present. This difference, expressed as a percentage of the theoretical unit weight, is the air content. Mixture proportions and specific gravities of the ingredients must be

Fig. 14-5. Volumetric air meter.

accurately known, otherwise results may be in error. Consequently, this method is suitable only where laboratory-type control is exercised. Significant changes in unit weight can be a convenient way to detect variability in air content.

A pocket-size air indicator (AASHTO T199) can be used for quick checks of air content, but it is not a substitute for the other more accurate methods (Fig. 14-6). A representative sample of mortar from the concrete is placed in the container. The container is then filled with alcohol and rolled with the thumb over the open end to remove the air from the mortar. The approximate air content is determined by comparing the drop in the level of the alcohol with a calibration chart. The test can be performed in a few minutes. It is especially useful in checking air contents in small areas near the surface that may have suffered reductions in air because of faulty finishing procedures.[*]

With any of the above methods, air-content tests should be started within 5 minutes after the sample has been obtained.

Recent studies into the effect of fly ash on the air-void stability of concrete have resulted in the development of the foam-index test. The test can be used to measure the relative air-entraining admixture requirements for concrete mixtures containing fly ash. The fly ash to be tested is placed in a widemouth jar along with the air-entraining admixture and shaken vigorously. Following a waiting period of 45 seconds, a visual determination of the stability of the foam or bubbles is made.[**]

Fig. 14-4. Pressure-type meter for determining air content.

[*]Reference 14-18.
[**]Reference 14-21.

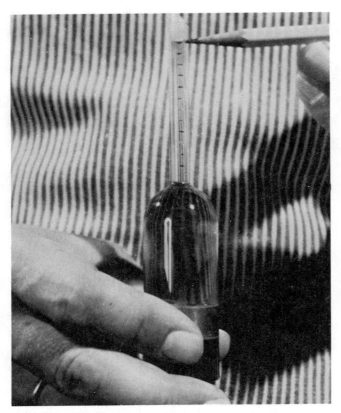

Fig. 14-6. Air indicator for checking approximate air content of concrete.

Fig. 14-7. Preparing standard test specimens for compressive strength of concrete.

Strength Specimens

Premolded specimens for strength tests should be made and cured in accordance with ASTM C 31 (field specimens) or ASTM C 192 (laboratory specimens). Molding of strength specimens should be started within 15 minutes after the sample is obtained.

The standard test specimen for compressive strength of concrete with a maximum aggregate size of 2 in. or smaller is a cylinder 6 in. in diameter by 12 in. high (Fig. 14-7). For larger aggregates, the diameter of the cylinder should be at least three times the maximum-size aggregate, and the height should be twice the diameter. While rigid metal molds are preferred, paraffined cardboard, plastic, or other types of disposable molds conforming to ASTM C 470 can be used. They should be placed on a smooth, level surface and filled carefully to avoid distortion of their shape.

Recently, 4-in.-diameter by 8-in.-high cylinder molds have been used with concrete containing up to 1 in. maximum-size aggregate.* The 4x8-in. cylinder is easier to cast, requires less sample, weighs considerably less than a 6x12-in. concrete cylinder and is therefore easier to handle, and requires less moist-curing storage space. Although 4-in.-diameter cylinders tend to break slightly higher than 6-in.-diameter cylinders, the difference is usually insignificant. Cylinders other than 6x12 in. should not be used unless specified.

Beams for the flexural strength test should be 6x6 in. in cross section for aggregates up to 2 in. For larger aggregates, the minimum cross-sectional dimension should be not less than three times the maximum size of aggregate. The length of beams should be at least three times the depth of the beam plus 2 in., or a total length of not less than 20 in. for a 6x6-in. beam.

Test cylinders to be rodded should be filled in three approximately equal layers with each layer rodded 25 times for 6-in.-diameter cylinders; beam specimens up to 8 in. deep should be filled in two equal layers with each layer rodded once with a ⅝ in. rod for each 2 sq in. of the specimen's top surface area. If the rodding leaves holes, the sides of the mold should be lightly tapped with a mallet or open hand. Beams over 8 in. deep and cylinders 12 to 18 in. deep to be vibrated should be filled in two layers; beams 6 to 8 in. deep to be vibrated can be filled in one layer. Concrete with a slump in excess of 3 in. should be rodded; concrete with a slump less than 1 in. should be vibrated; 1- to 3-in.-slump concrete can be rodded or vibrated. Internal vibrators should have a maximum width of no more than ⅓ the width of beams or ¼ the diameter of cylinders. Immediately after casting, the tops of the specimens should be (1) covered with an oiled glass or steel plate, (2) sealed with a plastic bag, or (3) sealed with a plastic cap.

The strength of a test specimen can be greatly affected by jostling, changes in temperature, and exposure to drying, particularly within the first 24 hours after casting. Thus, test specimens should be cast in locations where subsequent movement is unnecessary and

---

*References 14-15 and 14-23.

where protection is possible. Cylinders and test beams should be protected from rough handling at all ages.

Standard testing procedures require that specimens be cured under controlled conditions, either in the laboratory (Fig. 14-8) or in the field. Controlled laboratory curing in a moist room or in limewater gives a more accurate indication of the quality of the concrete as delivered. Specimens cured in the field in the same manner as the structure they represent may give a more accurate indication of the actual strength of concrete in the structure at the time of testing, but they give little indication of whether a deficiency is due to the quality of the concrete as delivered or to improper handling and curing. On some jobs, field-cured specimens are made in addition to those given controlled laboratory curing, especially when the weather is unfavorable, to determine when forms can be removed or when the structure can be put into use.*

In-place concrete strength development can also be evaluated by maturity testing (ACI 306 and ASTM C 1074), which was discussed in Chapter 12.

Fig. 14-8. Controlled moist curing in the laboratory for standard test specimens at a relative humidity of 95% to 100% and temperature of 73.4±3°F (ASTM C 511).

Time of Setting

The time of setting or rate of hardening is determined by ASTM C 403.

Accelerated Curing Tests

Accelerated strength tests can be used to expedite quality control of concrete in the production process and for the acceptance of structural concrete where adequate data correlated with the standard 28-day compressive strength test are available. Warm water (95 ± 5°F), boiling water, and autogenous accelerated curing methods used for such purposes are in ASTM C 684.

Chloride Content

The chloride content of concrete and its ingredients should be checked to make sure it is below the limit necessary to avoid corrosion of reinforcing steel. An approximation of the water-soluble chloride content of freshly mixed concrete, aggregates, and admixtures can be made using a method initiated by the National Ready Mixed Concrete Association (NRMCA).** The NRMCA method gives a quick approximation and should not be used to determine compliance. See Chapter 7 for water-soluble chloride-ion limitations for concrete.

Cement and Water Content and Water-Cement Ratio

Test methods are available for determining the cement and water content of freshly mixed concrete. These test results can assist in an estimate of strength and durability potential prior to the setting and hardening of the concrete and can affirm that the desired cement and water contents were obtained. ASTM test methods C 1078 and C 1079, based on the Kelly-Vail method, determine cement content and water content, respectively. In addition, the Willis-Hime test and Rapid Analysis Machine also measure cement content.† A combination of these test results can determine the water-cement ratio.

Mineral Admixture Content

Standard test methods are not available for determining the mineral admixture content of unhardened concrete. However, the presence of certain mineral admixtures such as fly ash can be determined by washing a sample of the concrete's mortar over a No. 325 sieve and viewing the residue retained with a stereo microscope (150 to 250X). Fly ash particles, for example, would appear as spheres of various colors. Sieving the mortar through the No. 100 or 200 sieve is helpful in removing sand grains.

Bleeding of Concrete

The bleeding properties of fresh concrete can be determined by two methods described in ASTM C 232. One method consolidates the specimen by tamping without further disturbance; the other method consolidates the specimen by vibration after which the specimen is vibrated intermittently throughout the test. The amount of bleed water at the surface is expressed as the volume of bleed water per unit area of exposed concrete or as a percentage of the net mixing water in the test specimen. The bleeding test is rarely used in the field. Bleeding was discussed in Chapter 1 under "Workability."

TESTING HARDENED CONCRETE

Premolded specimens (ASTM C 31, C 192, or C 873) or samples of hardened concrete obtained from the

---

*See "Strength Tests of Hardened Concrete" in this chapter and Reference 14-33 for more information.
**Reference 14-31.
†References 14-1, 14-7, 14-10, and 14-19.

construction (ASTM C 42, C 823, or C 873) can be used in tests on hardened concrete. Separate specimens should be obtained for different tests as specimen preconditioning for certain tests can make the specimen unusable for other tests.

Strength Tests of Hardened Concrete

Strength tests of hardened concrete can be performed on (1) cured specimens molded from samples of freshly mixed concrete, ASTM C 31 or C 192; (2) specimens cored or sawed from the hardened concrete in accordance with ASTM C 42; or (3) specimens made from cast-in-place cylinder molds, ASTM C 873 (Fig. 14-9). Cast-in-place cylinders can be used in concrete that is 5 to 12 in. in depth. For all methods, cylindrical samples should have a diameter at least three times the maximum size of the coarse aggregate in the concrete and a length as close to twice the diameter as possible. Correction factors are available in ASTM C 42 for samples with lengths of 1 to 2 times the diameter. Cores and cylinders with a height of less than 95% of the diameter before or after capping should not be tested.

Cores should not be taken until the concrete can be sampled without disturbing the bond between the mortar and the coarse aggregate. For horizontal surfaces, cores should be taken vertically and not near formed joints or edges. For vertical or sloped faces, cores should be taken perpendicular to the central portion of the concrete element. Coring through reinforcing steel should be avoided when possible. A pachometer (electromagnetic device) can be used to locate steel. Cores taken from structures that are normally wet or moist in service should be moist conditioned and tested moist, as described in ASTM C 42. Those from structures normally dry in service should be conditioned in an atmosphere approximating their service conditions and tested dry.

Test results are greatly influenced by the condition of the specimen. The ends of cylinders and cores for compression testing should be ground or capped in accordance with the requirements of ASTM C 617. Various commercially available materials can be used to cap compressive test specimens. Sulfur and granular materials can be used if the caps are allowed to harden at least two hours before the specimens are tested.* Caps should be made as thin as is practical. Reusable neoprene pads are being considered for standardization by ASTM.

Testing of specimens should be done in accordance with (1) ASTM C 39 for compressive strength (Fig. 1-6), (2) ASTM C 78 for flexural strength using third-point loading, (3) ASTM C 293 for flexural strength using center-point loading, and (4) ASTM C 496 for splitting tensile strength.

For both pavement thickness design and pavement mixture proportioning, the modulus of rupture (flexural strength) should be determined by the third-point-loading test. However, modulus of rupture by center-point loading (ASTM C 293) or cantilever loading can be used for job control if empirical relationships to

Fig. 14-9. Concrete cylinders cast in place in cylindrical molds provide means for determining the in-place compressive strength of concrete. The mold is filled in the normal course of concrete placement. The specimen is then cured in place and in the same manner as the rest of the concrete section (usually used in slabs). The specimen is removed from the concrete section and mold immediately prior to testing to determine the in-place concrete strength. This method is particularly applicable in cold-weather concreting, post-tensioning work, or any concrete work where a minimum strength must be achieved in place before construction can be continued.

third-point test results are determined before construction starts.

The moisture content of the specimen has considerable effect on the resultant strength. A saturated specimen will show lower compressive strength and higher flexural strength than those for companion specimens tested dry. This is important to consider when cores taken from hardened concrete in service are compared with molded specimens tested as they are taken from the moist-curing room.

The amount of variation in compressive-strength testing is far less than for flexural-strength testing. To avoid the extreme care needed in field flexural-strength testing to offset this disadvantage, compressive-strength tests can be used to monitor concrete quality if a laboratory-determined empirical relationship has been developed between the compressive and flexural strength of the concrete used.**

Air Content

The air-content and air-void-system parameters of hardened concrete can be determined by ASTM C 457. The hardened air-content test is performed to assure that the air-void system is appropriate for a particular

---

*In many prestressing plants, fast-setting compounds are used for capping compression cylinders to determine strength for detensioning. Such cylinders can be tested one-half hour after capping if the caps have attained strengths comparable to that expected of the cylinders.
**Reference 14-28.

environment. The test is also used to determine the effect different admixtures and methods of consolidation and placement have on the air-void system. The test can be performed on premolded specimens or samples removed from the structure. Using a polished section of a concrete sample, the air-void system is reviewed through a microscope. The information obtained from this test includes the volume of entrained air, its specific surface, and the spacing factor.

Density, Specific Gravity, Absorption, and Voids

The density, specific gravity, absorption, and voids content of hardened concrete can be determined in accordance with ASTM C 642 procedures. The boiling procedure of the method can render the specimens useless for certain additional tests, especially strength tests. The density can be obtained by multiplying the specific gravity by the unit weight of water (62.4 pcf).

Saturated, surface-dry (SSD) density is often required for specimens to be used in other tests. In this case, the density can be determined by soaking the specimen in water for 48 hours and then determining its weight in air (when SSD) and immersed in water. The SSD density is then calculated as follows:

$$D_{SSD} = \frac{W_1 \rho}{W_1 - W_2}$$

where

D_{SSD} is density in the SSD condition
W_1 is the SSD weight in air
W_2 is the weight immersed in water
ρ is the density of water, 62.4 pcf

The SSD density provides a close indication of the freshly mixed unit weight of concrete. The density of hardened concrete can also be determined by nuclear methods (ASTM C 1040).

Cement Content

The cement content of hardened concrete can be determined by ASTM C 85 and C 1084 standard methods or by the maleic acid or other nonstandard procedures.[*] Although not frequently performed, the cement-content tests are valuable in determining the cause of lack of strength gain or poor durability of concrete. Aggregate content can also be determined by these tests. The user of these test methods should be aware of certain admixtures and aggregate types that can alter test results. The presence of finely divided mineral admixtures would be reflected in the test results.

Mineral Admixture and Organic Admixture Content

The presence and amount of certain mineral admixtures, such as fly ash, can be determined by petrographic techniques (ASTM C 856). A sample of the mineral admixture used in the concrete is usually necessary as a reference to determine the type and amount of the mineral admixture present. The presence and possibly the amount of organic admixtures (such as water reducers) can be determined by infrared spectrophotometry.[**]

Chloride Content

Concern with chloride-induced corrosion of reinforcing steel has led to the need to monitor and limit the chloride content of reinforced concrete. Limits on the water-soluble chloride-ion content of hardened reinforced concrete are presented in Chapter 7. The water-soluble chloride-ion content of hardened concrete can be determined in accordance with procedures outlined in Federal Highway Administration report FHWA-RD-77-85.[†] Total chloride content can be determined by ASTM C 114 or AASHTO T260. ASTM test procedures for water-soluble chloride-ion content are currently being developed.

Petrographic Analysis

Petrographic analysis uses microscopic techniques described in ASTM C 856 to determine the constituents of concrete, concrete quality, and cause of inferior performance, distress, or deterioration. Estimating future performance and structural safety of concrete elements can be facilitated. Some of the items that can be reviewed by a petrographic examination include paste, aggregate, mineral admixture, and air content; frost and sulfate attack; alkali-aggregate reactivity; degree of hydration and carbonation; water-cement ratio; bleeding characteristics; fire damage; scaling; popouts; effect of admixture; and several other aspects.[††]

Volume and Length Change

Volume or length change limits are sometimes specified for certain concrete applications. Volume change is also of concern when a new ingredient is added to concrete to make sure there are no significant adverse effects. Length change due to drying shrinkage, chemical reactivity and forces other than intentionally applied forces and temperature changes can be determined by ASTM C 157 (water and air storage methods). Determination of early volume change of concrete before hardening can be performed using ASTM C 827. Creep can be determined in accordance with ASTM C 512. The static modulus of elasticity and Poisson's ratio of concrete in compression can be determined by methods of ASTM C 469 and dynamic values of these parameters can be determined by ASTM C 215.

Carbonation

The depth or degree of carbonation can be determined by petrographic techniques (ASTM C 856) through the

---

*References 14-8 and 14-9.
**References 14-6 and 14-30.
†Reference 14-12.
††Reference 14-30.

observation of calcium carbonate—the primary chemical product of carbonation. In addition, a phenolphthalein color test can be used to estimate the depth of carbonation by testing the pH of concrete (carbonation reduces pH). Upon application of the phenolphthalein solution to a freshly fractured surface of concrete, noncarbonated areas turn red or purple while carbonated areas remain colorless. The phenolphthalein indicator when observed against hardened paste changes color at a pH of 9.0 to 9.5. The pH of good quality noncarbonated concrete without admixtures is usually greater than 12.5.[*]

Durability

Durability refers to the ability of concrete to resist deterioration from the environment or service in which it is placed. Properly designed concrete should endure without significant distress throughout its service life. To meet project requirements, ensure or check durability, or determine the effect of certain ingredients or concreting procedures on durability, various tests can be performed. Freeze-thaw resistance can be determined in accordance with ASTM C 666, C 671, and C 682. Deicer scaling resistance can be determined by ASTM C 672. Corrosion protection and determining corrosion activity of reinforcing steel can be tested by ASTM C 876. Alkali-aggregate reactivity can be analyzed by ASTM C 227 (alkali-silica reaction), C 289, C 342, C 441 (effectiveness of mineral admixture inhibitors of alkali-silica reaction), and C 586 (alkali-carbonate reaction). Sulfate resistance can be evaluated by ASTM C 452 and C 1012. Abrasion resistance can be determined by ASTM C 418 (sandblasting), C 779 (revolving disk, dressing wheel, and ball-bearing methods), and C 944 (rotating cutter).

Moisture Content

The in-place moisture content or relative humidity of hardened concrete is sometimes useful to determine if concrete is dry enough for application of floor-covering materials and coatings. The moisture content must also be low enough to avoid spalling of concrete to be exposed to temperatures above the boiling point of water.

The most direct method for determining moisture content is to dry cut a specimen from the concrete element in question, place it in a moistureproof container, and transport it to a laboratory for test. After obtaining the specimen's initial weight, dry the specimen in an oven at about 220°F until constant weight is achieved. The difference between the two weights divided by the dry weight, times 100, yields the moisture content in percent. Moisture content or relative humidity meters are also available.

Although it can require several months of air drying, a relative humidity of about 80% or less is often required before floor-covering materials can be placed on concrete floors (see Chapter 13). Another useful test is the polyethylene-sheet test. A 4-ft-square sheet of clear polyethylene is taped to the floor; if no moisture condenses under it after 24 to 48 hours, the slab is considered dry enough for some coatings and flooring materials.[**]

Permeability

Various test methods are available for determining the permeability of concrete to various substances. Both direct and indirect methods are used. Resistance to chloride-ion penetration, for example, can be determined by ponding chloride solution on a concrete surface and, at a later age, determining the chloride content of the concrete at particular depths (AASHTO T259). The rapid chloride permeability (electrical resistance) test (AASHTO T277) correlates well with permeability and resistance to chloride-ion penetration of concrete. Various absorption methods are also used. Direct water permeability data can be obtained by using a method recommended by the American Petroleum Institute for rock. ASTM is in the process of developing a standard method for hydraulic permeability of concrete.[†]

Nondestructive Test Methods

Various nondestructive tests[††] can be used to evaluate the relative strength of hardened concrete. The most widely used are the rebound, penetration, pullout, and dynamic or vibration tests. Relatively new techniques being developed for testing the strength and other properties of hardened concrete include X-rays, gamma radiography, neutron moisture gages, magnetic cover meters, electricity, microwave absorption, and acoustic emissions. Each method has limitations, and caution should be exercised against acceptance of nondestructive test results as having a constant correlation to the traditional compression test, i.e., empirical correlations must be developed prior to use.

Rebound method. The Schmidt rebound hammer (Fig. 14-10) is essentially a surface-hardness tester that provides a quick, simple means of checking concrete uniformity. It measures the rebound of a spring-loaded plunger after it has struck a smooth concrete surface. The rebound number reading gives an indication of the compressive strength of the concrete.

The results of a Schmidt rebound hammer test (ASTM C 805) are affected by surface smoothness, size, shape, and rigidity of the specimen; age and moisture condition of the concrete; type of coarse aggregate; and carbonation of the concrete surface. When these limitations are recognized and the hammer is calibrated for the particular materials used in the concrete (Fig. 14-11), then this instrument can be useful for determining the relative compressive strength and uniformity of concrete in the structure.

Penetration method. The Windsor probe (ASTM

---

[*]References 14-3 and 14-5.
[**]Reference 14-27.
[†]See References 14-2, 14-4, 14-14, 14-16, 14-17, and 14-22 for more information.
[††]References 14-11, 14-13, 14-24, and 14-25.

Fig. 14-10. The rebound hammer gives an indication of the compressive strength of concrete.

A. Powder-actuated gun drives hardened alloy probe into concrete.

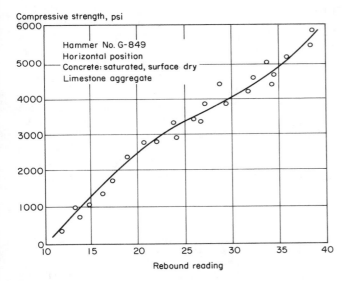

Fig. 14-11. Example of a calibration chart for an impact (rebound) test hammer.

B. Exposed length of probe is measured and relative compressive strength of the concrete then determined from calibration table.

Fig. 14-12. The Windsor-probe technique for determining the relative compressive strength of concrete.

C 803), like the rebound hammer, is basically a hardness tester that provides a quick means of determining the relative strength of the concrete. The equipment consists of a powder-actuated gun that drives a hardened alloy probe (needle) into the concrete (Fig. 14-12). The exposed length of the probe is measured and related by a calibration table to the compressive strength of the concrete.

The results of the Windsor-probe test will be influenced by surface smoothness of the concrete and the type and hardness of aggregate used. Therefore, a calibration table or curve for the particular concrete to be tested should be made, usually from cores or cast specimens, to improve accuracy.

Both the rebound hammer and the probe damage the concrete surface to some extent. The rebound hammer leaves a small indentation on the surface; the probe leaves a small hole and may cause minor cracking and small craters similar to popouts.

Pullout tests. A pullout test (ASTM C 900) involves casting the enlarged end of a steel rod in the concrete to be tested and then measuring the force required to pull it out (Fig. 14-13). The test measures the strength of the concrete—the measured strength being the direct shear strength of the concrete. This in turn is correlated with the compressive strength and thus a measurement of the in-place strength is made.

Fig. 14-13. Pullout test equipment being used to measure the strength of the concrete.

Dynamic or vibration tests. A dynamic or vibration (pulse velocity) test (ASTM C 597) is based on the principle that velocity of sound in a solid can be measured by (1) determining the resonant frequency of a specimen or (2) recording the travel time of short pulses of vibrations through a sample. High velocities are indicative of good concrete and low velocities are indicative of poor concrete.

Microseismic techniques employing low-frequency, mechanical energy can be used to detect, locate, and record discontinuities within solids. Modulus of elasticity as well as the presence and orientation of surface and internal cracking, can be determined. Fundamental transverse, longitudinal, and torsional frequencies of concrete specimens can be determined by ASTM C 215, a method frequently used in laboratory durability tests such as freezing and thawing (ASTM C 666).

Other tests. The use of X-rays for testing concrete properties is limited due to the costly and dangerous high-voltage equipment required as well as radiation hazards.

Gamma-radiography equipment can be used in the field to determine the location of reinforcement, density, and perhaps honeycombing in structural concrete units. ASTM C 1040 procedures use gamma radiation to determine the density of unhardened and hardened concrete in place.

Battery-operated magnetic detection devices like the pachometer or covermeter are available to measure the depth of reinforcement in concrete and to detect the position of rebars. Electrical-resistivity equipment is being developed to estimate the thickness of concrete pavement slabs.

A microwave-absorption method has been developed to determine the moisture content of porous building materials such as concrete. Acoustic-emission techniques show promise for studying load levels in structures and locating the origin of cracking.

Table 14-2 lists several nondestructive test methods along with main applications.*

Evaluation of Compression Test Results

The Building Code Requirements for Reinforced Concrete (ACI 318) states that the compressive strength of concrete can be considered satisfactory if the averages of all sets of three consecutive strength tests equal or exceed the specified 28-day strength and if no individual strength test (average of two cylinders) is more than 500 psi below the specified strength.**

If the strength of any laboratory-cured cylinder falls more than 500 psi below the specified strength, f'_c, the strength of the in-place concrete should be evaluated. The strength should also be evaluated if field-cured cylinders have a strength of less than 85% that of companion laboratory-cured cylinders. The 85% requirement may be waived if the field-cured strength exceeds f'_c by more than 500 psi.

When necessary, the in-place strength should be determined by testing three cores for each strength test where the laboratory-cured cylinders were more than 500 psi below f'_c. If the structure will be dry in service, prior to testing, the cores should be dried for 7 days at a temperature of 60 to 80°F and a relative humidity of less than 60%. The cores should be submerged in water for at least 40 hours before testing if the structure is to be used in a more than superficially wet state.

Nondestructive test methods are not a substitute for core tests (ASTM C 42). If the average strength of three cores is at least 85% of f'_c and if no single core is less than 75% of f'_c, the concrete in the area represented by the core is considered structurally adequate. If the results of properly made core tests are so low as to leave structural integrity in doubt, load tests as outlined in Chapter 20 of ACI 318 may be performed.†

---

*See Reference 14-25 for additional information on these and other test methods.
**In addition to the two 28-day cylinders, job specifications often require one or two 7-day cylinders and one or more "hold" cylinders. Seven-day cylinders monitor early strength gain. Hold cylinders are commonly used as backup in case the 28-day cylinders are damaged or do not come up to strength. For low 28-day breaks, the hold cylinder is tested at 56 days after standard curing.
†Refer to Chapter 7 and References 14-13, 14-20, and 14-29 for more information.

REFERENCES

14-1. Hime, W. G., and Willis, R. A., *A Method for the Determination of the Cement Content of Plastic Concrete,* Research Department Bulletin RX061, Portland Cement Association, 1955.

14-2. *Recommended Practice for Determining Permeability of Porous Media,* API RP 27, American Petroleum Institute, Washington, D.C., 1956.

Table 14-2. Nondestructive Evaluation (NDE) Methods for Concrete Materials

Concrete properties	Recommended NDE methods	Possible NDE methods
Strength	Penetration probe Rebound hammer Pullout methods	
General quality and uniformity	Penetration probe Rebound hammer Ultrasonic pulse velocity Gamma radiography	Ultrasonic pulse echo
Thickness		Radar Gamma radiography Ultrasonic pulse echo
Stiffness	Ultrasonic pulse velocity	Proof loading (load-deflection)
Density	Gamma radiography Ultrasonic pulse velocity	Neutron density gage
Rebar size and location	Covermeter (pachometer) Gamma radiography	X-ray radiography Ultrasonic pulse echo Radar
Corrosion state of reinforcing steel	Electrical potential measurement	
Presence of subsurface voids	Acoustic impact Gamma radiography Ultrasonic pulse velocity	Thermal inspection X-ray radiography Ultrasonic pulse echo Radar
Structural integrity of concrete structures	Proof loading (load-deflection)	Proof testing using acoustic emission

Adapted from Reference 14-25.

14-3. Verbeck, G. J., *Carbonation of Hydrated Portland Cement,* Research Department Bulletin RX087, Portland Cement Association, 1958.

14-4. Tyler, I. L., and Erlin, Bernard, *A Proposed Simple Test Method for Determining the Permeability of Concrete,* Research Department Bulletin RX133, Portland Cement Association, 1961.

14-5. Steinour, Harold H., *Influence of the Cement on Corrosion Behavior of Steel in Concrete,* Research Department Bulletin RX168, Portland Cement Association, 1964.

14-6. Hime, W. G.; Mivelaz, W. F.; and Connolly, J. D., *Use of Infrared Spectrophotometry for the Detection and Identification of Organic Additions in Cement and Admixtures in Hardened Concrete,* Research Department Bulletin RX194, Portland Cement Association, 1966.

14-7. Kelly, R. T., and Vail, J. W., "Rapid Analysis of Fresh Concrete," *Concrete,* The Concrete Society, Palladian Publications, Ltd., London, vol. 2, no. 4, April 1968, pages 140-145; vol. 2, no. 5, May 1968, pages 206-210.

14-8. Tabikh, A. A.; Balchunas, M. J.; and Schaefer, D. M., "A Method Used to Determine Cement Content in Concrete," *Concrete,* Highway Research Record Number 370, Transportation Research Board, National Research Council, Washington, D.C., 1971.

14-9. Clemena, Gerardo G., *Determination of the Cement Content of Hardened Concrete by Selective Solution,* PB-213 855, Virginia Highway Research Council, Federal Highway Administration, National Technical Information Service, U.S. Department of Commerce, Springfield, Virginia, 1972.

14-10. Forester, J. A.; Black, B. F.; and Lees, T. P., *An Apparatus for Rapid Analysis Machine,* Technical Report, Cement and Concrete Association, Wexham Springs, Slough, England, April 1974.

14-11. Malhotra, V. M., *Testing Hardened Concrete, Nondestructive Methods,* ACI Monograph No. 9, American Concrete Institute-Iowa State University Press, Detroit, 1976.

14-12. Clear, K. C., and Harrigan, E.T., *Sampling and Testing for Chloride Ion in Concrete,* FHWA-RD-77-85, Federal Highway Administration, Washington, D.C., August 1977.

14-13. *In-Place Concrete Strength Evaluation—A Recommended Practice,* NRMCA Publication 133-79, revised 1979, National Ready Mixed Concrete Association, Silver Spring, Maryland.

14-14. *Resistance of Concrete to Chloride Ion Penetration,* AASHTO T259-80, American Association of State Highway and Transportation Officials, Washington, D.C., 1980.

14-15. Forstie, Douglas A., and Schnormeir, Russell, "Development and Use of 4 by 8 Inch Concrete Cylinders in Arizona," *Concrete International,* American Concrete Institute, July 1981, pages 42-45.

14-16. Whiting, D., *Rapid Determination of the Chloride Permeability of Concrete,* FHWA-RD-81-119, Federal Highway Administration, 1981.

14-17. Pfeifer, D. W., and Scali, M. J., *Concrete Sealers for Protection of Bridge Structures,* NCHRP Report 244, Transportation Research Board, National Research Council, 1981.

14-18. *Standard Method of Test for Air Content of Freshly Mixed Concrete by the Chace Indicator,* AASHTO T199-82, American Association of State Highway and Transportation Officials, 1982.

14-19. "Rapid Analysis of Fresh Concrete," *Concrete Technology Today,* PL832B, Portland Cement Association, June 1983.

14-20. *Recommended Practice for Evaluation of Strength Test Results of Concrete,* ACI 214-77, reapproved 1983, ACI Committee 214 Report, American Concrete Institute.

14-21. Gebler, S. H., and Klieger, P., *Effects of Fly Ash on the Air-Void Stability of Concrete,* Research and Development Bulletin RD085T, Portland Cement Association, 1983.

14-22. *Rapid Determination of the Chloride Permeability of Concrete,* AASHTO T277-83I, American Association of State Highway and Transportation Officials, 1983.

14-23. Date, Chetan G., and Schnormeier, Russell H., "Day-to-Day Comparison of 4 and 6 Inch Diameter Concrete Cylinder Strengths," *Concrete International,* American Concrete Institute, August 1984, pages 24-26.

14-24. Malhotra, V. M., *In Situ/Nondestructive Testing of Concrete,* SP-82, American Concrete Institute, 1984.

14-25. Clifton, James R., "Nondestructive Evaluation in Rehabilitation and Preservation of Concrete and Masonry Materials," SP-85-2, *Rehabilitation, Renovation, and Preservation of Concrete and Masonry Structures,* SP-85, American Concrete Institute, 1985, pages 19-29.

14-26. *Permeability of Concrete and Its Control,* The Concrete Society, London, England, 1985.

14-27. Kosmatka, Steven H., "Floor-Covering Materials and Moisture in Concrete," *Concrete Technology Today,* PL853B, Portland Cement Association, September 1985.

14-28. Kosmatka, Steven H., "Compressive versus Flexural Strength for Quality Control of Pavements," *Concrete Technology Today,* PL854B, Portland Cement Association, 1985.

14-29. *Building Code Requirements for Reinforced Concrete,* ACI 318-83, revised 1986, ACI Committee 318 Report, American Concrete Institute.

14-30. Kosmatka, Steven H., "Petrographic Analysis of Concrete," *Concrete Technology Today,* PL862B, Portland Cement Association, 1986.

14-31. "Standard Practice for Rapid Determination of Water Soluble Chloride in Freshly Mixed Concrete, Aggregate and Liquid Admixtures," *NRMCA Technical Information Letter No. 437,* National Ready Mixed Concrete Association, March 1986.

14-32. Mor, Avi, and Ravina, Dan, "The DIN Flow Table," *Concrete International,* American Concrete Institute, December 1986.

14-33. *Manual of Aggregate and Concrete Testing,* revised 1987, American Society for Testing and Materials, Philadelphia.

CHAPTER 15
Special Types of Concrete

Special types of concrete are those with out-of-the-ordinary properties or those produced by unusual techniques. Concrete is by definition a composite material consisting essentially of a binding medium and aggregate particles, and it can take many forms. Table 15-1 lists many special types of concrete made with portland cement and some made with binders other than portland cement. In many cases the terminology of the listing describes the use, property, or condition of the concrete. Brand names are not given. A few of these concretes are discussed here.

STRUCTURAL LIGHTWEIGHT CONCRETE

Structural lightweight concrete is similar to normal-weight concrete except that it has a lower density. It is made with lightweight aggregates (all-lightweight concrete) or with a combination of lightweight and normal-weight aggregates. The term "sand lightweight" refers to lightweight concrete made with coarse lightweight aggregate and natural sand.

Structural lightweight concrete has an air-dry density in the range of 85 to 115 pcf and a 28-day compressive strength in excess of 2500 psi. Some job specifications allow air-dry densities up to 120 pcf. Normal-weight concrete containing regular sand, gravel, or crushed stone has a dry density in the range of 130 to 155 pcf. Structural lightweight concrete is used primarily to reduce the dead-load weight in concrete members such as floors in high-rise buildings.

Structural Lightweight Aggregates

Structural lightweight aggregates are usually classified according to their production process, because various

Table 15-1. Some Special Types of Concrete

Special types of concrete made with portland cement		
Air-entrained concrete	Heavyweight concrete	Prestressed concrete
Architectural concrete	High-early-strength concrete	Roller-compacted concrete
Cellular concrete	High-strength concrete	Sawdust concrete
Centrifugally cast concrete	Insulating concrete	Shielding concrete
Colloidal concrete	Latex-modified concrete	Shotcrete
Colored concrete	Low-density concrete	Shrinkage-compensating concrete
Controlled-density fill	Mass concrete	Silica-fume concrete
Cyclopean (rubble) concrete	Moderate-strength lightweight concrete	Soil-cement
Dry-packed concrete	Nailable concrete	Stamped concrete
Epoxy-modified concrete	No-slump concrete	Structural lightweight concrete
Exposed-aggregate concrete	Polymer-modified concrete	Superplasticized concrete
Ferrocement	Porous concrete	Terrazzo
Fiber-reinforced concrete	Pozzolan concrete	Tremie concrete
Fill concrete	Precast concrete	Vacuum-treated concrete
Flowing concrete	Prepacked concrete	Vermiculite concrete
Fly-ash concrete	Preplaced-aggregate concrete	White concrete
Gap-graded concrete		Zero-slump concrete
Special types of concrete not using portland cement		
Acrylic concrete	Furan concrete	Polyester concrete
Aluminum phosphate concrete	Gypsum concrete	Polymer concrete
Asphalt concrete	Latex concrete	Potassium silicate concrete
Calcium aluminate concrete	Magnesium phosphate concrete	Sodium silicate concrete
Epoxy concrete	Methyl methacrylate (MMA) concrete	Sulfur concrete

Most of the definitions of these types of concrete appear in *Cement and Concrete Terminology*, ACI 116R-78, ACI Committee 116, American Concrete Institute, 1978.

processes produce aggregates with somewhat different properties. Processed structural lightweight aggregates should meet the requirements of ASTM C 330, which includes

- Rotary kiln expanded clays, shales, and slates
- Sintering grate expanded shales and slates
- Pelletized or extruded fly ash
- Expanded slags

Structural lightweight aggregates can also be produced by processing other types of material, such as naturally occurring pumice and scoria.

Structural lightweight aggregates have densities significantly lower than normal-weight aggregates, ranging from 35 to 70 pcf compared to 75 to 110 pcf for normal-weight aggregates. They may absorb 5% to 20% water by weight of dry material. To control the uniformity of structural lightweight concrete mixtures, the aggregates are prewetted (but not saturated) prior to batching.

Compressive Strength

The compressive strength of structural lightweight concrete is usually related to the cement content at a given slump and air content rather than to a water-to-cement ratio. This is due to the difficulty in determining how much of the total mixture water is absorbed into the aggregate particles and thus not available for reaction with the cement. ACI 211.2, Standard Practice for Selecting Proportions for Structural Lightweight Concrete, provides guidance on the relationship between compressive strength and cement content. Typical compressive strengths range from 3000 psi to 5000 psi. High-strength concrete can also be made with structural lightweight aggregates.

In well-proportioned mixtures, the cement content and strength relationship is fairly constant for a particular source of lightweight aggregate. However, the relationship will vary from one aggregate source or type to another. When information on this relationship is not available from the aggregate manufacturer, trial mixtures with varying cement contents are required to develop a range of compressive strengths, including the strength specified. Fig. 15-1 shows the relationship between cement content and compressive strength. An example of a 4000-psi structural lightweight mixture with an air-dry density of about 112 pcf using a combination of natural sand and gravel and a particular lightweight rotary kiln expanded clay coarse aggregate is as follows (slump—3 in., air content—6%):

600 lb Type I portland cement
900 lb sand, ovendry
540 lb gravel (½ in. to #8), ovendry
600 lb lightweight aggregate (⅜ in. to #30), ovendry

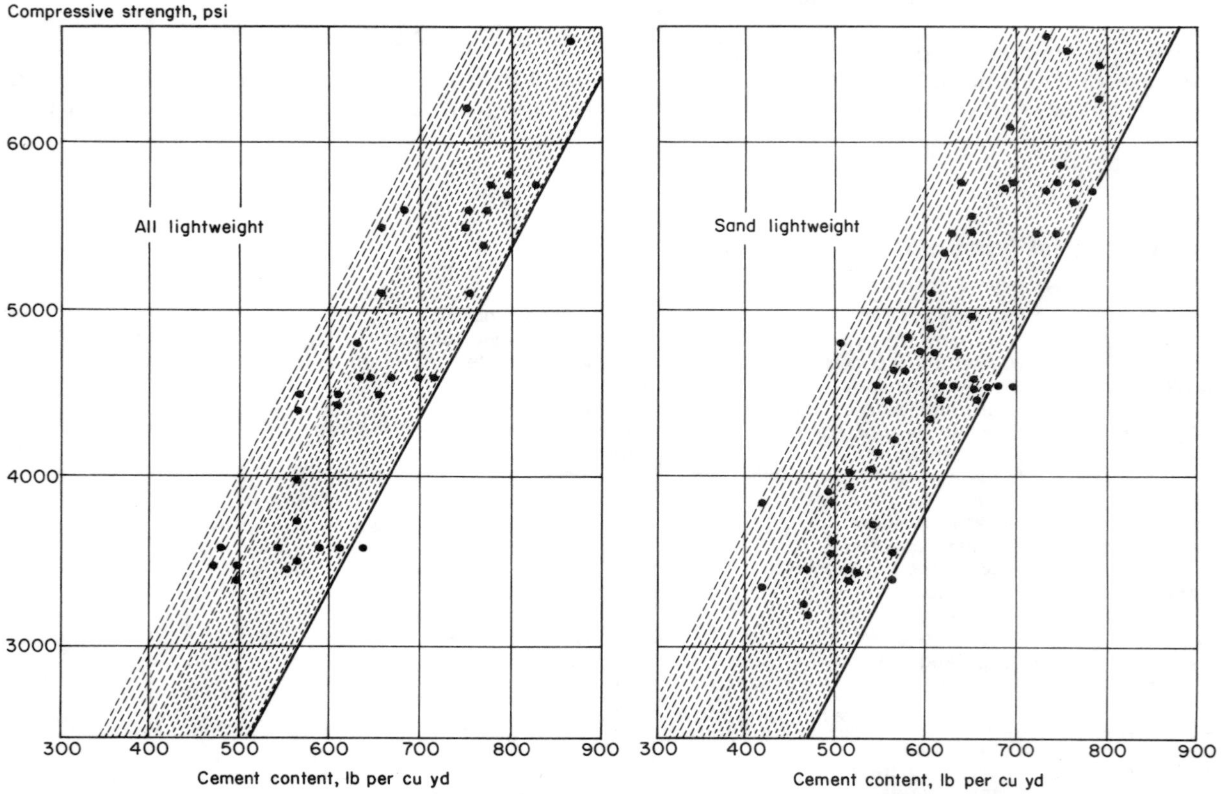

Fig. 15-1. Relationship between compressive strength and cement content of field structural lightweight concrete using lightweight fine aggregate and coarse aggregate or lightweight coarse aggregate and normal-weight fine aggregate (data points represent actual project strength results using a number of cement and aggregate sources). Reference 15-30.

290 lb mix water added
20 oz water-reducing admixture
2.5 oz air-entraining admixture
1 cu yd yield

Material proportions vary significantly for different materials and strength requirements.

Entrained Air

As with normal-weight concrete, entrained air in structural lightweight concrete ensures resistance to freezing and thawing and to deicer applications. It also improves workability, reduces bleeding and segregation, and may compensate for minor grading deficiencies in the aggregate.

The amount of entrained air should be sufficient to provide good workability to the plastic concrete and adequate freeze-thaw resistance to the hardened concrete. Air content is generally between 4.5% and 9%, depending on the maximum size of coarse aggregate used and the exposure conditions. Testing for air content should be performed by the volumetric method (ASTM C 173). The freeze-thaw durability is also significantly improved if structural lightweight concrete is allowed to dry before exposure to a freeze-thaw environment.

Specifications

Many suppliers of lightweight aggregates for use in structural lightweight concrete have information on suggested specifications and mixing proportions pertaining to their product. The usual specifications for structural concrete state a minimum compressive strength, a maximum density, a maximum slump, and an acceptable range in air content.

The contractor should also be concerned with the bleeding, workability, and finishing properties of freshly mixed structural lightweight concrete.

Mixing

In general, mixing procedures for structural lightweight concrete are similar to those for normal-density concrete; however, some of the more absorptive aggregates may require prewetting before use. Water added at the batching plant should be sufficient to produce the specified slump at the jobsite. Measured slump at the batch plant will generally be appreciably higher than the slump at the site.

Workability and Finishability

Structural lightweight concrete mixtures can be proportioned to have the same workability, finishability, and general appearance as a properly proportioned normal-density concrete mixture. Sufficient cement paste must be present to coat each particle, and coarse-aggregate particles should not separate from the mortar. Enough fine aggregate is needed to keep the freshly mixed concrete cohesive. If aggregate is deficient in minus No. 30 sieve material, finishability may be improved by using a portion of natural sand, by increasing cement content, or by using satisfactory mineral fines. Since entrained air improves workability, it should be used regardless of exposure.

Slump

Due to lower aggregate density, structural lightweight concrete does not slump as much as normal-weight concrete with the same workability. An air-entrained mixture with a slump of 2 to 3 in. can be placed under conditions that would require a slump of 3 to 5 in. for normal-weight concrete. It is seldom necessary to exceed slumps of 5 in. for normal placement of structural lightweight concrete. With higher slumps, the large aggregate particles tend to float to the surface, making finishing difficult.

Vibration

As with normal-weight concrete, vibration can be used effectively to consolidate lightweight concrete. Frequencies in excess of 115 Hz, about the same as those commonly used for normal-density concrete, are recommended. The length of time for proper consolidation varies, depending on mixture characteristics. Excessive vibration causes segregation by forcing large aggregate particles to the surface.

Internal vibration is recommended for all slabs thicker than 8 in. and for thinner slabs that contain reinforcing steel or conduit. The vibrator head should be completely immersed during vibration. For thick slabs, it is possible to insert the vibrator head vertically; for thinner slabs it should be slowly dragged through the concrete at an angle or even horizontally, and at a constant velocity.

On flat surfaces, a high-frequency vibratory screed is effective for consolidation and to facilitate finishing.

Placing, Finishing, and Curing

Structural lightweight concrete is generally easier to handle and place than normal-weight concrete. A slump of 2 to 4 in. produces best results for finishing. Greater slumps may cause segregation, delay finishing operations, and result in rough, uneven surfaces.

If pumped concrete is being considered, the specifier, suppliers, and contractor should all be consulted about performing a field trial using the pump and the mixture planned for the project. Adjustments to the mixture may be necessary, as the pumping pressure causes the aggregate to absorb more water, thus reducing the slump and increasing the density of the concrete. The water absorbed by the aggregate will eventually evaporate upon air drying of the concrete.

Finishing operations should be started earlier than for comparable normal-weight concrete, but finishing too early may be harmful. A minimum amount of floating and troweling should be done; aluminum or magnesium finishing tools are preferred.

The same curing practices should be used for lightweight concrete as for normal-weight concrete. The two methods commonly used in the field are water curing (ponding, sprinkling, or using wet coverings) and preventing loss of moisture from the exposed surfaces (covering with waterproof paper, plastic sheets, or sealing with liquid membrane-forming compounds). Generally, 7 days of curing are adequate for ambient air temperatures from 50°F to 70°F and 5 days for temperatures over 70°F.*

LOW-DENSITY AND MODERATE-STRENGTH LIGHTWEIGHT CONCRETES

Low-density concrete—also referred to as insulating concrete—is a lightweight concrete with an ovendry unit weight of 50 pcf or less. It is made with portland cement, water, air, and with or without aggregate and mineral admixtures. The ovendry unit weight ranges from 15 to 50 pcf and the 28-day compressive strength is generally between 100 and 1000 psi. Cast-in-place low-density concrete is used primarily for thermal and sound insulation, roof decks, fill for slab-on-grade subbases, leveling courses for floors or roofs, firewalls, and underground thermal conduit linings. Low-density concrete is also used in precast, reinforced-concrete floor, roof, and wall units.

Moderate-strength lightweight concrete weighs about 50 to 120 pcf ovendry and has a compressive strength of approximately 1000 to 2500 psi. It is made with portland cement, water, air, and with or without aggregate and mineral admixtures. At lower densities, it is used as fill for thermal and sound insulation of floors, walls, and roofs and is referred to as fill concrete. At higher densities it is used in cast-in-place walls, floors and roofs, and precast wall and floor panels.**

For discussion purposes, low-density and moderate-strength lightweight concretes can be grouped as follows:

Group I is made with expanded aggregates such as perlite, vermiculite, or expanded polystyrene beads. Ovendry concrete unit weights using these aggregates generally range between 15 and 50 pcf. This group is used primarily in low-density concrete. Some moderate-strength concretes can also be made from aggregates in this group.

Group II is made with aggregates manufactured by expanding, calcining, or sintering materials such as blast-furnace slag, clay, diatomite, fly ash, shale, or slate, or by processing natural materials such as pumice, scoria, or tuff. Ovendry concrete unit weights using these aggregates can range between 45 and 90 pcf. Aggregates in this group are used in moderate-strength lightweight concrete and some of these materials (expanded slag, clay, fly ash, shale, and slate) are also used in both moderate-strength and structural lightweight concrete (up to about 120 pcf air-dry).

Group III concretes are made by incorporating into a cement paste or cement-sand mortar a uniform cellular structure of air voids that is obtained with preformed foam (ASTM C 869), formed-in-place foam, or special foaming agents. This concrete is commonly referred to as cellular concrete. Ovendry unit weights

ranging between 15 to 120 pcf are obtained by substitution of air voids for some or all of the aggregate particles; air voids can consist of up to 80% of the volume. Cellular concrete can be made to meet the requirements of both low-density and moderate-strength lightweight concrete.

Aggregates used in Groups I and II should meet the requirements of ASTM C 332, Standard Specification for Lightweight Aggregates for Insulating Concrete. These aggregates have dry unit weights in the range of from 6 to 70 pcf down to 1 pcf for expanded polystyrene beads.

Mixture Proportions

Examples of mixture proportions for Group I and III concretes appear in Table 15-2. In Group I, air contents may be as high as 25% to 35%. The air-entraining agent can be prepackaged with the aggregate or added at the mixer. Because of the absorptive nature of the aggregate, the volumetric method (ASTM C 173) should be used to measure air content accurately.

Water requirements for insulating and fill concretes vary considerably, depending on aggregate characteristics, entrained air, and mixture proportions. An effort should be made to avoid excessive amounts of evaporable water in insulating concrete used in roof fills. Excessive water causes high drying shrinkage and cracks that may damage the waterproofing membrane. Accelerators containing calcium chloride should not be used where galvanized steel will remain in permanent contact with the concrete because of possible corrosion problems.

Mixture proportions for Group II concretes usually are based on volumes of dry, loose materials even when aggregates are moist as batched. Satisfactory proportions can vary considerably for different aggregates or combinations of aggregates. Mixture proportions ranging from 4 to 14 cu ft of aggregate per 100 lb of cement can be used in lightweight concretes that are made with pumice, expanded shale, and expanded slag. Some mixtures, such as those for no-fines concretes, are made without fine aggregate but with total void contents of 20% to 35%. Cement factors for Group II concretes range between 200 and 600 lb per cubic yard depending on air content, aggregate gradation, and mixture proportions.

No-fines concretes containing pumice, expanded slag, or expanded shale can be made with 250 to 290 lb of water per cubic yard, with total air voids of 20% to 35%, and a cement content of about 470 lb per cubic yard.

Workability

Because of their high air content, lightweight concretes weighing less than 50 pcf generally have excellent

---

*See References 15-17, 15-30, and 15-45 for more information.
**See References 15-32, 15-33, and other references for more information.

Table 15-2. Examples of Lightweight Insulating Concrete Mixtures

Type of concrete	Ratio: portland cement to aggregate by volume	Ovendry density, pcf	Type I portland cement, lb/cu yd	Water-cement ratio, by weight	28-day compressive strength, psi, 6x12-in. cylinders
Perlite*	1:4	30 to 38	610	0.94	400
	1:5	26 to 36	516	1.12	325
	1:6	22 to 34	414	1.24	220
	1:8	20 to 32	395	1.72	200
Vermiculite*	1:4	31 to 37	640	0.98	300
	1:5	28 to 31	498	1.30	170
	1:6	23 to 29	414	1.60	130
	1:8	20 to 21	300	2.08	80
Polystyrene:**					
0 lb sand	1:3.4	34‡	750	0.40	325
124 lb sand per cubic yard	1:3.1	39‡	750	0.40	400
261 lb sand per cubic yard	1:2.9	44‡	750	0.40	475
338 lb sand per cubic yard	1:2.5	48‡	800	0.40	550
Cellular*	—	39	884	0.57	350
(neat cement)	—	34	790	0.56	210
	—	28	668	0.57	130
	—	23	535	0.65	50
Cellular†	1:1	58	724	0.40	460
(sanded)††	1:2	78	630	0.41	820
	1:3	100	602	0.51	2190

*Reference 15-13.
**Source: Reference 15-24. The mix also included air entrainment and a water-reducing agent.
†Source: Reference 15-12.
††Dry-rodded sand weighing 100 pcf.
‡Air-dry density at 28 days, 50% relative humidity.

workability. Slumps of up to 10 in. usually are satisfactory for Group I and Group III concretes; appearance of the mix, however, may be a more reliable indication of consistency. Cellular concretes are handled as liquids and are poured or pumped into place without further consolidation.

Mixing and Placing

All concrete should be mechanically mixed to produce a uniform distribution of materials of proper consistency and required density. In batch-mixing operations, various sequences can be used for introducing the ingredients, but the preferred sequence is to introduce the required amount of water into the mixer, then add the cement, air-entraining or foaming agent, aggregate, preformed foam, and any other ingredients.

Excessive mixing and handling should be avoided because they tend to break up aggregate particles, thereby changing density and consistency. Segregation is not usually a problem (though it could be for Group II) because of the relatively large amounts of entrained air.

Pumping is the most common method of placement, but other methods can be used. Finishing operations should be kept to a minimum; smoothing with a darby is usually sufficient. Placement of insulating concretes should be done by workers experienced with these special concretes.

Periodic wet-density tests (ASTM C138) at the job-site can be performed to check the uniformity of the concrete. Variations in density generally should not exceed plus or minus 2 pcf. A close approximation of

the ovendry density can be determined from the freshly mixed unit weight.

Thermal Conductivity

ASTM C177, Test Method for Steady-State Heat Flux Measurements and Thermal Transmission Properties by Means of the Guarded-Hot-Plate Apparatus, is used to determine values of thermal conductivity. Fig. 15-2 shows an approximate relationship between thermal conductivity and density. The thermal conductivity of concrete increases with an increase in moisture content and density.

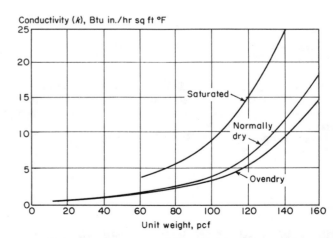

Fig. 15-2. Approximate relationship between unit weight and thermal conductivity of insulating structural lightweight and normal-weight concretes. Reference 15-10.

Strength

Strength requirements depend on the intended use of the concrete. For example, a compressive strength of 100 psi or even less may be satisfactory for insulation of underground steam lines. Roof-fill insulation requires sufficient early strength to withstand foot traffic. Compressive strengths of 100 to 200 psi are usually adequate for roof fills, but strengths up to 500 psi are sometimes specified. In general, the strength of insulating concrete is of minor importance. Compressive strength of lightweight insulating concrete should be determined by the methods specified in ASTM C 495 or C 513.

Table 15-2 and Fig. 15-3 give examples of the relationship between density and strength for lightweight insulating concretes. Fig. 15-4 shows examples for cellular concrete containing sand. Mixtures with strengths outside the ranges shown can be made by varying the mixture proportions. Strengths comparable to those at 28 days would be obtained at 7 days with high-early-strength cement. The relationships shown do not apply to autoclaved products.

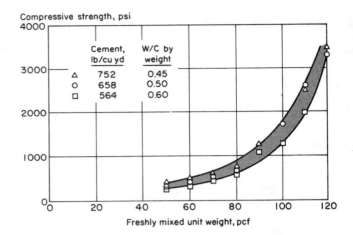

Fig. 15-4. Freshly mixed wet unit weight versus compressive strength for sanded cellular concretes. Compressive strength was determined with 6x12-in. cylinders that were cured for 21 days in a 100% moist room followed by 7 days in air at 50% relative humidity. References 15-11 and 15-33.

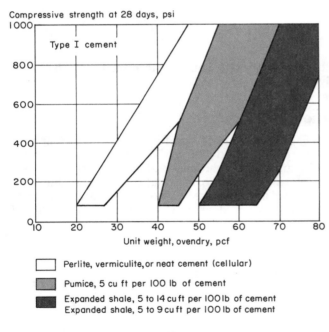

Fig. 15-3. Approximate relationship between ovendry unit weight and compressive strength of 6x12-in. cylinders tested in air-dry conditions for some insulating and fill concretes. For the perlite and vermiculite concretes, mix proportions ranged from 3 to 10 cu ft of aggregate per 100 lb of cement. All proportions are volumetric.

Resistance to Freezing and Thawing

Low-density and moderate-strength lightweight concretes normally are not required to withstand freeze-thaw exposure in a saturated condition. In service they are usually protected from the weather; thus little research has been done on their resistance to freezing and thawing.

Drying Shrinkage

The shrinkage of low-density or moderate-strength lightweight concrete is not usually critical when it is used for insulation or fill; however, excessive shrinkage can cause curling. In structural use, shrinkage should be considered. Moist-cured cellular concretes made without aggregates have high drying shrinkage. Moist-cured cellular concretes made with sand may shrink from 0.1% to 0.6%, depending on the amount of sand used. Autoclaved cellular concretes shrink very little on drying. Insulating concretes made with perlite or pumice aggregates may shrink 0.1% to 0.3% in six months of drying at 50% relative humidity; vermiculite concretes may shrink 0.2% to 0.45% during the same period. Drying shrinkage of insulating concretes made with expanded slag or expanded shale ranges from about 0.6% to 0.1% in six months.

Expansion Joints

Where insulating concrete is used on roof decks, a 1-in. expansion joint at the parapets and all roof projections is often specified. Its purpose is to accommodate expansion caused by the heat of the sun so that the insulating concrete can expand independently of the roof deck. Transverse expansion joints should be placed at a maximum of 100 ft in any direction for a thermal expansion of 1 in. per 100 lin ft. A fiberglass material that will compress to one-half its thickness under a stress of 25 psi is generally used to form the joints.

HEAVYWEIGHT CONCRETE

Heavyweight concrete, such as radiation-shielding concrete, is produced with special heavy aggregates and has a density of up to about 400 pcf.

Heavyweight concrete is used principally for radiation shielding but is also used for counterweights and other applications where high density is important. As a shielding material, heavyweight concrete protects against the harmful effects of X-rays, gamma rays, and neutron radiation.* Selection of concrete for radiation shielding is based on space requirements and on the type and intensity of radiation. Where space requirements are not important, normal-weight concrete will generally produce the most economical shield; where space is limited, heavyweight concrete will allow reductions in shield thickness without sacrificing shielding effectiveness.

Type and intensity of radiation usually determine the requirements for density and water content of shielding concrete. Effectiveness of a shield against gamma rays is approximately proportional to the density of the concrete—the heavier the concrete, the more effective the shield. On the other hand, an effective shield against neutron radiation requires both heavy and light elements. The hydrogen in water provides an effective light element in concrete shields. Some aggregates contain crystallized water, called fixed water, as part of their structure. For this reason, heavyweight aggregates with high fixed-water contents often are used if both gamma rays and neutron radiation are to be attenuated.

High-Density Aggregates

High-density aggregates such as barite, ferrophosphorus, goethite, hematite, ilmenite, limonite, magnetite, and steel punchings and shot are used to produce high-density concrete. Where high fixed-water content is desirable, serpentine (which is slightly heavier than normal-weight aggregate) or bauxite can be used (see ASTM C 637 and C 638).

Table 15-3 gives typical bulk density, specific gravity, and percentage of fixed water for some of these materials. The values are a compilation of data from a wide variety of tests or projects reported in the literature. Steel punchings and shot are used where concrete with a density of more than 300 pcf is required.

In general, selection of an aggregate is determined by physical properties, availability, and cost. Heavyweight aggregates should be reasonably free of fine material, oil, and foreign substances that affect either the bond of paste to aggregate particle or the hydration of cement. For good workability, maximum density, and economy, aggregates should be roughly cubical in shape and free of flat or elongated particles.

Additions

Boron additions such as colemanite, boron frits, and borocalcite are sometimes used to improve the neutron shielding properties of concrete. They may adversely affect setting and early strength of concrete; therefore, trial mixes should be made with the addition under field conditions to determine suitability. Admixtures such as pressure-hydrated lime can be used with coarse-sand sizes to minimize any retarding effect.

Properties of Heavyweight Concrete

The properties of heavyweight concrete in both the freshly mixed and hardened states can be tailored to meet job conditions and shielding requirements by proper selection of materials and mixture proportions.

Except for density, the physical properties of heavyweight concrete are similar to those of normal-weight concrete. Strength is a function of water-cement ratio; thus, for any particular set of materials, strengths comparable to those of normal-weight concretes can be achieved. Typical densities of concretes made with some commonly used high-density aggregates are shown in Table 15-3. Because each radiation shield has special requirements, trial mixtures should be made with job materials and under job conditions to determine suitable mixture proportions.

---

*See References 15-6 and 15-16.

Table 15-3. Physical Properties of Typical Heavyweight Aggregates and Concrete

Type of aggregate	Fixed water,* percent by weight	Aggregate specific gravity	Aggregate bulk density, pcf	Concrete density, pcf
Goethite	10-11	3.4-3.7	130-140	180-200
Limonite	8-9	3.4-4.0	130-150	180-210
Barite	0	4.0-4.6	145-160	210-230
Ilmenite	**	4.3-4.8	160-170	220-240
Hematite	**	4.9-5.3	180-200	240-260
Magnetite	**	4.2-5.2	150-190	210-260
Ferrophosphorus	0	5.8-6.8	200-260	255-330
Steel punchings or shot	0	6.2-7.8	230-290	290-380

*Water retained or chemically bound in aggregates.

**Test data not available. Aggregates may be combined with limonite to produce fixed-water contents varying from about ½% to 5%.

Proportioning, Mixing, and Placing

The procedures for selecting mixture proportions for heavyweight concrete are the same as those used for normal-weight concrete. However, additional mixture information and sample calculations are given in ACI 211.1. Following are the most common methods of mixing and placing high-density concrete:

Conventional methods of mixing and placing often are used, but care must be taken to avoid overloading the mixer, especially with very heavy aggregates such as steel punchings. Batches should be reduced to about 50% of the rated mixer capacity. Because some heavy aggregates are quite friable, excessive mixing should be avoided to prevent aggregate breakup with resultant detrimental effects on workability and bleeding.

Preplaced aggregate methods can be used for placing normal and high-density concrete in confined areas and around embedded items to minimize segregation of coarse aggregate, especially steel punchings or shot. The method also reduces drying shrinkage and produces concrete of uniform density and composition. The coarse aggregates are preplaced in the forms and grout made of cement, sand, and water is then pumped through pipes to fill the voids in the aggregate.

Pumping of heavyweight concrete through pipelines may be advantageous in locations where space is limited. Heavyweight concretes cannot be pumped as far as normal-weight concretes because of their higher densities.

Puddling is a method whereby 2 in. or more of mortar is placed in the forms and then covered with a layer of coarse aggregate that is rodded or internally vibrated into the mortar. Care must be taken to ensure uniform distribution of aggregate throughout the concrete.

HIGH-STRENGTH CONCRETE

High-strength concrete is generally defined as concrete with a compressive strength of 6000 psi or greater. Concrete having strengths of 20,000 psi has been used in buildings. The specified strength of concrete has been based traditionally on 28-day test results. However, in high-rise structures requiring high-strength concrete, the process of construction is such that the structural elements in lower floors are not fully loaded for periods of a year or more. For this reason, compressive strengths based on 56- or 90-day test results are commonly specified in order to achieve significant economy in material costs.

With use of low-slump or no-slump rich mixes, high-compressive-strength concrete is produced routinely under careful control in precast and prestressed concrete plants. These stiff mixes are placed in ruggedly built forms and consolidated by prolonged vibration or shock methods. However, cast-in-place concrete uses more fragile forms that do not permit the same compaction procedures, hence more workable concretes are necessary to achieve the required compaction and avoid segregation and honeycomb. Superplasticizing admixtures are often added to the concrete mix to produce workable and sometimes flowable mixtures.

Materials Selection

Production of high-strength concrete may or may not require purchase of special materials. The producer must know the factors affecting compressive strength and know how to vary those factors for best results. Each variable should be analyzed separately in developing the mixture design. When an optimum or near optimum is established for each variable, it should be incorporated as the remaining variables are studied. An optimum mixture design is then developed keeping in mind the economic advantages of using locally available materials.

Cement. Selection of portland cement for high-strength concrete should not be based only on mortar-cube tests but should also include comparative strengths of concretes at 28, 56, and 90 days. A cement that yields the highest concrete compressive strength at extended ages (90 days) is obviously preferable. For high-strength concrete, the cement should produce a minimum 7-day mortar-cube strength of approximately 4200 psi.

Trial mixtures should be made with cement contents between 650 and 950 lb per cubic yard for each cement—amounts will vary depending on target strengths. Other than decreases in sand content as cement content increases, the trial mixtures should be as nearly identical as possible with the slump between 3 to 4 in.

Pozzolans. Fly ash or silica fume (see Chapter 6) are often mandatory in the production of high-strength concrete as the strength gain obtained with these pozzolans cannot be attained by using additional cement alone. However, the fly ash or silica fume must be used as an addition to the regular amount of cement, not as a partial substitute for it. These pozzolans are usually added at dosage rates of 5% to 20% by weight of cement. The water to cement plus pozzolan ratio should be adjusted so that equal workability becomes the basis of comparison. For each set of materials, there will be an optimum cement-plus-pozzolan content at which strength does not continue to increase with greater amounts and the mixture becomes too sticky to handle properly. Ground, granulated blast-furnace slag can also be used in the production of high-strength concrete, although its use for this purpose is small in the United States. Also certain blended cements can be used in place of portland cement plus a pozzolan as they already contain these pozzolanic materials.

Aggregates. Careful attention must be given to aggregate size, shape, surface texture, mineralogy, and cleanness. For each source of aggregate and concrete strength level there is an optimum-size aggregate that will yield the most compressive strength per pound of cement (Fig. 15-5). To find the optimum size, trial batches should be made with ¾ in. and smaller coarse aggregates and varying cement contents. Many studies have found that ⅜-in. to ½-in. maximum-size aggregates appear to give optimum strength. Maximum sizes of

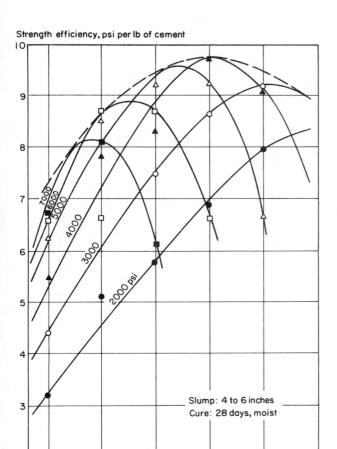

Strength efficiency, psi per lb of cement

Maximum-size aggregate, in.

Slump: 4 to 6 inches
Cure: 28 days, moist

Fig. 15-5. Maximum-size aggregate for strength efficiency envelope. Reference 15-7.

¾ in. and 1 in. are also used successfully. Fig. 15-6 shows the effect of maximum aggregate size on compressive strength of concrete at 28 and 90 days for several types of concrete. The effect of maximum size in rich mixes is more significant at 90 days than at 28 days.

In high-strength concretes, the strength of the aggregate itself and the bond or adhesion between the cement paste and aggregate become important factors. Tests have shown that crushed-stone aggregates produce higher compressive strength in concrete than gravel aggregates using the same size aggregate and the same cement content, probably due to a superior aggregate-to-paste bond when using rough, angular, crushed material.

Coarse aggregates used in high-strength concrete should be clean, that is, free from detrimental coatings of dust and clay. Removing dust is important since it may affect the quantity of fines and consequently the water demand. Clay may affect the aggregate-paste bond. Washing of coarse aggregates may be necessary.

The quantity of coarse aggregate in high-strength concrete should be the maximum consistent with required workability. Because of the high percentage of cementitious material in high-strength concrete, an increase in coarse-aggregate content beyond values recommended in standards for normal-strength mixtures is necessary and allowable.

Because of the high amount of cementitious material in high-strength concrete, the role of the fine aggregate (sand) in providing workability and good finishing characteristics is not as crucial as in conventional strength mixes. Sand with a fineness modulus (FM) of about 3.0—considered a coarse sand—has been found to be satisfactory for producing good workability and high compressive strength. Finer sand, say with an FM of 2.5 to 2.7, may produce lower-strength, sticky mixtures.

Admixtures. Most ready mix producers feel that the use of chemical admixtures such as water reducers, retarding water reducers, high-range water reducers (superplasticizers), or a combination of these, are necessary to make use efficiently of the large amount of cementitious material in high-strength concrete and to help obtain the lowest practical water-cement ratio. Chemical admixture efficiency must be evaluated by comparing strengths of trial batches. Also, factors such as cement and pozzolan compatibility, water reduction, setting times, workability, and admixture dosage rate and time of addition must be determined by trial batches.

The use of air-entraining admixtures is not necessary or desirable in high-strength concrete that is protected from the weather, such as interior columns and shearwalls of high-rise buildings. However, for projects such as bridges, concrete piles, piers, or parking structures, where durability in a freeze-thaw environment is required, entrained air is mandatory. Because air entrainment decreases concrete strength of rich mixtures, testing to establish optimum air contents may be required.

Proportioning

The trial mixture approach is best for selecting proportions for high-strength concrete. To obtain high strength, it is necessary to use the lowest possible water-cement ratio (generally 0.30 to 0.40) and a high cement content.

The unit strength in pounds per square inch obtained for each pound of cement used in a cubic yard of concrete can be plotted as strength efficiency, as shown in Fig. 15-5. The figure confirms that smaller-size aggregates result in the most efficient use of cement as cement content is increased above about 700 lb per cubic yard.

The water requirement of concrete increases as the fine aggregate content is increased for any given size of coarse aggregate. Because of the high cement content of these concretes, the fine aggregate content can be kept low. However, even with well-graded aggregates, a low water-cement ratio may result in concrete that is not sufficiently workable for the job. If a superplasticizer is not already being used, this may be the time to consider one. A slump of 4 in. will generally provide adequate workability.

A typical mixture design for 9000 psi concrete at 56 days (8000 psi at 28 days) being delivered by several ready mix producers is as follows:

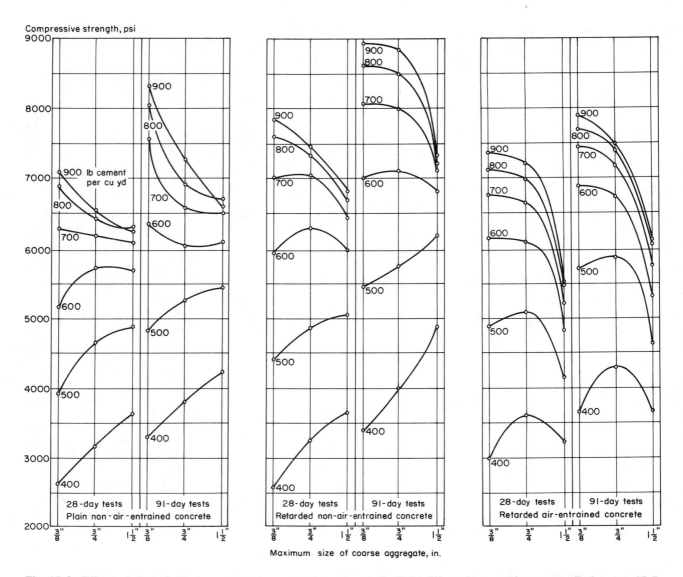

Fig. 15-6. Effect of size of coarse aggregate on compressive strength in different types of concrete. Reference 15-5.

Type I cement	846 lb per cu yd
Fly ash	100 lb per cu yd
Sand	1060 lb per cu yd
Stone, ½ in. maximum	1710 lb per cu yd
Water, net	317 lb per cu yd
Water reducer-retarder	38 oz per cu yd

An example of a high-strength concrete mixture using silica fume is illustrated in Table 15-4. The strength levels developed in laboratory trial batches may be difficult to achieve in the field. Therefore, all final mixtures proposed for use should be tested in full-size batches under typical job conditions.

Placing, Consolidation, and Curing

There must be close liaison between the contractor and the concrete producer so that the concrete can be discharged rapidly after arrival at the jobsite. Final adjustment of the concrete should be supervised by the concrete producer's technicians at the site, by a concrete laboratory, or by a consultant familiar with the performance and use of high-strength concrete.

Delays in delivery and placing must be eliminated and sometimes it may be necessary to reduce batch sizes if placing procedures are slower than anticipated. Rigid surveillance must be exercised at the jobsite to control any addition of retempering water. The contractor must be prepared to receive the concrete and understand the consequences of exceeding the specified slump and water-cement ratio.

Consolidation is very important in achieving the potential strengths of high-strength concrete. Concrete must be vibrated as quickly as possible after placement in the forms. High-frequency vibrators should be small enough to allow sufficient clearance between the vibrating head and reinforcing steel. Overvibration of workable normal-strength concrete often results in seg-

Table 15-4. Example of High Strength Silica Fume Concrete Mixture

Mix proportions and properties	
Type I cement	1000 lb per cu yd
Silica-fume admixture*	200 lb per cu yd
Sand (2.90 FM)	905 lb per cu yd
Limestone (½ in. maximum)	1682 lb per cu yd
Water	266 lb per cu yd
Water- (cement + silica fume) ratio	0.22
Slump (average)	4.7 in.
Air content (average)	1.5%

Age, days	Compressive strength, psi**	Flexural strength, psi
0.5	6,260	—
1	9,430	—
3	11,300	1615
7	13,730	1790
14	14,920	—
28	16,170	2070
60	16,260	—
90	16,700	—
128	18,020	—
365	18,360	—

*Includes high-range water reducer.
**4x8-in. cylinders, moist-cured.
Adapted from Reference 15-47.

regation, loss of entrained air, or both. High-strength concrete, on the other hand, will generally be relatively stiff and contain little air. Consequently, inspectors should be more concerned with undervibration rather than overvibration.

If high-strength concrete is to achieve its potential strength, curing is of utmost importance. Providing adequate moisture and favorable temperature conditions is necessary for a prolonged period, particularly when 56- or 90-day concrete strengths are specified.

Quality Control

A comprehensive quality-control program is required at both the concrete plant and the site to guarantee consistent production and placement of high-strength concrete. Inspection of concreting operations from stockpiling of aggregate through completion of curing is important. Closer production control than is normally obtained on most projects is necessary. Also, routine sampling and testing of all materials is particularly necessary to control uniformity of the concrete.

In testing high-strength concrete, some changes and more attention to detail are required. For example, cardboard cylinder molds, which can cause lower strength-test results, should be replaced with reuseable steel or plastic molds or disposable tin molds. Capping of cylinders must be done with great care using only high-strength capping compounds as per ASTM C 670. ACI 318 requires that the average strength of all sets of three consecutive tests of concrete delivered to a project equal or exceed the specified value, with no individual test (average of two cylinders) falling more than 500 psi below this value. To meet these requirements, it is necessary to aim at an average strength

higher than the specified minimum. The level of the design strength depends on the control exercised over the variables that influence the strength of the concrete. The less effective the control exercised, the higher will be the average strength required to meet specification requirements.

The production of concrete with a high compressive strength involves achieving a low variance in test results, because in most cases it will not be possible to produce concrete with an average strength significantly higher than the specified strength. The lower standard of deviation necessary on high-strength concrete projects can only be achieved by strict vigilance in all aspects of quality control on the part of the producer and quality testing on the part of the laboratory.*

HIGH-EARLY-STRENGTH CONCRETE

High-early-strength concrete is concrete that achieves its specified strength at an earlier age than normal concrete. The time period in which a specified strength should be achieved may range from a few hours (or even minutes) to several days. High-early-strength can be attained by using traditional concrete ingredients and concreting practices, although sometimes special materials or techniques are needed.

High early strength can be obtained by using one or a combination of the following depending on the age at which the specified strength must be achieved and on job conditions:

1. Type III high-early-strength cement
2. High cement content (600 to 1000 lb/cu yd)
3. Low water-cement ratio (0.2 to 0.45 by weight)
4. Higher freshly mixed concrete temperature
5. Higher curing temperature (see Chapter 11)
6. Chemical admixtures
7. Silica fume (see Table 15-4)
8. Steam or autoclave curing (see Chapter 10)
9. Insulation to retain heat of hydration
10. Regulated-Set cement or other special cements

High-early-strength concrete is used for prestressed concrete to allow for early stressing; precast concrete for rapid production of elements; high-speed cast-in-place construction; rapid form reuse; cold-weather construction; rapid repair to reduce traffic downtime; fast-track paving; and several other uses. In fast-track paving, use of high-early-strength mixtures allows traffic to open within 24 hours after concrete is placed. A fast-track concrete mixture used for a bonded concrete overlay on a major highway consisted of 640 lb of Type III cement, 70 lb of Class C fly ash, 6½% air, a water reducer, and a water to cement plus fly ash ratio of 0.4. Strength data for this 1½-in.-slump concrete was as follows:**

*See References 15-8, 15-14, 15-18, 15-23, 15-42, 15-50, and 15-57 for more information.
**Reference 15-57.

187

Age	Compressive strength, psi	Flexural strength (center-point loading), psi	Bond strength, psi
4 hours	252	126	120
6 hours	1020	287	160
8 hours	1883	393	200
12 hours	2546	494	225
18 hours	2920	574	250
24 hours	3467	604	302
7 days	4960	722	309
14 days	5295	825	328
28 days	5900	830	359

Several state highway departments have used high-early-strength patching concrete capable of carrying traffic within 4 hours. The patching mixes had a low water-cement ratio and contain 550 to 750 lb of Type III cement and an accelerator. Insulation was used to retain the heat of hydration.*

MASS CONCRETE

Mass concrete is defined by ACI 116 as "Any large volume of cast-in-place concrete with dimensions large enough to require that measures be taken to cope with the generation of heat and attendant volume change to minimize cracking." Mass concrete includes not only low-cement-content concrete used in dams and other massive structures but also moderate- to high-cement-content concrete in structural members that require special considerations to handle heat of hydration and temperature rise.

In mass concrete, temperature rise is caused by heat of hydration as shown in Fig. 15-7. As the interior concrete increases in temperature, the surface concrete may be cooling and contracting. This causes tensile stress and cracks at the surface if the temperature

differential is too great. The width and depth of cracks depends upon the temperature gradient between the hot internal concrete and the cooler concrete surface.

A definite member size beyond which a concrete structure should be classified as mass concrete is not readily available. However, ACI 211.1 states that "Many large structural elements may be massive enough that heat generation should be considered, particularly when the minimum cross-sectional dimensions of a solid concrete member approach or exceed 2 to 3 ft or when cement contents above 600 lb per cubic yard are being considered." Temperature rise in mass concrete is related to the initial concrete temperature (Fig. 15-8), ambient temperatures, size of the concrete element (volume to surface ratio), and the amount of reinforcement. Small concrete members are of little concern as the generated heat is rapidly dissipated.

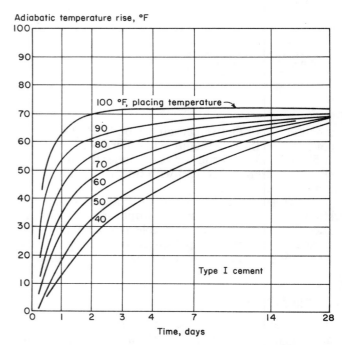

Fig. 15-8. The effect of concrete-placing temperature on temperature rise in mass concrete with 376 lb of cement per cubic yard. Higher placing temperatures accelerate temperature rise. Reference 15-52.

To avoid cracking, the internal concrete temperature for dams and other nonreinforced mass concrete structures of relatively low compressive strength should not be allowed to rise more than 20°F to 25°F above the mean annual ambient temperature.** Internal concrete temperature gain can be controlled through the use of (1) a low cement content—200 to 450 lb per cubic yard, large aggregate size—3 in. to 6 in., and high coarse aggregate content—up to 80% of total aggregate; (2) low-heat-of-hydration portland or blended cement;

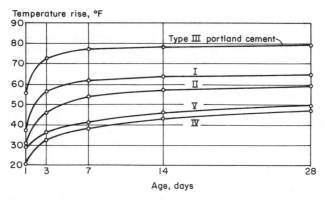

Fig. 15-7. Temperature rise for mass concrete with 4½-in. maximum-size aggregate and 376 lb of cement per cubic yard. Specimens were 17x17-in. cylinders, sealed and cured in adiabatic calorimeter rooms. The temperature rise for higher cement content concretes would be proportionately higher. Reference 15-29.

*See References 15-3, 15-47, 15-57, 15-58, and ASTM C 928 for more information.
**Reference 15-51, page 10.

(3) pozzolans—the heat of hydration of a pozzolan is approximately 25% to 50% that of cement; (4) reductions in the initial concrete temperature to approximately 50°F by cooling the concrete ingredients—see Chapter 11; (5) cooling the concrete through the use of embedded cooling pipes; (6) steel forms for rapid heat dissipation; (7) water curing; and (8) low lifts—5 ft or less—during placement.

In massive structures of high volume-to-surface ratio an estimate of the adiabatic temperature rise can be made using the equation

$$T = \frac{CH}{S}$$

where

T = temperature rise in degrees Fahrenheit of the concrete due to heat generation of cement under adiabatic conditions

C = proportion of cement in the concrete, by weight

H = heat generation due to hydration of cement, Btu per pound

S = specific heat of concrete, Btu per pound per degree Fahrenheit

Example: Assume a concrete weighing 4000 lb per cubic yard contains 300 lb of Type II portland cement and 100 lb of pozzolan per cubic yard. Assume the heat of hydration of the pozzolan is 40% that of the cement.

$$C = \frac{300 + (0.4 \times 100)}{4000} = 0.085$$

From Fig. 15-9, the heat of hydration of cement, H, at seven days is 137 Btu per pound. Assume that the specific heat of concrete, S, is equal to 0.24 (specific heat of concrete ranges from 0.20 to 0.28 Btu per pound). The temperature rise,

$$T = \frac{CH}{S} = \frac{0.085 \times 137}{0.24} = 48.5°F \text{ at 7 days}$$

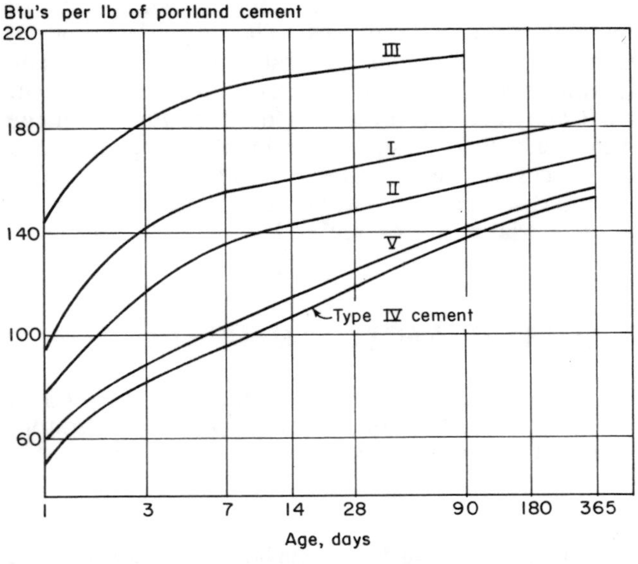

Btu's per lb of portland cement

Fig. 15-9. Typical heat of hydration curves for various types of cement. References 15-9 and 15-56.

Therefore, if this concrete is placed at 70°F, it can be expected to have a temperature of approximately 118.5°F (70 + 48.5°F) at the interior of the member at 7 days if there is no heat loss.* For massive structures, a temperature differential of 48.5°F (assuming the surface temperature drops to 70°F) is sufficient to cause cracking.**

Massive structural reinforced concrete with high cement contents (500 to 1000 lb per cu yard) cannot use many of the placing techniques and controlling factors mentioned earlier to maintain low temperatures to control cracking. For these concretes (often used in mat foundations and power plants), a good technique is to (1) place the entire concrete section in one continuous pour, (2) avoid external restraint from adjacent concrete elements, and (3) control internal differential thermal strains by preventing the concrete from experiencing excessive temperature differential between the internal concrete and the surface. This is done by keeping the concrete warm through use of insulation (tenting, quilts, or sand on polyethylene sheeting). Studies and experience have shown that the maximum temperature differential between the interior and exterior concrete should not exceed about 36°F to avoid surface cracking. Internal cracking is also reduced.† Some sources indicate that the maximum temperature differential (MTD) for concrete containing granite or limestone (low-thermal-coefficient aggregates) should be 45°F and 56°F, respectively.†† However, an MTD of 36°F should be assumed unless tests on the actual concrete mix to be used show that higher MTD values are allowable.

By reducing the temperature differential to 36°F or less, the concrete will cool slowly to ambient temperature with little or no surface cracking as long as it is not restrained by continuous reinforcement crossing the interface of adjacent or opposite sections of hardened concrete. Restrained concrete will crack due to eventual thermal contraction after the cool down. Unrestrained concrete will not crack if proper procedures are followed and the temperature differential is monitored and controlled. If there is any concern over excess temperature differentials in a concrete member, the element should be considered as mass concrete and appropriate precautions taken.

Fig. 15-10 illustrates the relationship between temperature rise, cooling, and temperature differentials for a section of mass concrete. As can be observed, if the forms (which are providing adequate insulation in this case) are removed too early, cracking will occur once the difference between interior and surface concrete temperatures exceeds the critical temperature differential of 36°F.

The maximum temperature rise can be estimated by methods previously mentioned or by an approxima-

---

*References 15-9 and 15-56.
**The temperature rise can also be calculated through the use of equations and tables in References 15-46 and 15-52, which take into account the effect of volume-to-surface ratios and placement temperatures on temperature rise.
†References 15-21 and 15-22.
††Reference 15-31.

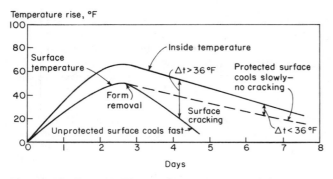

Fig. 15-10. Potential for surface cracking after form removal, assuming a critical temperature differential, Δt, of 36°F. No cracking should occur if concrete is cooled slowly and Δt is less than 36°F. References 15-22 and 15-56.

tion if the concrete contains 500 to 1000 lb of cement per cubic yard and the least dimension of the member is 6 ft or more. This approximation (under normal, not adiabatic conditions) would be 12.8°F for every 100 lb of cement per cubic yard. For example, the maximum temperature of such an element made with concrete having 900 lb of cement per cubic yard and cast at 60°F would be about

$$60°F + (12.8°F \times \frac{900}{100}) \text{ or } 175°F$$

The slow rate of heat exchange between concrete and its surroundings is caused by its low thermal conductivity. Heat escapes from concrete at a rate that is inversely proportional to the square of its least dimension. A 6-in.-thick wall cooling from both sides will take approximately 1½ hours to dissipate 95% of its developed heat. A 5-ft-thick wall would take an entire week to dissipate the same amount of developed heat.*

PREPLACED AGGREGATE CONCRETE

Preplaced aggregate concrete is concrete produced by placing coarse aggregate in a form and later injecting a cement-sand grout, usually with admixtures, to fill the voids. Properties of the resulting concrete are similar to those of comparable concrete placed by conventional methods; however, considerably less thermal and drying shrinkage can be expected because of the point-to-point contact of aggregate particles.

Coarse aggregates should meet requirements of ASTM C 33. In addition, most specifications limit both the maximum and minimum sizes; for example, 3-in. maximum and ½-in. minimum. Aggregates are generally graded to produce a void content of 35% to 40%. Fine aggregate used in the grout is generally graded to a fineness modulus of between 1.2 and 2.0, with nearly all of the material passing a No. 16 sieve.

Although the preplaced aggregate method has been used principally for restoration work and in the construction of reactor shields, bridge piers, and underwater structures, it has also been used in buildings to produce unusual architectural effects. Since the forms

are completely filled with coarse aggregate prior to grouting, a dense and uniform exposed-aggregate facing is obtained when the surface is sandblasted, tooled, or retarded and wire-brushed at an early age.

Tests for preplaced aggregate concrete are given in ASTM C 937 through C 943. Preplaced aggregate concrete is discussed in more detail in ACI 304, Guide for Measuring, Transporting, and Placing Concrete.

NO-SLUMP CONCRETE

No-slump concrete is defined by ACI 116 as concrete with a consistency corresponding to a slump of ¼ in. or less. Such concrete, while very dry, must be sufficiently workable to be placed and consolidated with the equipment to be used on the job. The methods referred to here do not necessarily apply to mixtures for concrete masonry units or for compaction by spinning techniques.

Many of the basic laws governing the properties of higher-slump concretes are applicable to no-slump concrete. For example, the properties of hardened concrete depend principally on the ratio of water to cement, provided consolidation is suitable.

Measurement of the consistency of no-slump concrete differs from that for higher-slump concrete because the slump cone is impractical for the drier consistencies. ACI 211.3, Standard Practice for Selecting Proportions for No-Slump Concrete, describes three methods for measuring the consistency of no-slump concrete: the Vebe apparatus, the compacting-factor test, and the Thaulow drop table. In the absence of the above test equipment, workability can be adequately judged by a trial mixture that is placed and compacted with equipment and methods to be used on the job.

Intentionally entrained air is recommended for no-slump concrete where durability is required. The amount of air-entraining admixture usually recommended for higher-slump concretes will not produce air contents in no-slump concretes as high as those in the higher-slump concretes. The lower volume of entrained air, however, generally provides adequate durability for no-slump concretes, that is, sufficient small air voids are present. This departure from the usual methods of designing and controlling entrained air is necessary for no-slump concretes.

For a discussion of water requirements and computation of trial mixtures, see ACI 211.3.

ROLLER-COMPACTED CONCRETE

Roller-compacted concrete is a lean, no-slump, almost dry concrete that is compacted in place by vibratory roller or plate compaction equipment (Fig. 15-11). It is

---

*Reference 15-26, page 26. For more information on mass concrete, see References 15-4, 15-9, 15-15, 15-21, 15-22, 15-26, 15-31, 15-41, 15-46, 15-52, 15-53, 15-56, and 15-59.

Fig. 15-11. Vibratory rollers are used to compact roller-compacted concrete.

a mixture of aggregate, cement, and water; supplementary cementing materials such as fly ash also have been used. Cement contents range from 100 to 600 lb per cubic yard. Mixing is done with conventional batch mixers, continuous mixers, or in some instances tilting-drum truck mixers.

Roller-compacted concrete is continuing to develop as a fast and economical method of constructing large gravity dams, off-highway pavement projects such as container-handling facilities and dry-land log-sorting areas, airport aprons, county roads, and as subbases for conventional highway and street pavements. Compressive strengths of 1000 to 4500 psi have been obtained from roller-compacted concrete in dam projects. Pavement projects have had design compressive strengths of about 5000 psi with field strengths in the range of 5000 to 10,000 psi.*

Roller-compacted concrete must be placed in layers thin enough to allow complete compaction by the construction equipment available. Optimum layer thicknesses ranging from 8 to 12 in. are placed and consolidated with conventional earthmoving or paving equipment (graders, rollers, and so forth). On multiple-layer projects it is essential that a construction procedure be adopted that will ensure good bond between layers. The method used to spread roller-compacted concrete is a major factor in the control of production.

ACI 207.5 discusses mixture proportions, physical properties, mixing, and construction procedures for roller-compacted concrete used in massive placements.

SOIL-CEMENT

Soil-cement is a mixture of pulverized soil or granular material, cement, and water. Some other terms applied to soil-cement are "cement-treated base or subbase," "cement stabilization," "cement-modified soil," and "cement-treated aggregate." The mixture is compacted to high density, and as the cement hydrates the material becomes hard and durable.

Soil-cement is primarily used as base course for roads, streets, airports, and parking areas. A bituminous or portland cement concrete wearing course is usually placed over the base. Soil-cement is also used as a subbase for concrete pavements, as slope protection for earth dams and embankments, reservoir and ditch linings, and foundation stabilization.

The soil material in soil-cement can be almost any combination of sand, silt, clay, and gravel or crushed stone. Local granular materials (such as slag, caliche, limerock, and scoria) plus a wide variety of waste materials (such as cinders, ash, and screenings from quarries and gravel pits) can be used to make soil-cement. Also, old granular-base roads, with or without their bituminous surfaces, can be recycled to make soil-cement.

Soil-cement should contain sufficient portland cement to resist deterioration from freeze-thaw and wet-dry cycling and sufficient moisture for maximum compaction. Cement contents range from 130 to 430 lb per cubic yard. The soil, cement, and water is mixed in a central mixing plant or in place using transverse shaft mixers or traveling pugmills. The mixture is placed and compacted with conventional road-building equipment to 96% to 100% of maximum density (ASTM D 558).

Depending on the soil used, the 7-day compressive strengths of saturated specimens at the minimum cement content meeting soil-cement criteria are generally between 300 and 800 psi. Soil-cement continues to gain strength with age; compressive strengths in excess of 2500 psi have been obtained after many years of service.**

SHOTCRETE

Shotcrete is mortar or concrete that is pneumatically projected onto a surface at high velocity (Fig. 15-12). Developed in 1911, its concept is essentially unchanged even in today's use. The relatively dry mixture is consolidated by the impact force and can be placed on vertical or horizontal surfaces without sagging. Shotcrete is used for both new construction and repair work. It is especially suited for curved or thin concrete structures and shallow repairs. The hardened properties of shotcrete are very operator dependent. Shotcrete has a unit weight and compressive strength similar to normal- and high-strength concrete. Aggregate sizes up to ¾ in. can be used.

Shotcrete can be made by a dry or wet process. In the dry process, a premixed blend of cement and damp aggregate is propelled through a hose by compressed

*Reference 15-55.
**Reference 15-25. Also controlled-density, or flowable, fill is essentially fluid, low-strength soil-cement used for pipe bedding, backfill, and subbase in place of compacted soil (see Reference 15-59).

Fig. 15-12. Shotcrete.

air to a nozzle. Water is added to the cement and aggregate mixture at the nozzle and the intimately mixed ingredients are projected onto the surface. In the wet process, all the ingredients are premixed. Compressed air conveys the mixture through the hose to the nozzle. Additional compressed air is added at the nozzle to increase the velocity, at which time the mixture is projected onto the surface. Guidelines for the use of shotcrete are described in ACI 506R-85, Guide to Shotcrete (Reference 15-49).

SHRINKAGE-COMPENSATING CONCRETE

Shrinkage-compensating concrete, using an expansive cement or expansive admixture added to portland cement, expands after setting and during hardening to an amount equal to or slightly greater than the amount of drying shrinkage expected in a normal concrete mix. Shrinkage-compensating concrete is used in concrete slabs, pavements, structures, and repair work to minimize drying shrinkage cracks.

Reinforcing steel in the structure restrains the concrete and goes into tension as the shrinkage compensating concrete expands. Upon shrinking due to drying contraction caused by moisture loss in hardened concrete, the tension generally is relieved and as long as the resulting tension in the concrete does not exceed the tensile strength of the concrete, no cracks should result. Shrinkage-compensating concrete can be proportioned, batched, placed, and cured similarly to normal concrete with some precautions necessary to assure the expected expansion. Additional information can be found in Chapter 2 and in ACI 223, Standard Practice for the Use of Shrinkage-Compensating Concrete.

POROUS CONCRETE

Porous (no fines) concrete contains a narrowly graded coarse aggregate, little to no fine aggregate, and insuffi-

cient cement paste to fill voids in the coarse aggregate. This low water-cement ratio, low-slump concrete resembling popcorn is primarily held together by cement paste at the contact points of the coarse aggregate particles, producing a concrete with a high volume of voids (20% to 35%) and a high permeability that allows water to flow through it easily.

Porous concrete is used in hydraulic structures as drainage media, and in parking lots, pavements, and airport runways to reduce storm water run off. It also recharges the local groundwater supply by allowing the water to penetrate through the concrete to the ground below. Porous concretes have also been used in tennis courts and greenhouses.

As a paving material, porous concrete is raked or slipformed into place with conventional paving equipment and then roller compacted. Vibratory screeds or hand rollers can be used for smaller work. In order to maintain porous properties, the surfaces should not be closed up or sealed; therefore, troweling and finishing are not desired. The compressive strength of different mixes can range from 500 to 4000 psi. Drainage rates commonly range from 2 to 18 gallons per minute per square foot.

Porous concrete is used in building construction (particularly walls) for its thermal insulating properties. For example, a 10-in.-thick porous-concrete wall can have an R value of 5 compared to 0.75 for normal concrete. Porous concrete is also lightweight, 100 to 120 pcf, and has low shrinkage properties.*

WHITE AND COLORED CONCRETE

White Concrete

White portland cement is used to produce white concrete, a widely used architectural material. It is also used in mortar, plaster, stucco, terrazzo, and portland cement paint. White portland cement is manufactured to conform to ASTM C150 even though that specification does not specifically mention white portland cement.

White concrete is made with aggregates and water that contain no materials that will discolor the concrete. White or light-colored aggregates can be used. Oil that could stain concrete should not be used on the forms. Care must be taken to avoid rust stains from tools and equipment. Curing materials that could cause stains must be avoided. Slabs should be cured with nonstaining waterproofed paper and the paper should be overlapped and sealed at the seams with a nonstaining material.

Colored Concrete

Colored concrete can be produced by using colored aggregates or by adding color pigments (ASTM C979) or both. When colored aggregates are used, they should

---

*References 15-20 and 15-39.

be exposed at the surface of the concrete. This can be done by casting against a form that has been treated with a retarder. Unhydrated paste at the surface is later brushed or washed away. Other methods involve removing the surface mortar by sandblasting, waterblasting, bushhammering, grinding, or acid washing. If surfaces are to be washed with acid, a delay of approximately two weeks after casting is necessary. Colored aggregates may be natural aggregates such as quartz, marble, and granite, or they may be ceramic materials.

Pigments for coloring concrete should be pure mineral oxides ground finer than cement and be insoluble in water, free of soluble salts and acids, colorfast in sunlight, resistant to alkalies and weak acids, and virtually free of calcium sulfate. Mineral oxides occur in nature and are also produced synthetically; synthetic pigments generally give more uniform results.

The amount of color pigments added to a concrete mixture should not be more than 10% of the weight of the cement. The amount required depends on the type of pigment and the color desired. For example, a dose of pigment equal to 1.5% by weight of cement may produce a pleasing pastel color, but 7% may be needed to produce a deep color. White portland cement will produce cleaner, brighter colors and is recommended in preference to gray cement, except for black or dark gray colors.

To maintain uniform color, do not use calcium chloride, and proportion all materials carefully by weight. To prevent streaking, the dry cement and color pigment must be thoroughly blended before they are added to the mix. Mixing time should be longer than normal to ensure uniformity.

In air-entrained concrete, the addition of pigment may require an adjustment in the amount of air-entraining admixture to maintain the desired air content.

Dry-shake method. Slabs or precast panels that are cast horizontally can be colored by the dry-shake method. Dry coloring materials consisting of mineral oxide pigment, white portland cement, and specially graded silica sand or other fine aggregate are marketed ready for use by various manufacturers.

After the slab has been floated once, two-thirds of the dry coloring material should be broadcast evenly by hand over the surface. The required amount of coloring material can usually be determined in pounds per square foot from previously cast sections. After the material has absorbed water from the fresh concrete, it should be floated into the surface. Immediately after, the rest of the material should be applied at right angles to the initial application, so that a uniform color is obtained. The slab should again be floated to work the remaining material into the surface.

Other finishing operations may follow depending on the type of finish desired. Curing should begin immediately after finishing—taking precautions to prevent discoloring the surface.

POLYMER-PORTLAND CEMENT CONCRETE

Polymer-portland cement concrete (PPCC), also called polymer-modified concrete, is basically normal portland cement concrete to which a polymer or monomer has been added during mixing. Thermoplastic and elastomeric latexes are the most commonly used polymers in PPCC. Epoxies and other polymers are also commonly used. In general, latex improves ductility, durability, adhesive properties, resistance to chloride-ion ingress, shear bond, and tensile and flexural strength of concrete and mortar. Latex-modified concretes (LMC) also have excellent freeze-thaw, abrasion, and impact resistance. Some LMC materials can also resist certain acids, alkalies, and organic solvents. Polymer-portland cement concrete is primarily used in concrete patching and overlays.*

FERROCEMENT

Ferrocement is a special type of reinforced concrete composed of closely spaced layers of continuous relatively thin metallic or nonmetallic mesh or wire embedded in mortar. It is constructed by hand plastering, shotcreting, laminating (forcing the mesh into fresh mortar), or a combination of these methods.

The mortar mixture generally has a sand-cement ratio of 1.5 to 2.5 and a water-cement ratio of 0.35 to 0.5. Reinforcement makes up about 5% to 6% of the ferrocement volume. Fibers and admixtures may also be used to improve the mortar quality. Polymers or cement-based coatings are often applied to the finished surface to reduce porosity.

Ferrocement is considered easy to produce in a variety of shapes and sizes; however, it is labor intensive. Ferrocement is used to construct shell roofs, swimming pools, tunnel linings, silos, tanks, prefabricated houses, barges, boats, sculptures, and thin panels or sections usually less than 1 in. thick.**

FIBER-REINFORCED CONCRETE

Fiber-reinforced concrete is conventional concrete to which discontinuous discrete fibers are added during mixing. The fibers, made from steel, plastic, glass, and natural (cellulose) and other materials, are available in a variety of shapes (round, flat, crimped, and deformed) and sizes with typical lengths of 0.25 to 3 in. and thicknesses ranging from 0.0002 to 0.030 in.†

Steel fibers have been shown to significantly improve concrete flexural strength, impact strength, toughness, fatigue strength, and resistance to cracking. Other types of fibers show varied results. If the fibers are not added to the mix in the proper batching sequence or if the volume percentage of fibers is too high, fibers may clump together or ball up during mixing. Fiber contents up to 4% or 5% by volume of concrete or mortar can be used; however, 1% to 2% is the practical upper limit for field placement of most fibers.

---

*See Reference 15-54 for more information.
**Reference 15-35.
†See References 15-61 and 15-62 for more information.

Steel-fiber-reinforced concrete can be placed by most conventional methods including pumping, as long as the mix is not too wet. Overly wet mixes that are pumped can segregate. Steel fibers can also be used in slurry-infiltrated fiber concrete in which a cement slurry is poured into a bed of fibers (6% to 18% volume fiber content). Steel, glass, and polypropylene plastic fibers can also be used in fiber-reinforced shotcrete.

Steel-fiber-reinforced concrete is primarily used in pavements, overlays, patching, hydraulic structures, thin shells, and precast products. Glass fibers are primarily used in spray-up thin panel applications, which are referred to as GFRC—glass-fiber-reinforced concrete. Additional information can be obtained in ACI 544.1R, State-of-the-Art Report on Fiber Reinforced Concrete. Tests for fiber-reinforced concrete are described in ASTM C 995 (inverted-slump test), ASTM C 1018 (flexural toughness), and ACI 544.2R, Measurement of Properties of Fiber Reinforced Concrete.

REFERENCES

15-1. Valore, R. C., Jr., "Cellular Concretes, Parts 1 and 2," *Proceedings of the American Concrete Institute,* vol. 50, American Concrete Institute, Detroit, May 1954, pages 773-796; June 1954, pages 817-836.

15-2. Valore, R. C., Jr., "Insulating Concrete," *Proceedings of the American Concrete Institute,* vol. 53, American Concrete Institute, November 1956, pages 509-532.

15-3. Klieger, Paul, *Early-High-Strength Concrete for Prestressing,* Research Department Bulletin RX091, Portland Cement Association, 1958.

15-4. Carlson, Roy W., and Thayer, Donald P., "Surface Cooling of Mass Concrete to Prevent Cracking," *Journal of the American Concrete Institute,* American Concrete Institute, August 1959, page 107.

15-5. Walker, Stanton, and Bloem, Delmar L., "Effects of Aggregate Size on Properties of Concrete," *Proceedings of the American Concrete Institute,* vol. 57, no. 3, American Concrete Institute, September 1960, pages 283-298; and Discussion by N. G. Zoldner, March 1961, pages 1245-1248.

15-6. *Concrete for Radiation Shielding,* Compilation No. 1, American Concrete Institute, 1962.

15-7. Cordon, William A., and Gillespie, Aldridge H., "Variables in Concrete Aggregates and Portland Cement Paste which Influence the Strength of Concrete," *Proceedings of the American Concrete Institute,* vol. 60, no. 8, American Concrete Institute, August 1963, pages 1029-1050; and Discussion, March 1964, pages 1981-1998.

15-8. Mather, K., "High-Strength, High-Density Concrete," *Proceedings of the American Concrete Institute,* vol. 62, American Concrete Institute, August 1965, pages 951-960.

15-9. Townsend, C. L., "Control of Cracking in Mass Concrete Structures," *Engineering Monograph No. 34,* U.S. Department of Interior, Bureau of Reclamation, Denver, 1965.

15-10. Brewer, Harold W., *General Relation of Heat Flow Factors to the Unit Weight of Concrete,* Development Department Bulletin DX114, Portland Cement Association, 1967.

15-11. McCormick, Fred C., "Rational Proportioning of Preformed Foam Cellular Concrete," *Journal of the American Concrete Institute,* American Concrete Institute, February 1967, pages 104-110.

15-12. Gustaferro, A. H.; Abrams, M. S.; and Litvin, Albert, *Fire Resistance of Lightweight Insulating Concretes,* Research and Development Bulletin RD004B, Portland Cement Association, 1970.

15-13. Reichard, T. W., "Mechanical Properties of Insulating Concretes," *Lightweight Concrete,* American Concrete Institute, 1971, pages 253-316.

15-14. Freedman, Sidney, *High-Strength Concrete,* IS176T, Portland Cement Association, 1971.

15-15. Tuthill, Lewis H., and Adams, Robert F., "Cracking Controlled in Massive, Reinforced Structural Concrete by Application of Mass Concrete Practices," *Journal of the American Concrete Institute,* American Concrete Institute, August 1972, page 481.

15-16. *Concrete for Nuclear Reactors,* SP-34, 3 volumes, 73 papers, American Concrete Institute, 1972, 1766 pages.

15-17. *Structural Lightweight Concrete,* IS032T, Portland Cement Association, revised 1986.

15-18. Perenchio, W. F., *An Evaluation of Some of the Factors Involved in Producing Very-High-Strength Concrete,* Research and Development Bulletin RD014, Portland Cement Association, 1973.

15-19. Klieger, Paul, "Proportioning No-Slump Concrete," *Proportioning Concrete Mixes,* SP-46, American Concrete Institute, 1974, pages 195-207.

15-20. Malhotra, V. M., "No-Fines Concrete—Its Properties and Applications," *Journal of the American Concrete Institute,* American Concrete Institute, November 1976.

15-21. FitzGibbon, Michael E., "Large Pours for Reinforced Concrete Structures," Current Practice Sheets No. 28, 35, and 36, *Concrete,* Cement and Concrete Association, Wexham Springs, Slough, England, March and December 1976 and February 1977.

15-22. Fintel, Mark, and Ghosh, S. K., "Mass Reinforced Concrete Without Construction Joints," presented at the Adrian Pauw Symposium on

Designing for Creep and Shrinkage, Fall Convention of the American Concrete Institute, Houston, Texas, November 1978.

15-23. Perenchio, William F., and Klieger, Paul, *Some Physical Properties of High-Strength Concrete,* Research and Development Bulletin RD056T, Portland Cement Association, 1978.

15-24. Hanna, Amir N., *Properties of Expanded Polystyrene Concrete and Applications for Pavement Subbases,* Research and Development Bulletin RD055P, Portland Cement Association, 1978.

15-25. *Soil-Cement Construction Handbook,* EB003S, Portland Cement Association, 1979.

15-26. *Mass Concrete for Dams and Other Massive Structures,* ACI 207.1R-70, reaffirmed 1980, ACI Committee 207 Report, American Concrete Institute.

15-27. *Standard Practice for Selecting Proportions for No-Slump Concrete,* ACI 211.3-75, revised 1980, ACI Committee 211 Report, American Concrete Institute.

15-28. *Roller-Compacted Concrete,* ACI 207.5R-80, ACI Committee 207 Report, American Concrete Institute, 1980.

15-29. *Concrete Manual,* 8th ed., U.S. Bureau of Reclamation, revised 1981.

15-30. *Standard Practice for Selecting Proportions for Structural Lightweight Concrete,* ACI 211.2-81, ACI Committee 211 Report, American Concrete Institute, 1981.

15-31. Bamforth, P. B., "Large Pours," letter to the editor, *Concrete,* Cement and Concrete Association, February 1981.

15-32. *Guide for Cast-in-Place Low-Density Concrete,* ACI 523.1R-67, revised 1982, ACI Committee 523 Report, American Concrete Institute.

15-33. *Guide for Cellular Concretes Above 50 pcf and for Aggregate Concretes Above 50 pcf with Compressive Strengths Less Than 2500 psi,* ACI 523.3R-75, revised 1982, ACI Committee 523 Report, American Concrete Institute.

15-34. *State-of-the-Art Report on Fiber Reinforced Concrete,* ACI 544.1R-82, ACI Committee 544 Report, American Concrete Institute, 1982.

15-35. *State-of-the-Art Report on Ferrocement,* ACI 549R-82, ACI Committee 549 Report, American Concrete Institute, 1982.

15-36. *Standard Practice for the Use of Shrinkage-Compensating Concrete,* ACI 223-83, ACI Committee 223 Report, American Concrete Institute, 1983.

15-37. *Building Code Requirements for Reinforced Concrete,* ACI 318-83, ACI Committee 318 Report, American Concrete Institute, 1983.

15-38. *Recommended Practice for Measuring, Mixing, Transporting, and Placing Concrete,* ACI 304-73, reaffirmed 1983, ACI Committee 304 Report, American Concrete Institute.

15-39. "Porous Concrete Slabs and Pavement Drain Water," *Concrete Construction,* Concrete Construction Publications, Inc., Addison, Illinois, September 1983, pages 685 and 687-688.

15-40. *Guide for Specifying, Mixing, Placing, and Finishing Steel Fiber Reinforced Concrete,* ACI 544.3R-84, ACI Committee 544 Report, American Concrete Institute, 1984.

15-41. *Standard Practice for Selecting Proportions for Normal, Heavyweight, and Mass Concrete,* ACI 211.1-81, revised 1984, ACI Committee 211 Report, American Concrete Institute.

15-42. *State-of-the-Art Report on High-Strength Concrete,* ACI 363R-84, ACI Committee 363 Report, American Concrete Institute, 1984.

15-43. *State-of-the-Art Report on Fiber Reinforced Shotcrete,* ACI 506.1R-84, ACI Committee 506 Report, American Concrete Institute, 1984.

15-44. *Fiber Reinforced Concrete,* SP-81, American Concrete Institute, 1984, 460 pages.

15-45. *Guide for Structural Lightweight Aggregate Concrete,* ACI 213R-79, reaffirmed 1984, ACI Committee 213 Report, American Concrete Institute.

15-46. *Control of Cracking in Concrete Structures,* ACI 224R-80, revised 1984, ACI Committee 224 Report, American Concrete Institute.

15-47. Wolsiefer, John, "Ultra High-Strength, Field Placeable Concrete with Silica Fume Admixture," *Concrete International,* American Concrete Institute, April 1984, pages 25-31.

15-48. Van Geem, Martha G.; Litvin, Albert; and Musser, Donald W., *Insulative Lightweight Concrete for Building Walls,* presented at the Annual Convention and Exposition of the American Society of Civil Engineers in Detroit, October 1985.

15-49. *Guide to Shotcrete,* ACI 506R-85, ACI Committee 506 Report, American Concrete Institute, 1985.

15-50. *Very High Strength Cement-Based Materials,* Materials Research Society, Pittsburgh, Pennsylvania, 1985.

15-51. *Standard Practice for Curing Concrete,* ACI 308-81, revised 1986, ACI Committee 308 Report, American Concrete Institute.

15-52. *Effect of Restraint, Volume Change, and Reinforcement on Cracking of Massive Concrete,* ACI 207.2R-73, reapproved 1986, ACI Committee 207 Report, American Concrete Institute.

15-53. *Cooling and Insulating Systems for Mass Concrete,* ACI 207.4R-80, revised 1986, ACI Committee 207 Report, American Concrete Institute.

15-54. *Guide for the Use of Polymers in Concrete,* ACI 548.1R-86, ACI Committee 548 Report, American Concrete Institute, 1986.

15-55. Hansen, Kenneth D., "A Pavement for Today and Tomorrow," *Concrete International,* American Concrete Institute, February 1987, pages 15-17.

15-56. *Concrete for Massive Structures,* IS128T, Portland Cement Association, 1987.

15-57. Knutson, Marlin, and Riley, Randall, "Fast-Track Concrete Paving Opens Door to Industry Future," *Concrete Construction,* Concrete Construction Publications, Inc., January 1987, pages 4-13.

15-58. Popovics, Sandor; Rajendran, N.; and Penko, Michael, "Rapid Hardening Cements for Repair of Concrete," *ACI Materials Journal,* American Concrete Institute, January-February 1987, pages 64-73.

15-59. *Use of Fly Ash in Concrete,* ACI 226.3R-87, ACI Committee 226 Report, American Concrete Institute, 1987.

15-60. Godfrey, K. A., Jr., "Concrete Strength Record Jumps 36%," *Civil Engineering,* American Society of Civil Engineers, New York, October 1987.

15-61. *Fiber Reinforced Concrete,* SP039T, Portland Cement Association, 1991.

15-62. Panarese, William C., "Fiber: Good for the Concrete Diet?," *Civil Engineering,* American Society of Civil Engineers, New York, May 1992, pages 44-47.

Appendix

ASTM STANDARDS

American Society for Testing and Materials* documents related to aggregates, cement, and concrete that are relevant to or referred to in the text are listed as follows:

C 29-87   Test Method for Unit Weight and Voids in Aggregate

C 31-87   Practice for Making and Curing Concrete Test Specimens in the Field

C 33-86   Specification for Concrete Aggregates

C 39-86   Test Method for Compressive Strength of Cylindrical Concrete Specimens

C 40-84   Test Method for Organic Impurities in Fine Aggregates for Concrete

C 42-85   Method of Obtaining and Testing Drilled Cores and Sawed Beams of Concrete

C 70-79   Test Method for Surface Moisture in Fine Aggregate

C 78-84   Test Method for Flexural Strength of Concrete (Using Simple Beam with Third-Point Loading)

C 85-66   Test Method for Cement Content of Hardened Portland Cement Concrete

C 87-83   Test Method for Effect of Organic Impurities in Fine Aggregate on Strength of Mortar

C 88-83   Test Method for Soundness of Aggregates by Use of Sodium Sulfate or Magnesium Sulfate

C 91-87   Specification for Masonry Cement

C 94-86   Specification for Ready-Mixed Concrete

C 109-86   Test Method for Compressive Strength of Hydraulic Cement Mortars (Using 2-in. or 50-mm Cube Specimens)

C 114-85   Methods for Chemical Analysis of Hydraulic Cement

C 115-86   Test Method for Fineness of Portland Cement by the Turbidimeter

C 117-87   Test Method for Materials Finer than 75-μm (No. 200) Sieve in Mineral Aggregates by Washing

C 123-83   Test Method for Lightweight Pieces in Aggregate

C 125-86   Definitions of Terms Relating to Concrete and Concrete Aggregates

C 127-84   Test Method for Specific Gravity and Absorption of Coarse Aggregate

C 128-84   Test Method for Specific Gravity and Absorption of Fine Aggregate

C 131-81   Test Method for Resistance to Degradation of Small-Size Coarse Aggregate by Abrasion and Impact in the Los Angeles Machine

C 136-84   Method for Sieve Analysis of Fine and Coarse Aggregates

C 138-81   Test Method for Unit Weight, Yield, and Air Content (Gravimetric) of Concrete

C 141-85   Specification for Hydraulic Hydrated Lime for Structural Purposes

C 142-78   Test Method for Clay Lumps and Friable Particles in Aggregates

C 143-78   Test Method for Slump of Portland Cement Concrete

C 150-86   Specification for Portland Cement

C 151-84   Test Method for Autoclave Expansion of Portland Cement

C 156-80   Test Method for Water Retention by Concrete Curing Materials

C 157-86   Test Method for Length Change of Hardened Hydraulic-Cement Mortar and Concrete

C 171-69   Specification for Sheet Materials for Curing Concrete

C 172-82   Method of Sampling Freshly Mixed Concrete

C 173-78   Test Method for Air Content of Freshly Mixed Concrete by the Volumetric Method

C 174-87   Test Method for Measuring Length of Drilled Concrete Cores

C 177-85   Test Method for Steady-State Heat Flux Measurements and Thermal Transmission Properties by Means of the Guarded-Hot-Plate Apparatus

C 183-83   Methods of Sampling and Acceptance of Hydraulic Cement

C 184-83   Test Method for Fineness of Hydraulic Cement by the 150-μm (No. 100) and 75-μm (No. 200) Sieves

C 185-85   Test Method for Air Content of Hydraulic Cement Mortar

C 186-86   Test Method for Heat of Hydration of Hydraulic Cement

C 187-86   Test Method for Normal Consistency of Hydraulic Cement

C 188-84   Test Method for Density of Hydraulic Cement

C 190-85   Test Method for Tensile Strength of Hydraulic Cement Mortars

C 191-82   Test Method for Time of Setting of Hydraulic Cement by Vicat Needle

C 192-81   Method of Making and Curing Concrete Test Specimens in the Laboratory

C 204-84   Test Method for Fineness of Portland Cement by Air Permeability Apparatus

C 215-85   Test Method for Fundamental Transverse, Longitudinal, and Torsional Frequencies of Concrete Specimens

C 219-84   Terminology Relating to Hydraulic Cement

---

*ASTM, 1916 Race Street, Philadelphia, Pa. 19103

METRIC CONVERSION FACTORS

The following list provides the conversion relationship between U.S. customary units and SI (International System) units. The proper conversion procedure is to multiply the specified value on the left (primarily U.S. customary values) by the conversion factor exactly as given below and then round to the appropriate number of significant digits desired. For example, to convert 11.4 ft to meters: 11.4 × 0.3048 = 3.47472, which rounds to 3.47 meters. Do not round either value before performing the multiplication, as accuracy would be reduced. A complete guide to the SI system and its use can be found in ASTM E 380, Metric Practice.

To convert from	to	multiply by
Length		
inch (in.)	micron (μ)	25,400 E*
inch (in.)	centimeter (cm)	2.54 E
inch (in.)	meter (m)	0.0254 E
foot (ft)	meter (m)	0.3048 E
yard (yd)	meter (m)	0.9144
Area		
square foot (sq ft)	square meter (sq m)	0.09290304 E
square inch (sq in.)	square centimeter (sq cm)	6.452 E
square inch (sq in.)	square meter (sq m)	0.00064516 E
square yard (sq yd)	square meter (sq m)	0.8361274
Volume		
cubic inch (cu in.)	cubic centimeter (cu cm)	16.387064
cubic inch (cu in.)	cubic meter (cu m)	0.00001639
cubic foot (cu ft)	cubic meter (cu m)	0.02831685
cubic yard (cu yd)	cubic meter (cu m)	0.7645549
gallon (gal) Can. liquid	liter	4.546
gallon (gal) Can. liquid	cubic meter (cu m)	0.004546
gallon (gal) U.S. liquid**	liter	3.7854118
gallon (gal) U.S. liquid	cubic meter (cu m)	0.00378541
fluid ounce (fl oz)	milliliters (ml)	29.57353
fluid ounce (fl oz)	cubic meter (cu m)	0.00002957
Force		
kip (1000 lb)	kilogram (kg)	453.6
kip (1000 lb)	newton (N)	4,448.222
pound (lb) avoirdupois	kilogram (kg)	0.4535924
pound (lb)	newton (N)	4.448222
Pressure or stress		
kip per square inch (ksi)	megapascal (MPa)	6.894757
kip per square inch (ksi)	kilogram per square centimeter (kg/sq cm)	70.31
pound per square foot (psf)	kilogram per square meter (kg/sq m)	4.8824
pound per square foot (psf)	pascal (Pa)†	47.88
pound per square inch (psi)	kilogram per square centimeter (kg/sq cm)	0.07031
pound per square inch (psi)	pascal (Pa)†	6,894.757
pound per square inch (psi)	megapascal (MPa)	0.00689476
Mass (weight)		
pound (lb) avoirdupois	kilogram (kg)	0.4535924
ton, 2000 lb	kilogram (kg)	907.1848
grain	kilogram (kg)	0.0000648

To convert from	to	multiply by
Mass (weight) per length		
kip per linear foot (klf)	kilogram per meter (kg/m)	0.001488
pound per linear foot (plf)	kilogram per meter (kg/m)	1.488
Mass per volume (density)		
pound per cubic foot (pcf)	kilogram per cubic meter (kg/cu m)	16.01846
pound per cubic yard (lb/cu yd)	kilogram per cubic meter (kg/cu m)	0.5933
Temperature		
degree Fahrenheit (°F)	degree Celsius (°C)	$t_C = (t_F - 32)/1.8$
degree Fahrenheit (°F)	degree Kelvin (°K)	$t_K = (t_F + 459.7)/1.8$
degree Kelvin (°K)	degree Celsius (C°)	$t_C = t_K - 273.15$
Energy and heat		
British thermal unit (Btu)	joule (J)	1055.056
calorie (cal)	joule (J)	4.1868 E
Btu/°F · hr · ft²	W/m² · °K	5.678263
kilowatt-hour (kwh)	joule (J)	3,600,000. E
British thermal unit per pound (Btu/lb)	calories per gram (cal/g)	0.55556
British thermal unit per hour (Btu/hr)	watt (W)	0.2930711
Power		
horsepower (hp) (550 ft-lb/sec)	watt (W)	745.6999 E
Velocity		
mile per hour (mph)	kilometer per hour (km/hr)	1.60934
mile per hour (mph)	meter per second (m/s)	0.44704
Permeability		
darcy	centimeter per second (cm/sec)	0.000968
feet per day (ft/day)	centimeter per second (cm/sec)	0.000352

*E indicates that the factor given is exact.
**One U.S. gallon equals 0.8327 Canadian gallon.
†A pascal equals 1.000 newton per square meter.

Note:
One U.S. gallon of water weighs 8.34 pounds (U.S.) at 60°F.
One cubic foot of water weighs 62.4 pounds (U.S.).
One milliliter of water has a mass of 1 gram and has a volume of one cubic centimeter.
One U.S. bag of cement weighs 94 lb.

The prefixes and symbols listed below are commonly used to form names and symbols of the decimal multiples and submultiples of the SI units.

Multiplication Factor	Prefix	Symbol
$1,000,000,000 = 10^9$	giga	G
$1,000,000 = 10^6$	mega	M
$1,000 = 10^3$	kilo	k
$1 = 1$	—	—
$0.01 = 10^{-2}$	centi	c
$0.001 = 10^{-3}$	milli	m
$0.000001 = 10^{-6}$	micro	μ
$0.000000001 = 10^{-9}$	nano	n

Index